Fuzzy Systems Modeling in Environmental and Health Risk Assessment

Fuzzy Systems Modeling in Environmental and Health Risk Assessment

Edited by

Boris Faybishenko
Lawrence Berkeley National Laboratory

Rehan Sadiq
University of British Columbia

Ashok Deshpande (Deceased)

WILEY

Published by John Wiley & Sons, Inc., Hoboken, New Jersey.

Published simultaneously in Canada.

A catalogue record for this book is available from the Library of Congress

Hardback: 9781119569473; ePub ISBN: 9781119569480; ePDF ISBN: 9781119569497;
oBook ISBN: 9781119569503

Cover Design: Wiley
Cover Image: © peterschreiber.media/Shutterstock,
josefkubes/Shutterstock

Set in 9.5/12.5pt STIXTwoText by Integra Software Services Pvt. Ltd, Pondicherry, India

Contents

List of Contributors

Editors

Boris Faybishenko
Lawrence Berkeley National Laboratory, Earth and Environmental Sciences Area, Energy Geosciences Division, Berkeley, California, USA

Rehan Sadiq
School of Engineering, University of British Columbia (Okanagan), Kelowna, BC, Canada

Ashok Deshpande
Berkeley Initiative in Soft Computing (BISC), Special Interest Group Environmental Management Systems (EMS), University of California, Berkeley, California, USA; and College of Engineering, Pune, India

Contributors

Farid Wajdi Akashah
Department of Building Surveying, Faculty of Built Environment, University of Malaya, Kuala Lumpur, Malaysia

Majed Alinizzi
Qassim University, Buraydah, Qassim, Saudi Arabia

Saleem S. AlSaleem
Qassim University, Buraydah, Qassim, Saudi Arabia

Péter Böröcz
Széchenyi István University, Győr, Hungary

Ádám Bukovics
Széchenyi István University, Győr, Hungary

Adrienn Buruzs
Széchenyi István University, Győr, Hungary

Gyan Chhipi-Shrestha
School of Engineering, University of British Columbia (Okanagan), Kelowna, BC, Canada

R.A. Christian
SV National Institute of Technology, Surat, India

Roberta Dyck
School of Engineering, University of British Columbia (Okanagan), Kelowna, BC, Canada

Péter Földesi
Széchenyi István University, Győr, Hungary

Rajesh Gupta
Visvesvaraya National Institute of Technology, Nagpur, India

James Hager
School of Engineering, University of British Columbia (Okanagan), Kelowna, BC, Canada

Husnain Haider
Qassim University, Buraydah, Qassim, Saudi Arabia

Kasun Hewage
School of Engineering, University of British Columbia (Okanagan), Kelowna, BC, Canada

Guangji Hu
School of Engineering, University of British Columbia (Okanagan), Kelowna, BC, Canada

Abdullah Ibrahim
Faculty of Engineering Technology, University Malaysia Pahang, Pahang, Malaysia

Namrata Jariwala
SV National Institute of Technology, Surat, India

Muhammad Nomani Kabir
Department of Computer Science & Engineering, Trust University, Barishal, Ruiya, Bangladesh

Shichang Kang
Chinese Academy of Sciences, State Key Laboratory of Cryospheric Sciences, Beijing, China

Hirushie Karunathilake
School of Engineering, University of British Columbia (Okanagan), Kelowna, BC, Canada

Manjot Kaur
School of Engineering, University of British Columbia (Okanagan), Kelowna, BC, Canada

László T. Kóczy
Széchenyi István University, Győr, Hungary

Emmi Matern
School of Engineering, University of British Columbia (Okanagan), Kelowna, BC, Canada

Haroon R. Mian
School of Engineering, University of British Columbia (Okanagan), Kelowna, BC, Canada

Lindell Ormsbee
Kentucky Water Resources Research Institute, University of Kentucky, Lexington, KY, USA

Rachid Ouache
School of Engineering, University of British Columbia (Okanagan), Kelowna, BC, Canada; and Faculty of Chemical Engineering and Natural Resources, University of Malaysia, Pahang, Malaysia

Tharindu Prabatha
School of Engineering, University of British Columbia (Okanagan), Kelowna, BC, Canada

Thais Ayres Rebello
School of Engineering, University of British Columbia (Okanagan), Kelowna, BC, Canada

Kedar Rijal
Tribhuvan University, Department of Environmental Science, Kathmandu, Nepal

Manuel Rodriguez
Laval University, École Supérieure d'Aménagement du Territoire, Canada

Rajeev Ruparathna
Department of Civil and Environmental Engineering, University of Windsor, Windsor, ON, Canada

Nurdin Said
Faculty of Chemical Engineering and Natural Resources, University of Malaysia, Pahang, Malaysia

Sana Saleem
School of Engineering, University of British Columbia (Okanagan), Kelowna, BC, Canada

Kalyani Salla
Modern College of Arts, Commerce and Science, Pune, India

Chhatra Mani Sharma
Tribhuvan University, Central Department of Environmental Science, Kathmandu, Nepal

Pushpinder Singh
Lovely Professional University, Department of Mathematics, Chaheru, Phagwara, Punjab, India

Sunil K. Sinha
Virginia Tech, Blacksburg, VA, USA

Devna Singh Thapa
Kathmandu University, Kathmandu, Nepal

Venkata U.K. Vadapalli
School of Engineering, University of British Columbia (Okanagan), Kelowna, BC, Canada

Anmol Vishwakarma
Virginia Tech, Blacksburg, VA, USA

Hao Xu
Virginia Tech, Blacksburg, VA, USA

Jyoti Yadav
Department of Computer Science, Savitribai Phule Pune University, Pune, India

Foreword

Ashok Deshpande, who is one of the pioneers in the application of fuzzy set techniques in health science, along with his co-authors Rehan Sadiq and Boris Faybishenko have edited a ground-breaking volume on fuzzy systems modeling for environmental management and human health risk assessment. Zadeh's original view on the potential role of fuzzy sets was that it could be applied to enable systemic thinking and analysis that had been used so successfully in solving engineering problems to aid in the solution of human-focused problems. The editors of this volume have taken an important step toward achieving this goal. The interconnected areas of environmental management and human health are replete with the kinds of soft and subjective concepts, goals, and criteria that fuzzy set theory was developed to model. The authors contributing to this volume have provided numerous well-thought-out applications that show conclusively that the framework and mathematics of fuzzy sets and fuzzy logic have much to contribute to important problems related to human physical well-being. As the world becomes more and more enamored by the advances in artificial intelligence, it is very refreshing to see the latest technological ideas being applied to such complex, but real, problems as disease risk assessment among solid waste workers, bathing in swimming pools, and the like.

What is notable about the papers in this volume is that, while the authors are modeling processes and concepts not easily modeled by standard methodologies, the actual mathematics being used is not very difficult. This is a hallmark of fuzzy set methods. Because of their inherent simplicity and directness, the use of fuzzy methods enables engineers and environmental professionals to easily explain to the responsible manager what they have done.

I am very happy to have had the opportunity to write this foreword as it has taken me to worlds in which I don't usually travel.

Ronald R. Yager
New York, 2022

Preface

Boris Faybishenko and Rehan Sadiq

This special issue is dedicated to the memory of Professor Lotfi A. Zadeh, the father of fuzzy logic, and to the memory of Dr. Ashok Deshpande, who initiated this volume on using fuzzy logic for environmental problems, and solicited most of the chapters.

Since the pioneering work of Lotfi Zadeh [1], fuzzy logic (FL) analysis has been applied successfully in many scientific and real-life engineering situations, such as electrical, mechanical, civil, chemical, aerospace, agricultural, biomedical, computer, environmental, geological, industrial, as well as by mathematicians, computer software developers and researchers, natural scientists (biology, chemistry, earth science, and physics), medical researchers, social scientists (economics, management, political science, and psychology), public policy analysts, business analysts, and jurists [e.g., 2]. The methods of FL have proven to be extremely useful when dealing with environmental data, because environmental management and risk assessment activities are often based on limited, imprecise, or uncertain observations and numerical simulations. Environmental data are usually characterized by both highly aleatoric and epistemic uncertainties, which are intrinsically present in environmental problems. Aleatory uncertainty indicates that the system (or media) variables or parameters can be characterized by their probability distributions. Aleatory uncertainty is a naturally occurring phenomenon, and is not because of a *lack of information*. Therefore, aleatory uncertainty is irreducible. Examples of aleatory uncertainty are such measured variables as river discharge, precipitation, and traffic data, which can be described by probability distributions if sufficient information is available. If there is no sufficient information, these variables can be described using fuzzy numbers. Epistemic uncertainty indicates the lack of, or limited, knowledge about the real system or its models. This type of uncertainty is reducible, because if we provide more measurements or improve the models, we reduce the level of uncertainty. Moreover, a combination of human judgment, behavior, and emotions plays a central role in the management of environmental systems. FL methods can manipulate both the aleatory and epistemic types of uncertainties, and can be used to handle imprecision or uncertainty by means of using various measures of possibility. It is also important to indicate that FL and fuzzy systems modeling are among the primary components of the scientific field of soft computing, which is an emerging discipline rooted in a group of technologies that aim at exploiting the tolerance for imprecision and uncertainty in achieving solutions to complex problems [3, 4].

This volume is a collection of 16 articles representing a snapshot of the current state of the fields of the application of FL for solving water and air risk assessment, healthcare decision-making, the management of water distribution networks, and the optimization of water treatment and waste management systems. The case studies included in this book are just the tip of the iceberg regarding the potential and usefulness of FL-based techniques and methods in environmental system modeling. We are conscious there is a great need and potential for applying fuzzy set theory and FL-based methods in other facets of environmental management systems.

This volume is expected to serve scientists, researchers, and practitioners who are interested in FL and soft computing theories and their applications in a variety of environmental sciences and engineering problems. The editors and contributors to the current volume made concerted efforts in the development of new applications of FL concepts to solve real-world problems. This volume symbolizes a milestone of transitioning from a purely scientific method to practical environmental engineering and science applications. We believe that the interested reader will get in touch with the authors to further improve the practical applications of FL-based modeling in addressing environmental challenges.

References

1 Zadeh, L.A. (1965). Fuzzy sets. *Inform Control* 8 (2): 338–353.

2 Singh, H., Gupta, M.M., Meitzler, T., Hou, Z.-G., Garg, K.K., Solo, A.M.G., and Zadeh, L.A. (2013). Real-life applications of Fuzzy logic. *Advances in Fuzzy Systems* Article ID 581879. https://doi.org/10.1155/2013/581879.

3 Bouchon-Meunier, B., Yager, R.R., and Zadeh, L.A. (eds.) (1995). *Advances in Fuzzy Systems: Applications and Theory: Volume 4: Fuzzy Logic and Soft Computing*. https://doi.org/10.1142/2829.

4 Sadiq, R. and Tesfamariam, S. (2009). Environmental decision-making under uncertainty using intuitionistic fuzzy analytic hierarchy process (IF-AHP). *Stoch Environ Res Risk Assess* 23: 75–91. https://doi.org/10.1007/s00477-007-0197-z.

Introduction

Boris Faybishenko[1], Rehan Sadiq[2], and Ashok Deshpande[3]

[1] *Lawrence Berkeley National Laboratory, Earth and Environmental Sciences Area, Energy Geosciences Division, Berkeley, California, USA*
[2] *School of Engineering, University of British Columbia (Okanagan), Kelowna, BC, Canada*
[3] *Berkeley Initiative in Soft Computing (BISC), Special Interest Group Environmental Management Systems (EMS), University of California, Berkeley, California, USA; and College of Engineering, Pune, India*

Environmental pollution has been the greatest problem facing humanity for many years, and it is the leading cause of morbidity and mortality. Humankind's activities, such as urbanization, industrialization, mining, exploration, and organic and radioactive contamination, are major reasons of global environmental pollution in both developed and developing nations [e.g. 1, 2]. These problems call for immediate action on initiating pollution abatement strategies. However, predictions and environmental decision-making are often limited due to the uncertainty, vagueness, or ambiguity of observational data [e.g. 3–7]. The methods of statistical mechanics have embraced two-valued logic-based probability theory, wherein a random variable is used as the basis of probability computations. However, the standard probability theory is not designed to deal with imprecise probabilities that pervade real-world uncertainties. Fuzzy set theory is an alternative approach to modeling these uncertainties.

Since the publication of the first scientific paper on fuzzy sets by Lotfi A. Zadeh [8], fuzzy systems modeling has been applied successfully in many scientific and technological fields and has proven useful when dealing with environmental problems. The theory of fuzzy sets is based on the notion of relative graded membership, as inspired by the processes of human perception and cognition. Fuzzy logic (FL) can deal with information arising from computational perception and cognition that is uncertain, imprecise, vague, partially true, or without sharp boundaries. FL allows for the inclusion of vague human assessments in computing problems. Also, it provides an effective means for conflict resolution of multiple criteria and for the better assessment of options [4, 9, 10].

The main idea of the application of FL is the presentation of all system parameters and variables as a matter of degree or partial belief, which will allow one to produce acceptable, definitive outputs in response to incomplete, ambiguous, distorted, or imprecise inputs. Despite FL being successfully applied in various sectors, new applications are constantly being found for FL methods and fuzzy sets.

It is a deep-seated tradition in science to employ the conceptual structure of bivalent logic and probability theory as a basis for the formulation of concepts. What is widely unrecognized is that, in reality, most concepts are fuzzy in nature rather than bivalent, and thus it is generally not possible to accurately formulate most real-world problems within the conceptual structure of bivalent logic and probability theory. Thus, the techniques based on FL are more applicable for environmental system modeling wherein an expert's perception takes center stage of the modeling process.

Fuzzy Systems Modeling in Environmental and Health Risk Assessment, First Edition. Edited by Boris Faybishenko, Rehan Sadiq, and Ashok Deshpande.

The fuzzy set of a concept is defined by a distribution function of the degree of belief (DoB) in a qualitative parameter (the concept) over a range of variations in a quantitative or less-qualitative parameter (the scale). The concept may be determined by different scaling parameters, and each parameter on its own is not necessarily unique. So, the form of a fuzzy set depends almost entirely on the scale selected. In the environmental field, regulators, health authorities, epidemiologists, politicians, environmentalists, engineers, and the general public often define a concept, such as contamination, in different ways. The only term which is more or less unequivocally understood by all interested groups is the final risk often associated with dollar value. The proper definition and scaling of fuzzy sets can provide a common language through which experts from different disciplines can communicate during the entire process of risk assessment. Uncertainty in the input information propagates (but is neither magnified nor dampened) in a fuzzy way, so that the output remains fuzzy and can then be translated into either quantitative risk values or qualitative linguistic expressions. In this book, we examine a real risk assessment case scenario using fuzzy arithmetic.

There are several emerging and complex environmental issues wherein fuzzy sets and FL can be applied. FL-based techniques are well suited for problem-solving in a number of environmental study areas, including but not limited to climate change adaptation, water resources management, air quality management, watershed management, wildlife management, flood control and management, water quality management, environmental risk assessment, wildfire control and management, emerging pollutants control, and socio-environmental issues. The complex environmental issues in these areas can be approached, analyzed, managed, and resolved effectively using FL-based methods. The developed methods are able to measure, evaluate, and analyze complex issues characterized by uncertain, imprecise, ambiguous, and subjective features.

It is important to note that life cycle assessment (LCA) and environmental impact assessment (EIA) are commonly used for evaluating environmental impacts. LCA is used to quantify environmental impacts throughout a product's, system's, or service's entire life. It has also become a measure for evaluating the environmental performance of management strategies and industrial products. Similarly, EIA is applied to evaluate the environmental impacts of any development project before it is implemented, such as a road construction project. Both tools involve extensive modeling. Environmental modeling using imprecise or vague data may consequently lead to inaccurate or uncertain predictions. Expert knowledge is often used to address the issue of a lack of quantitative data. However, expert knowledge is intrinsically associated with uncertainties due to the vagueness and ambiguity of human thoughts. FL can be used to reduce such uncertainties. The modeling of environmental systems is primarily conducted using the principles of either Newtonian mechanics based on closed form solutions with very little uncertainty or statistical methods, such as various optimization techniques, which are based on bivalent probability theory. Moreover, quantitative environmental data can be converted to fuzzy data using fuzzy inference systems (FIS). The hybrid methodology combining FL and LCA or EIA can be used to reduce data-related uncertainty in EIA.

In today's world, data are being generated at an unprecedented speed. This will continue to increase, too. The use of data-driven methods – such as machine learning, artificial intelligence (e.g. artificial neural networks), data mining, case-based reasoning, pattern recognition – is becoming prevalent in systematic decision-making, especially in making "smart" decisions. Such decision-making systems often use automated data collection using sensors and employ the Internet of Things (IoT) for rapid decision-making. The accuracy of such data-driven methods can be improved by incorporating expert knowledge and other qualitative data, and fuzzy set theory and FL can be a powerful enhancement to the intelligent models [11]. The complexity, vagueness, and ambiguity in data can be properly approximated using fuzzy concepts. For example, an adaptive neurofuzzy inference system (ANFIS) integrates both neural networks and the fuzzy inference concept, which can capture the benefits of both techniques. The FIS of ANFIS uses a group of *if–then rules* that have the capability to approximate highly nonlinear and complex relationships that are indiscernible by conventional mathematical techniques. FL principles enable the intelligent/smart models to conduct approximate reasoning like human brains.

The IoT assists in rapid and automated decision-making even in real time using sensor-generated data transmitted via the Internet. The biggest challenge in such a situation is data reliability. Sensors often generate faulty data, which may adversely affect the accuracy of modeling outcomes. However, the outliers flagged based on the conventional definition of an outlier can be correct data points in a highly complex and irregular system. On the other hand, the data points within a conventional data range can be faulty, too, due to instrumental error. In such situations, FL principles incorporating expert knowledge can be very useful in identifying faulty data. The fuzzy if–then rules are very practical and easier to comprehend and use. Moreover, the use of fuzzy concepts in neural networks reduces computational complexity, which further reduces the computational time in modeling complex systems. The FL-enhanced neural networks also require less cloud space for data storage, ultimately enhancing the efficacy of the developed models and decision-making systems. FL theory and methods can benefit the application of neural networks and data mining techniques. For example, Talpur et al. [12] provide a review of deep neural networks (DNNs) and reasoning aptitude from FIS. This study revealed that the proposed deep neural fuzzy systems' (DNFS) architectures performed better than nonfuzzy models, with an overall accuracy of 81.4%. The novel hybridization of DNN and FL was an effective way to reduce uncertainty using fuzzy if–then rules. The study also showed that the DNFS networks presented in the literature have integrated DNN with typical FIS, although satisfactory results can be obtained using a new generation of FIS, termed fractional FIS (FFIS), and the Mamdani complex FIS (M-CFIS). Dynamic neural networks are suggested in the replacement of static DNNs to facilitate dynamic learning for solving highly nonlinear problems. Thus, deep learning methods along with FL techniques provide better interpretability of the network, while solving complex real-world problems using fuzzy if–then inference rules. The introduction of fuzzy layers to the deep learning architecture can help exploit the powerful aggregation properties expressed through fuzzy methodologies, and can be used to represent fuzzified intermediate, or hidden, layers. For example, FL can be used to automatically cluster information into categories which improve performance by decreasing sensitivity to noise and outliers. For example, Prince et al. [13] propose introducing fuzzy layers into the deep learning architecture to exploit the powerful aggregation properties expressed through fuzzy methodologies, such as the Choquet and Sugeno fuzzy integrals.

The following is a summary of the individual chapters included in the current volume.

1 Theoretical Considerations

In Chapter 1, "Fuzzy Logic and Fuzzy Set Theory: Overview of Mathematical Preliminaries," Yadav provides basic definitions of FL and fuzzy set theory that are applicable to environmental management and risk assessment activities, taking into account that observations and results of numerical simulations are commonly uncertain. In particular, the author describes the general ideas of FL and fuzzy set theory, arithmetic operation rules, rules of computing with words (CW), and methods of fuzzy relational calculus: fuzzy equivalence and fuzzy tolerance relation, similarity measures or value assignments, cosine amplitude method, max–min method, monotone measures, Dempster–Shafer theory, fuzzy C-means (FCM) clustering, and basic concepts of the Bellman–Zadeh approach.

2 Fuzzy Logic for Environmental Risk Assessment

In Chapter 2, "Fuzzy-based Integrated Risk Assessment of Methylmercury in Lake Phewa, Nepal," Chhipi-Shrestha et al. study the distribution of mercury (Hg) entering a lake via processes of air deposition and run-off. In lake water, Hg is converted into highly neurotoxic methylmercury (MeHg), primarily by bacteria, and the MeHg is then bioaccumulated and biomagnified along the food chain. The data and models used in the risk assessment may be associated with uncertainty that can be incorporated into the analysis using a fuzzy approach. The objective of this research was to assess the integrated human health and ecological risks of MeHg in Lake Phewa, Nepal using a fuzzy approach. The measured MeHg in fish tissue was used to estimate the concentration of MeHg in the water by applying a bioaccumulation factor. A questionnaire survey was conducted to estimate fish consumption rates among different occupations: fishermen, local people, hotel owners, government staff, army/police, and others (visitors). An integrated risk assessment framework proposed by the World Health Organization (WHO) was applied to estimate the health risk to humans and the ecological risk to fishes. The results showed that fish consumption contributed approximately 90% or higher to overall human health risk and the remaining risk was attributed to rice consumption. The higher health risk was primarily due to a very high fish consumption rate. Moreover, the ecological risk to fish was within acceptable levels. The risk estimate using the fuzzy approach was used to approximate uncertainty and to build confidence in decisions for risk assessors.

In Chapter 3, "A Fuzzy Approach to Analyze Data Uncertainty in the Life Cycle Assessment of a Drinking Water System: A Case Study of the City of Penticton (CA)" Rebello et al. perform an LCA and evaluate the effect of the uncertainty in input data on the final environmental impacts of water treatment and distribution systems in the City of Penticton, Canada. Data uncertainty was evaluated using a fuzzy analysis coupled with the weighted product method (WPM) to estimate life cycle impact. Additionally, a comparison between the life cycle impact and ReCiPe endpoint single score is presented to understand how sensitive those indicators are to data uncertainties. The results indicate that the most sensible life cycle stage is water treatment and distribution use, with variations from 3 to 15%, mainly resulting from the chemical analysis in the usage phase. Additionally, the WPM methodology presented a deviation from 60 to 378% when considering the individual life cycle stages. The total LCA presented a higher robustness to the changes, with 1–8% variations in the midpoint categories; however, the WPM disparities ranged from 80 to 410% in the midpoint categories.

Ever-increasing pollution levels due to rapid urbanization and industrialization, especially in many developing countries with a minimal focus on adequate pollution abatement strategies, have resulted in widespread damage to the natural environment. It is, therefore, important to classify environmental quality. The practice en vogue for classifying air or water quality for variety of usage is by computing the air quality index (AQI) or water quality index (WQI). Why compute a numeric AQI or WQI, and then describe air/water quality linguistically? The human brain does not compute numbers. Why not describe air or water quality, for the defined usage, straightway in linguistic terms with some linguistic degree of certainty attached to each linguistic description?

Chapter 4, "Environmental Quality Assessment Using Fuzzy Logic," by Yadav and Rijal, presents two research case studies. The first one is on the fuzzy air quality description in the Pimpri-Chinchwad Municipal Corporation (PCMC) monitoring location, and the other one relates to the linguistic classification of water quality with a degree of certainty in the PCMC area, India.

In Chapter 5, "Assessing Spatiotemporal Water Quality Variations in Polluted Rivers with Uncertain Flow Variations: An Application of Triangular Type-2 Fuzzy Sets," Haider et al. address an important problem of dealing with high pollution loads in rivers by large cities. Rivers also experience extreme spatial and temporal flow variations due to the overexploitation of freshwater resources, poor management practices, and climate change impacts. Water quality data are usually scarce in developing countries due to the absence of planned periodic monitoring programs. Consequently, water quality parameters (i.e. biochemical oxygen demand, dissolved oxygen, unionized ammonia, and coliforms) are significantly dependent on river discharge. The morphology of rivers also influences water quality parameters because of changes in the water residence time. For a river, a WQI is a robust assessment tool that indicates the overall water quality based on the water source, such as natural freshwater, single point source loads, and cumulative loads from several outfalls along the river length. Other factors that are taken into account in the derivation of the WQI are seasonal variations in river flow, hydrodynamics, and pollution loads. Type-2 fuzzy sets, an extension of ordinary fuzzy sets, directly model the uncertainties by providing an additional degree of freedom. Type-2 fuzzy sets improve the specific kind of interface that exists due to increasing uncertainties associated with imprecision in knowledge and vagueness in information due to limited water quality data. Using the triangular type-2 fuzzy sets approach, the WQI developed in the present work effectively overcomes these uncertainties and has proved to be a more reliable water quality assessment measure for highly polluted rivers with extreme flow variations.

In Chapter 6, "Optimal Ranking of Air Quality Monitoring Stations and Thermal Power Plants in a Fuzzy Environment," Yadav and Salla apply the Zadeh–Deshpande (ZD) FL-based formalism for linguistic description of the air quality criteria of pollutants, and the Bellman–Zadeh method for optimal ranking on the basis of risk for thermal power plants. The air quality classification obtained for 12 cities using the ZD formalism is described linguistically using the concept of degree of certainty (DC). The air quality assessment is provided for 26 thermal power plants in India.

3 Fuzzy Logic Application in Healthcare Decision-making

Environmental professionals and medical practitioners have made significant contributions using statistical methods to analyze medical data and associated air and water pollution parameters in order to determine the cause–effect relationship between air/water-borne diseases and air/water-pollution parameters. Multiple studies have been carried out and are primarily centered on exposure to exhaust pollution, and epidemiological studies associated with various respiratory diseases. Chapter 7, "Evaluation of Health Effects Due to Environmental Pollution Based on Belief and Possibility," by Yadav and Rijal, includes two case studies that reveal a strong correlation between polluted-water and water-borne diseases, as well as polluted-air and air-borne diseases. The perceptions of experienced medical practitioners are modeled to assess the collective DoB for all possible combinations of air/water-borne diseases, and evidence theory and fuzzy relational calculus without the need of having a sizeable parametric data are also used. The objective of this chapter is, therefore, to critically evaluate the vast armamentarium available on these facets of air- and water-pollution studies.

Chapter 8, "Respiratory Disease Risk Assessment Among Solid Waste Workers Using Fuzzy Rule Based System Approach," by Jariwala and Christian, deals with the prediction of respiratory disease, which is considered one of the major causes of mortality. Because the symptoms of respiratory disease require a long time to manifest, an early evaluation of an individual risk prior to a medical diagnosis or a laboratory test is essential. Medical science considers the criteria of odds ratio and relative risk for the study of disease occurrence, wherein comparison is made between exposed and nonexposed groups. Reportedly, solid waste workers are at high risk of developing respiratory diseases. The present study aims at determining an individual worker's risk of respiratory diseases under selected parameters. The factors responsible for the development of respiratory diseases were identified and measured in a group of solid waste workers from the city of Surat in India. An individual worker's risk without any test or diagnosis is evaluated on a scale from 0 to 1 by means of applying a fuzzy rule based system (FRBS). The calculated risk value ranges from 0.08 to 0.92. The FRBS modeling results are validated based on the comparison with the pulmonary function test.

Chapter 9, "Risk Analysis for Indoor Swimming Pools: A Fuzzy-based Approach," by Saleem et al., addresses the risk analysis in indoor swimming pools using a fuzzy-based approach to incorporate the uncertainty. The likelihood and consequences are defined using fuzzy numbers to identify the fuzzy risk as the product of both likelihood and consequence. The resultant defuzzified risk will be categorized on a linguistic scale from *very low* to *low*, *medium*, *high*, and *very high* to estimate the formation risk level of the different pool types. The model created is based on FL, which gives a risk assessor the ability to solve complex problems plagued with uncertainty and vagueness. In this chapter, formation risk is studied through the notions of likelihood and consequences, for which membership functions are established. The method adopted allows for a realistic preliminary assessment of the risk of formation in an indoor swimming pool. This method can be used by swimming pool facility managers to evaluate the risk in all pool types to ensure that safety measurements are satisfactorily based on the given data. This method can be used as a preliminary risk assessment tool, which can highlight critical situations and the need for more in-depth and complete analysis.

4 Fuzzy Logic Applied to the Management of Water Distribution Networks

In Chapter 10, "Fuzzy Parameters in the Analysis of Water Distribution Networks," Gupta and Ormsbee present the results of a fuzzy analysis of water distribution networks (WDNs) to determine how the uncertainties in independent or basic parameters (such as nodal demands and pipe roughness coefficients) are dependent on such parameters as pipe flows, pipe velocities, and available pressure heads. The chapter shows how fuzzy analysis can be applied to identify the vulnerable zone in WDNs, as well as how to conduct a reliability-based design of WDNs under the uncertainty of various parameters. Several methodologies have been suggested for fuzzy analysis. These methodologies are categorized as: (i) optimization-based methodologies and (ii) analysis-based methodologies. The membership functions obtained by both optimization and analysis-based methodologies are compared using two types of networks. Methods for obtaining the approximate fuzzy membership functions are also discussed.

In Chapter 11, "Selection of Wastewater Treatment for Small Canadian Communities: An Integrated Fuzzy AHP and Grey Relational Analysis Approach," Hu et al. describe an integrated fuzzy analytic hierarchy process (F-AHP) and grey relational analysis (GRA), which can be used to facilitate the selection of appropriate wastewater treatment (WWT) alternatives for small communities. Seven commonly used WWT technology alternatives were assessed for a hypothetical small community in Canada. The assessment was based on the holistic evaluation of technical, economic, social, and environmental criteria, with each criterion composed of several subindices. The weights of criteria and subindices were determined using F-AHP to address nonprobabilistic uncertainties, such as vagueness and ambiguities in human thoughts resulting from the subjective weighting process. The weighted criteria were then aggregated and, based on the aggregation results, alternatives were ranked using GRA. The results from the integrated approach show that constructed wetland, stabilized pond, and extended aeration lagoon are the top three appropriate WWT technologies for small Canadian communities. It was also found that the fuzzy-based approach and the non-fuzzy-based approach generated different rankings for the alternatives, indicating that fuzzy uncertainties could affect the decision-making process.

Vishwakarma and Sinha, in Chapter 12, "Fuzzy Logic Applications for Water Pipeline Risk Analysis," state that civil infrastructures are considered critical systems providing economic, social, and environmental benefits for society. However, given the issue of aging water infrastructures, the cost of maintaining a sustainable level of service for a growing population is increasing. Consequently, there is an urgent need to advance the processes that drive current water pipeline renewal (repair, rehabilitation, and replacement) decision support systems. Fortunately, with the developments in computational techniques and data collection through advanced sensors, it is now possible to support prioritization of critical water pipelines with data and knowledge. This chapter explains a state-of-the-art approach to assessing the consequence of failure (CoF) built on a comprehensive list of parameters and utilizing a novel hierarchical FIS to support better criticality-based renewal decisions for water pipelines.

This chapter can also help water utilities to build data-driven decision support systems within their asset management programs. All data and models shown in this chapter are part of the PIPE*i*D or PIPEline Infrastructure Database. This platform is envisioned as a national database platform for advanced asset management addressing the major management levels (e.g. strategic, tactical, and operational) that will assist water utilities of all sizes in sustaining targeted levels of service with acceptable risk. It will also provide secure access to the aggregated data, models, and tools that will enable the synthesis, analysis, query, and visualization of the data for decision support.

In Chapter 13, "Fuzzy Logic Applications for Water Pipeline Performance Analysis," Xu and Sinha present the results of studies to improve decision-making related to the problems of aging drinking water pipes. Multiple statistical models have been developed to support advanced asset management. Many models have been prepared for specific materials or conditions, and fail to be used holistically for a water distribution system. To address this issue, a comprehensive understanding of performance parameters for different pipe materials is needed. PIPE*i*D was created to help in understanding the critical drinking water pipeline infrastructure. It is envisioned as a national database platform for advanced asset management addressing the major management levels, including strategic, tactical, and operational, that will assist water utilities of all sizes in sustaining targeted levels of service with acceptable risk. It will also provide secure access to aggregated data, models, and tools that will enable the synthesis, analysis, querying, and visualization of the data for decision support. With the help of PIPE*i*D, it becomes possible to incorporate the many previous models and develop a systematic and comprehensive approach to the statistical modeling of drinking water pipeline performance prediction. The research provides knowledge and insights about water pipe condition and performance. The research questions are related to the core tasks (i.e. assessing the pipe performance index/rating and developing pipe performance curves). This chapter explains the background knowledge needed for the development of the FL-based performance index for water pipelines.

5 Using Fuzzy Logic for the Optimization of Water Treatment and Waste Management

In Chapter 14, "Developing a Fuzzy-based Model for Regional Waste Management," Karunathilake et al. describe a decision-making method for an integrated waste management plan that becomes complicated due to inherent variabilities and uncertainties associated with the waste inputs, processes, and the external environment, as well as the lack of data and high human involvement. It is shown that fuzzy multicriteria decision-making techniques can be used to develop a robust waste management planning model to customize regional needs and conditions. Different waste treatment modes, such as landfilling, composting, material recovery and recycling, and waste-to-energy conversion can be used to process municipal solid waste in a regional waste management strategy. This chapter uses the fuzzy technique for order of preference by similarity to ideal solution (TOPSIS) method to rank and identify the best-performing technology in a particular mode of treatment, and a fuzzy optimization model is developed to identify the best combination of technologies for a region.

In Chapter 15, "Development of a Fuzzy-based Risk Assessment Model for Process Engineering," Ouache et al. state that ample evidence is available on the impacts of engineered systems on ecological and social risks, and there is an urgent need to assess the risks involved in operational management. The risk matrix has been commonly used to prioritize the risks associated with the prevention and remediation of environmental damage. Despite the popularity of using the risk matrix, data uncertainty is the main challenge associated with the risk matrix. The published literature identifies imprecision of a risk level and absence of data as added challenges. A FL-based risk matrix model (RMM) has been developed to overcome these challenges. The proposed RMM adopts a unique approach, integrating frequency of the risk consequence with the impact of the corresponding consequence. The frequency of risk consequence is based on four factors: (i) the frequency of risk events, (ii) exposure to risk events, (iii) the probability of failure on demand of the safeguards, and (iv) the vulnerability of safeguards. The consequence impacts involve the impact on humans, the environment, properties, and the reputation of the industry. A Simulink-MATLAB model was developed to facilitate the RMM computation. A case study is used to demonstrate the application of this model by using a case study of a reboiler oven in the petroleum industry. The study revealed that the results of the new RMM can be more reliable than traditional risk matrices.

In Chapter 16, "Application of Fuzzy Theory to Investigate the Effect of Innovation Power in the Emergence of an Advanced Reusable Packaging System," Böröcz et al. present a novel technique to analyze the role of subjective factors, such as innovation in the economy, which influence the design of reusable packaging systems in a given industrial region. In modern supply chains, companies and packaging engineers have to make decisions on determining adequate packaging with optimal waste. The decision-making process is usually based on available data and information, taking into account the need to minimize the environmental impact. Fuzzy systems analysis is used to show that a willingness to innovate is an indispensable requirement of the appearance of advanced packaging, but most of the time it depends on the synergic effect of local production factors and regional peculiarities.

The case studies presented in this volume may serve as models for the application of soft computer modeling of various types of pollution – air, water, soil, noise, radioactive, light, and thermal – which are the primary causes that affect our environment. All these types of pollution are interlinked and influence each other. However, discussing each of these issues in depth is beyond the scope of this volume.

References

1 Linkov, I., Bridges, T.S., Jamil, S., Kiker, G.A., Seager, T.P., and Varghese, A. (2004). *Multi-criteria Decision Analysis: Framework for Applications in Remedial Planning for Contaminated Sites*, 15–54. Amsterdam: Kluwer.

2 Ukaogo, P.O., Ewuzie, U., and Onwuka, C.V. (2020). Environmental pollution: causes, effects, and the remedies. In: *Microorganisms for Sustainable Environment and Health* (ed. P. Chowdhary et al.). Elsevier.

3 Anaokar, G.S. and Khambete, A.K. (2016). Application of fuzzy logic in environmental engineering for determination of air quality index. *International Journal of Engineering Technology, Management and Applied Sciences* 4 (2). 109–116.

4 Boclin, A. and Mello, R. (2006). A decision support method for environmental impact assessment using a fuzzy logic approach. *Ecological Economics* 58: 170–181. doi:10.1016/j.ecolecon.2005.06.007.

5 Deshpande, A.W. and Raje, D.V. (2003). Fuzzy logic applications to environment management systems: case studies. doi:10.1109/INDIN.2003.1300356. IEEE.

6 Deshpande, A.W., Raje, D.V., and Khanna, P. (1996). Fuzzy description of river water quality. *Paper for International Conference-EUFIT*.

7 Ghomshei, M.M. and Meech, J.A. (2000). Application of fuzzy logic in environmental risk assessment: some thoughts on fuzzy sets. *Cybernetics and Systems: An International Journal* 31: 317–332.

8 Zadeh, L.A. (1965). Fuzzy sets. *Information and Control* 8 (3): 338–353.

9 Vairal, K.L., Kulkarni, S.D., and Basotia, V. (2020). Fuzzy logic and its applications in some area: a mini review. *Journal of Engineering Sciences* 11 (8). 85–96.

10 Zabeo, A., Semenzin, E., Torresan, S., Gottardo, S., Pizzo, L. et al. (2010). Fuzzy logic based IEDSSs for environmental risk assessment and management. *International Congress on Environmental Modelling and Software* 209. https://scholarsarchive.byu.edu/iemssconference/2010/all/209 (accessed 4 October 2022).

11 Sànchez-Marrè, M., Gibert, K., Cabello, A., and Sem, F. (2012). A methodology for the characterization of intelligent environmental decision support systems. In: *International Congress on Environmental Modelling and Software Managing Resources of a Limited Planet, Sixth Biennial Meeting, Leipzig* (ed. G.R. Seppelt, A.A. Voinov, S. Lange, and D. Bankamp). https://scholarsarchive.byu.edu/cgi/viewcontent.cgi?article=1756&context=iemssconference (accessed 4 October 2022).

12 Talpur, N., Abdulkadir, S.J., Alhussian, H. et al. (2022). A comprehensive review of deep neuro-fuzzy system architectures and their optimization methods. *Neural Computing & Applications*. doi: 10.1007/s00521-021-06807-9.

13 Prince, S.R., Price, S.R., and Anderson, D.T. (2019). Introducing fuzzy layers for deep learning. *IEEE International Conference on Fuzzy Systems (FUZZ-IEEE)*, New Orleans, LA, USA, 1–6, doi:10.1109/FUZZ-IEEE.2019.8858790.

Part I

Theoretical Considerations

1

Fuzzy Logic and Fuzzy Set Theory: Overview of Mathematical Preliminaries

Jyoti Yadav

Department of Computer Science, Savitribai Phule Pune University, Pune, India

1.1 Introduction

The modeling of epistemic knowledge is a necessity for systems dealing with some sort of artificial reasoning. There exist several formalisms that mathematically model someone's degrees of belief. The theory of evidence, or Dempster–Shafer theory (DST), provides a method for combining evidences from different sources without prior knowledge of their distributions. In this model, it is possible to assign probability values to sets of possibilities rather than to single events only, and it is not needed to divide all the probability values among the events, once the remaining probability is assigned to the environment and not to the remaining events, thus modeling more naturally certain classes of problems. Uncertainty in the form of vagueness, imprecision, and ignorance is captured with the help of monotone measures. CW is a methodology for reasoning, computing, and decision-making with information described in natural language, which is basically a system for description of perception. Conventional systems of computation do not have the capability to deal with linguistic valuations.

1.2 Fuzzy Logic and Fuzzy Set Theory

1.2.1 Basic Definitions

Fuzzy logic: FL is a branch of fuzzy set theory. It is a multivalued logic with degree of truth values (or membership values) that can be any real numbers between 0 and 1 for variables. FL is different from the traditional binary logic, which only has two-valued logic: completely true (1) or completely false (0). Furthermore, FL allows for using linguistic variables to present the membership values to imprecise concepts such as *low*, *average*, and *high*.

Fuzzy set: If X is a universal set and A is a subset of X, then the ordered paired set $\tilde{A} = \{(x, \mu_{\tilde{A}}(x)) : x \in X\}$ is defined as a fuzzy set in X, where $\mu_{\tilde{A}} : X \to [0,1]$ is a membership function and x is a variable in X.

Degree of membership: The membership function $\mu_{\tilde{A}}$ assigns a real number in interval [0, 1] for all x in X, known as degree of membership $\mu_{\tilde{A}}(x)$. $\mu_{\tilde{A}}(x)$ denotes the degree of association of x to $\tilde{A}$. The higher value of $\mu_{\tilde{A}}(x)$, the higher degree of association of x to $\tilde{A}$.

Fuzzy Systems Modeling in Environmental and Health Risk Assessment, First Edition. Edited by Boris Faybishenko, Rehan Sadiq, and Ashok Deshpande.

α-cut: If $\tilde{A}$ is a fuzzy set in X and $\alpha \in [0,1]$ is a real number, then the α-cut or parametric form of fuzzy set $\tilde{A}$ is defined as $A^{\alpha} = \{x \in X, \mu_{\tilde{A}}(x) \geq \alpha\}$.

Normalized fuzzy set: A fuzzy set $\tilde{A}$ is said to be a normalized fuzzy set if the largest membership grade (i.e. $\sup_{x \in X}\{\mu_{\tilde{A}}(x)\}$) is equal to 1.

Convex fuzzy set: A fuzzy set $\tilde{A}$ is said to be a convex fuzzy set if and only if $\mu_{\tilde{A}}(\alpha x_1 + (1-\alpha)x_2) \geq \text{Minimum}\{\mu_{\tilde{A}}(x_1), \mu_{\tilde{A}}(x_2)\}$, $\forall\, x_1, x_2 \in X$ and $\alpha \in [0,1]$.

Fuzzy number: A convex and normalized fuzzy set $\tilde{A}$ is said to be a fuzzy number if and only if the membership function $\mu_{\tilde{A}}$ is piecewise continuous in X.

Non-negative fuzzy number: A fuzzy number is called a non-negative fuzzy number if and only if the membership value is equal to zero (i.e. $\mu_{\tilde{A}}(x) = 0$), $\forall\, x < 0$.

Trapezoidal fuzzy number: A fuzzy number $\tilde{A} = (a,b,c,d)$, which is defined on the universal set of real numbers $\mathbb{R}$, is said to be a trapezoidal fuzzy number if its membership function, $\mu_{\tilde{A}}(x)$, is defined as

$$\mu_{\tilde{A}}(x) = \begin{cases} \dfrac{(x-a)}{(b-a)}, & \text{for } a \leq x \leq b \\ 1, & \text{for } b \leq x \leq c \\ \dfrac{(x-d)}{(c-d)}, & \text{for } c \leq x \leq d \\ 0, & \text{otherwise} \end{cases}$$

A trapezoidal fuzzy number $\tilde{A} = (a,b,c,d)$ can be regarded as a triangular fuzzy number if $b = c$ and will be denoted by $\tilde{A} = (a,b,b,d) = (a,b,d)$ or $\tilde{A} = (a,c,c,d) = (a,c,d)$.

Non-negative trapezoidal fuzzy number: A trapezoidal fuzzy number $\tilde{A} = (a,b,c,d)$ is said to be a non-negative fuzzy number if and only if $a \geq 0$.

Equal trapezoidal fuzzy numbers: Two trapezoidal fuzzy numbers $A^{\alpha} = [a + (b-a)\alpha,\, d - (d-c)\alpha], 0 \leq \alpha \leq 1$ $\tilde{A} = (a_1,a_2,a_3,a_4)$ and $\tilde{B} = (b_1,b_2,b_3,b_4)$ are said to be equal if and only if $a_1 = b_1, a_2 = b_2, a_3 = b_3$ and$a_4 = b_4$.

1.2.2 Arithmetic Operation Rules

Let $\tilde{A} = (a_1,a_2,a_3,a_4)$ and $\tilde{B} = (b_1,b_2,b_3,b_4)$ be two non-negative trapezoidal fuzzy numbers, then the arithmetic operation rules can be defined as follows:

1) Addition

$$\tilde{A} \oplus \tilde{B} = (a_1 + b_1, a_2 + b_2, a_3 + b_3, a_4 + b_4)$$

2) Subtraction

$$\tilde{B} = (a_1 - b_4, a_2 - b_3, a_3 - b_2, a_4 - b_1)$$

3) Multiplication

$$\tilde{A} \otimes \tilde{B} = \left(a_1 \times b_1, a_2 \times b_2, a_3 \times b_3, a_4 \times b_4\right)$$

4) Scalar multiplication

$$\lambda\tilde{A} = \begin{Bmatrix} (\lambda a_1, \lambda a_2, \lambda a_3, \lambda a_4), & \text{if } \lambda \geq 0 \\ (\lambda a_4, \lambda a_3, \lambda a_2, \lambda a_1), & \text{if } \lambda \leq 0 \end{Bmatrix}$$

5) Inverse

$$\tilde{A}^{-1}=(a_1,a_2,a_3,a_4)^{-1}=\left(\frac{1}{a_4},\frac{1}{a_3},\frac{1}{a_2},\frac{1}{a_1}\right)$$

6) Division

$$\frac{\tilde{A}}{\tilde{B}}=\left(\frac{a_1}{b_4},\frac{a_2}{b_3},\frac{a_3}{b_2},\frac{a_4}{b_1}\right)$$

1.3 Computing with Words (CW)

Words mean different things to different people, and so are uncertain. We, therefore, need a fuzzy set model for a word that has the potential to capture their uncertainties. Computing means manipulation of numbers and symbols. In contrast, computing with words, or CW for short, is a methodology in which the objects of computation are words and propositions drawn from a natural language (e.g. *small*, *large*, *far*, *heavy*, *not very likely*, etc.). CW is inspired by the remarkable human capability to perform a wide variety of physical and mental tasks without any measurements and any computations. Familiar examples of such tasks are parking a car, driving in heavy traffic, riding a bicycle, understanding speech, and summarizing a story. Underlying this remarkable capability is the brain's crucial ability to manipulate perceptions – perceptions of distance, size, weight, color, speed, time, direction, force, number, truth, likelihood, and other characteristics of physical and mental objects. The manipulation of perceptions plays a key role in human recognition, decision, and execution processes. As a methodology, CW provides a foundation for a computational theory of perceptions – a theory which may have an important bearing on how humans make, and machines might make, perception-based rational decisions in an environment of imprecision, uncertainty, and partial truth. There are two major imperatives for CW. First, CW is a necessity when the available information is too imprecise to justify the use of numbers and, second, when there is a tolerance for imprecision which can be exploited to achieve tractability, robustness, low solution cost, and better rapport with reality. When perceptions are described in words, manipulation of perceptions is reduced to CW.

Figure 1.1 depicts the principal levels of complexity in CW. Linguistic natural language (LNL) deals with level 1 complexities and natural language (NL) deals with level 2 complexities. The intersection of LNL and NL is the fuzzy relation. FL is level 1 complexity and FL needs CW, but CW may or may not need FL.

1.3.1 Fuzzy Relational Calculus

A fuzzy relation R is a mapping from the Cartesian space $X \times Y$ to the interval [0, 1], where the strength of the mapping is expressed by the membership function of the relation for ordered pairs from the two universes, or $\mu_R(x,y)$. Relations can be used to represent *similarity*. Suppose R is a fuzzy relation on the Cartesian space $X \times Y$, S is a fuzzy relation on $Y \times Z$, and T is a fuzzy relation on $X \times Z$; then fuzzy max–min composition is defined in terms of the function-theoretic notation in the following manner

$$\mu_T(x,z)=\vee_{y\in Y}(\mu_R(x,y)\wedge\mu_S(y,z)) \tag{1.1}$$

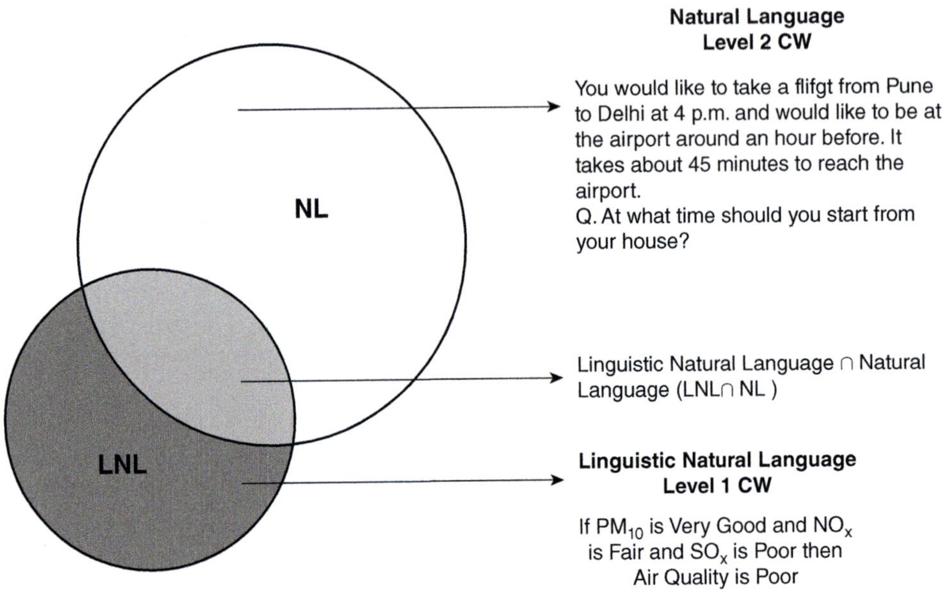

Figure 1.1 Computing with words: perception-based modeling.

1.3.1.1 Fuzzy Equivalence and Fuzzy Tolerance Relation

A fuzzy relation R on a single universe X is also a relation from X to X. It is a fuzzy equivalence relation or a similarity relation if the following properties for matrix relations define it.

Reflexivity

$$\mu_R(x_i, x_i) = 1 \tag{1.2}$$

$$\text{Symmetry}\ \mu_R(x_i, x_j) = \mu_R(x_j, x_i) \tag{1.3}$$

Transitivity

$$\mu_R(x_i, x_j) = \lambda_1 \ \&\ \mu_R(x_j, x_k) = \lambda_2 \rightarrow \mu_R(x_i, x_k) = \lambda \text{ where } \lambda \geq \min[\lambda_1, \lambda_2] \tag{1.4}$$

A fuzzy tolerance relation was observed while examining the similarity matrix. In order to defuzzify the fuzzy tolerance relation to obtain ultimately the classical relation, it is necessary to transform the relation to the fuzzy equivalence relation using a transitivity closure operation expression. A fuzzy tolerance relation can be transformed into a fuzzy equivalence relation by at most (n^{-1}) compositions.

$$R_1^{(n-1)} = R_1 \circ R_1 \circ R_1 \circ R_1 \ldots\ldots\ldots\ldots \circ R_1 = R \tag{1.5}$$

Fuzzy relations describe the interactions between variables. While an equivalence relation clearly groups elements that are equivalent under the relation into disjoint classes, the interpretation of a similarity relation can be approached in two different ways. First, it can be considered to effectively group elements into crisp sets whose members are similar to each other to some specified degree. Obviously, when this degree is equal to 1, the grouping is an equivalence class. Alternatively, however, we may wish to consider the degree of similarity that the elements of X have to some specified element $x \in X$. Thus, for each $x \in X$, a similarity class can be defined as a fuzzy set in which the membership grade of any particular element represents the similarity of that element to the element x. If all the elements in the class are similar to x to the degree of 1 and similar to all elements outside the set to the degree of 0 then the grouping again becomes an equivalence class. We know every fuzzy relation R can be uniquely represented in terms of its α-cuts by the formula $\bigcup \alpha_{\alpha \in (0,1]} R$.

It is easily verified that if R is a similarity relation then each α-cut, ${}^{\alpha}R$, is a crisp equivalence relation. Thus, we may use any similarity relation R and by taking an α-cut ${}^{\alpha}R$ for any value $\alpha \in (0, 1)$, create a crisp equivalence relation that represents the presence of similarity between the elements to the degree α. Each of these equivalence relations forms a partition of X. These partitions are nested in the sense that $\pi\,({}^{\alpha}R)$ is a refinement of $\pi\,({}^{\beta}R)$ if and only $\alpha \geq \beta$.

1.3.2 Similarity Measures or Value Assignments

An appropriate question regarding relations is: Where do the membership values that are contained in a relation come from? The answer to this question is that there are at least seven different ways to develop the numerical values that characterize a relation. Fuzzy relations can be assembled from linguistic knowledge, expressed as if–then rules. Such knowledge may come from experts, from polls, or from consensus building. One of the most prevalent forms of determining the values in relations is through manipulations of data. The more robust a dataset, the more accurate the relational entities are in establishing relationships among elements of two or more datasets.

1.3.2.1 Cosine Amplitude Method

This similarity metric method makes use of a collection of data samples, n data samples in particular. If these data samples are collected, they form a data array, $\boldsymbol{X} = \{\boldsymbol{x}_1, \boldsymbol{x}_2, ..., \boldsymbol{x}_n\}$. Each of the elements, $\boldsymbol{x}_i$, in the data array $\boldsymbol{X}$ is itself a vector of length m, i.e. $\boldsymbol{x}_i = \{\boldsymbol{x}_{i1}, \boldsymbol{x}_{i2}, ..., \boldsymbol{x}_{im}\}$. Hence, each of the data samples can be thought of as a point in m-dimensional space, where each point needs m coordinates for a complete description. Each element of a relation, r_{ij}, results from a pair-wise comparison of two data samples, say $\boldsymbol{x}_i$ and $\boldsymbol{x}_j$, where the strength of the relationship between data sample $\boldsymbol{x}_i$ and data sample $\boldsymbol{x}_j$ is given by the membership value expressing that strength, i.e. $r_{ij} = \mu_R(x_i, y_j)$. The relation matrix will be of size $n \times n$ and, as will be the case for all similarity relations, the matrix will be reflexive and symmetric, hence a tolerance relation. The cosine amplitude method calculates r_{ij} in the following manner, and guarantees, as do all the similarity methods, that $0 \leq r_{ij} \leq 1$

$$r_{ij} = \frac{\left|\sum_{k=1}^{n} x_{ik} x_{jk}\right|}{\sqrt{\left(\sum_{k=1}^{m} x_{ik}^2\right)\left(\sum_{k=1}^{m} x_{jk}^2\right)}} \text{ where } i, j = 1, 2, \ldots, n \tag{1.6}$$

When two vectors are colinear (most similar), their dot product is unity; when the two vectors are at right angles to one another (most dissimilar), their dot product is zero.

1.3.2.2 Max–Min Method

This method is computationally simpler than cosine amplitude method. It is found through simple max and min operations on pairs of the data points, x_{ij}, and is given by

$$r_{ij} = \frac{\sum_{k=1}^{m} min(x_{ik}, x_{jk})}{\sum_{k=1}^{m} max(x_{ik}, x_{jk})} \text{ where } i, j = 1, 2, 3, \ldots, n \tag{1.7}$$

1.3.3 What Is Fuzzy Logic?

One of the powerful definitions of FL is approximate reasoning with fuzzy sets. FL can also be termed *human centric logic* wherein human perceptions take center stage.

FL has two different meanings. In a narrow sense, FL is a logical system, which is an extension of multivalued logic. However, in a wider sense, FL is almost synonymous with the theory of fuzzy sets, a theory which relates to classes of objects with unsharp boundaries in which membership is a matter of degree. Even in its narrower definition, FL differs both in concept and substance from traditional multivalued logical systems. What might be added is that the basic concept underlying FL is that of a linguistic variable, that is a variable whose values are words rather than numbers. In effect, much of FL may be viewed as a methodology for CW rather than numbers. Although words are inherently less precise than numbers, their use is closer to human intuition. Furthermore, CW exploits the tolerance for imprecision and thereby lowers the cost of solution.

Another basic concept in FL, which plays a central role in most of its applications, is that of a fuzzy if–then rule or, simply, fuzzy rule. Although rule-based systems have a long history of use in artificial intelligence (AI), what is missing in such systems is a mechanism for dealing with fuzzy consequents and fuzzy antecedents. In FL, this mechanism is provided by the calculus of fuzzy rules. The calculus of fuzzy rules serves as a basis for what might be called the fuzzy dependency and command language (FDCL). Although FDCL is not used explicitly in the toolbox, it is effectively one of its principal constituents. In most of the applications of FL, a FL solution is, in reality, a translation of a human solution into FDCL.

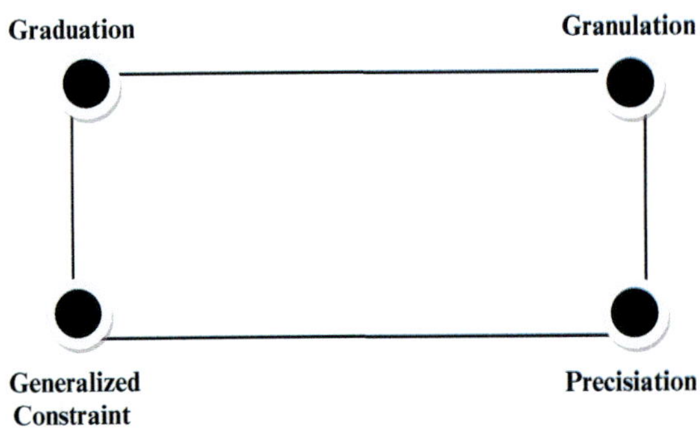

Figure 1.2 Four cornerstones of FL.

Figure 1.2 shows the four cornerstones of FL. The concept of precisiation coined by Professor Lotfi Zadeh [1] refers to support of a fuzzy set or *mm* precisiation. The first "m" stands for *man* and the second "m" stands for *machine*.

The output of FL is fuzzy or imprecise. It is, therefore, important to get final output as crisp; therefore, the concept of defuzzification.

1.3.3 Defuzzification Methods

A variety of defuzzification methods have been developed to convert a fuzzy number to a crisp numeric value, such as the centroid method, mean of maxima, bisector of area, weighted average method, center of sums, and max–min aggregation ranking method are available for defuzzification. In this chapter, two of the most commonly used defuzzification methods, the centroid method and max–min aggregation ranking method, are introduced.

1) The centroid method: For a trapezoidal fuzzy number $\tilde{A}=(a,b,c,d)$, the centroid is calculated as the defuzzified value (υ)

$$\upsilon=\frac{\int_a^b \frac{x-a}{m-a}xdx+\int_m^n xdx+\int_n^b \frac{b-x}{b-n}xdx}{\int_a^m \frac{x-a}{m-a}dx+\int_m^n dx+\int_n^b \frac{b-x}{b-n}dx}=\frac{1}{3}\frac{(b+n)^2-bn-(a+m)^2+am}{b+n-m-a}$$

2) The max–min aggregation ranking method: In the max–min aggregation ranking method, the max and min fuzzy sets are firstly defined as

$$f_{\max}(x)=\begin{cases}x; 0<x<1\\ 0; \text{otherwise}\end{cases}$$

$$f_{\min}(x)=\begin{cases}1-x; 0<x<1\\ 0; \text{otherwise}\end{cases}$$

The left and right utility scores (U_L and U_R) of a fuzzy number $\tilde{A}$ can then be calculated as

$$U_L\left(\tilde{A}\right)=\underbrace{\sup}_{x}\left[f_{\tilde{A}}\cap f_{max}\left(x\right)\right]$$

$$U_R\left(\tilde{A}\right)=\underbrace{\sup}_{x}\left[f_{\tilde{A}}\cap f_{\min}\left(x\right)\right]$$

where $f_{\tilde{A}}$ is the fuzzy membership function of $\tilde{A}\left(a,b,c,d\right)$. The graphic illustration of the ranking method is shown in Figure 1.3. Based on the two utility scores, the final ranking (υ) can be obtained as

$$\upsilon=\frac{\left[U_R\left(\tilde{A}\right)+1-U_L\left(\tilde{A}\right)\right]}{2}$$

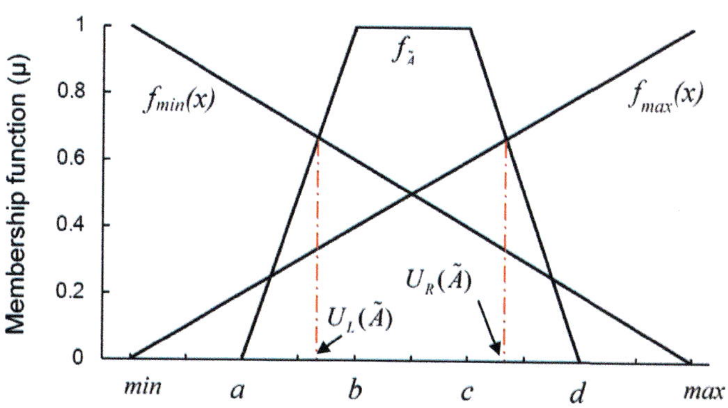

Figure 1.3 Graphic illustration of the min–max aggregation ranking method.

If $\tilde{A}$ and $\tilde{B}$ are two fuzzy numbers, then

1) $\tilde{A} \prec \tilde{B}$ if $v(\tilde{A}) < v(\tilde{B})$
2) $\tilde{A} \approx \tilde{B}$ if $v(\tilde{A}) = v(\tilde{B})$
3) $\tilde{A} \succ \tilde{B}$ if $v(\tilde{A}) > v(\tilde{B})$

where $v(\tilde{A})$ and $v(\tilde{B})$ are the crisp values from the defuzzification of $\tilde{A}$ and $\tilde{B}$, respectively.

1.4 Monotone Measures

Fuzzy sets provide us with an intuitively pleasing method of representing one form of uncertainty. Two prevalent forms of uncertainty are those arising from vagueness and from imprecision. *Vagueness* describes certain kinds of uncertainty associated with linguistic information or intuitive information. An example of vague information is, "The air quality is *good*." Imprecision can be associated with quantitative or countable data as well as noncountable data. As an example, one might say that the length of a bridge span is *long*. Imprecision can be used to quantify random variability in quantitative uncertainty and to describe lack of knowledge for descriptive entities (e.g. good air quality). Vagueness is usually related to nonmeasurable issues with no upper boundary. For example, "I will visit you."

Probability theory, evidence theory, and possibility theory are related under an umbrella theory termed monotone measures. A monotone measure describes the vagueness or imprecision in the assignment of an element "*a*" to two or more crisp sets. In a monotone measure we describe vagueness or imprecision in assigning this element to any of the crisp sets on the power set. This notion is not random: the crisp sets have no uncertainty about them. The uncertainty is about the assignment. This uncertainty is usually associated with evidence to establish an assignment. The evidence can be completely lacking (the case of total ignorance) or the evidence can be complete (the case of a probability assignment). Hence, the difference between a monotone measure and a fuzzy set in a universe of elements is that, in the former, the imprecision is in the assignment of an element to one of two or more crisp sets, and in the latter, the imprecision is in the prescription of the boundaries of a set.

1.4.1 Dempster–Shafer Evidence Theory

The theory of evidence is a branch of mathematics that is concerned with combining evidence to calculate the probability of an event. The DST is well known for its usefulness to express uncertain judgments of experts. The DST was introduced in 1967 by Arthur Dempster [2] and developed in 1976 by Glenn Shafer [3]. According to Glenn Shafer, the DST is a generalization of the Bayesian theory of subjective probability.

The DST can be viewed as a method for reasoning under epistemic uncertainty. Reasoning under epistemic uncertainty refers to logically arriving at decisions based on available knowledge. The most important part of this theory is Dempster's rule of combination which combines evidence from two or more sources to form inferences. According to Shafer, "the Dempster–Shafer theory is based on two ideas: the idea of obtaining degrees of belief for one question from subjective probabilities for a related question, and Dempster's rule for combining such degrees of belief when they are based on independent items of evidence." In traditional probability theory, evidence is associated with only one possible event. In DST, evidence can be associated with multiple possible events (e.g. sets of events). The significant innovation of this framework is that it allows for the allocation of a probability mass to sets or intervals. This is a potentially valuable tool for evaluation when knowledge is obtained from expert elicitation. An important aspect of this theory is the combination of evidence obtained from multiple sources and the modeling of conflict between them.

Helton [4] defines two prevalent forms of uncertainty as:

- **Aleatory Uncertainty**: It is a type of uncertainty which results from the fact that a system can behave in random ways. It is the inherent variation associated with the physical system or the environment under consideration. It is also known as stochastic uncertainty, type A uncertainty, irreducible uncertainty, and variability or objective uncertainty (e.g. dice throws).
- **Epistemic Uncertainty**: It is a type of uncertainty which results from the lack of knowledge or information in any phase or activity of the modeling process. A gain of information about the system or environmental factors can lead to a reduction in epistemic uncertainty. It is also known as subjective uncertainty, type B uncertainty, reducible uncertainty, state of knowledge uncertainty, and ignorance.

Traditionally, probability theory has been used to characterize both types of uncertainty. However, the recent criticisms of the probabilistic characterization of uncertainty claim that traditional probability theory is not capable of capturing epistemic uncertainty (Figure 1.4). The application of traditional probabilistic methods to epistemic or subjective uncertainty is often known as Bayesian probability. A probabilistic analysis requires that an analyst has information on the probability of all events. In situations where very little information is available to evaluate a probability or the information is nonspecific, ambiguous, or conflicting DST could be used as the framework for representing uncertainty.

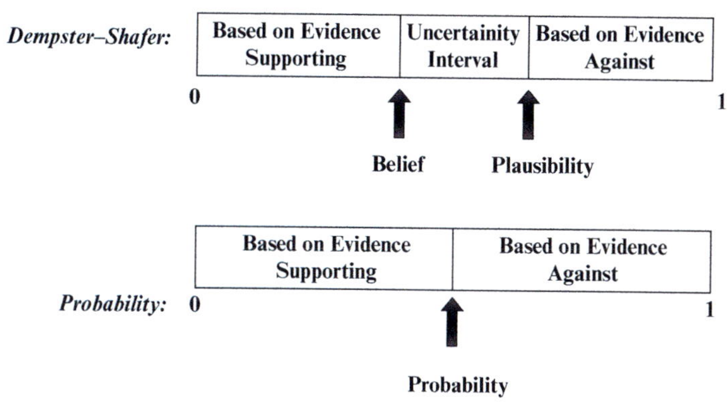

Figure 1.4 Dempster–Shafer evidence theory and probability theory.

There are special forms of monotone measures. A form associated with preconceived notions is called a *belief measure*. A form associated with information that is possible, or plausible, is called a *plausibility measure*. Specific forms of belief measures and plausibility measures are known as certainty and possibility measures, respectively.

The intersection of belief measures and plausibility measures (i.e. where belief equals plausibility) will be shown to be a probability. Monotone measures are defined by weaker axioms than probability theory, thus subsuming probability measures as specific forms of monotone measures. Basically, a belief measure is a quantity, denoted *Bel(A)*, that expresses the degree of support, or evidence, for a collection of elements defined by one or more of the crisp sets existing on the power set of a universe. The plausibility measure of this collection A is defined as the "complement of the belief of the complement of A" or is given by

$$Pl(A) = 1 - Bel(\bar{A}) \tag{1.8}$$

Since belief measures are quantities that measure the degree of support for a collection of elements or crisp sets in a universe, it is entirely possible that the belief measure of a set A plus the belief measure of $\bar{A}$ will not be equal to unity. When this sum equals 1, we have the condition where the belief measure is a probability, that is the evidence supporting set A can be described probabilistically. The difference between the sum of these two quantities $[Bel(A) + Bel(\bar{A})]$ and 1 is called the *ignorance*, which is given by

$$\text{Ignorance} = 1 - \left[Bel(A) + Bel(\bar{A})\right] \tag{1.9}$$

When the ignorance equals 0, we have the case where the evidence can be described by probability measures. Let's say we have evidence about a certain prospect in our universe of discourse, evidence of some set occurring or some set being realized, and we have no evidence (zero evidence) of the complement of that event. In probability theory we must assume, because of the excluded middle axioms, that if we know the probability of A then the probability of $\bar{A}$ is also known, because we have in all cases involving probability measures, $prob(A) + prob(\bar{A}) = 1$. This constraint of the excluded middle axioms is not a requirement in evidence theory. The probability of $\bar{A}$ also has to be supported with some sort of evidence. If there is no evidence (zero degree of support) for $\bar{A}$ then the degree of ignorance is large.

This distinction between evidence theory and probability theory is important. Monotone measures are very useful in quantifying uncertainty that is difficult to measure or that is linguistic in nature.

For example, in assessing structural damage in buildings and bridges after an earthquake or hurricane, evidence theory has proven quite successful because what we have are non-quantitative estimates from experts: the information concerning damage is not about how many inches of displacement or micro inches per inch of strain the structure might have undergone but rather expert judgment concerning the suitability of the structure for habitation or its intended function. These kinds of judgments are not quantitative: they are qualitative. We begin by assigning a value of membership to each crisp set existing in the power set of a universe, signifying the degree of evidence or belief that a particular element from the universe, say x, belongs in any of the crisp sets on the power set. We will label this membership $g(A)$, where it is a mapping between the power set and the unit interval

$$g : P(X) \to \left[0,1\right] \tag{1.10}$$

where $P(X)$ is the power set of all crisp subsets on the universe, X. So, the membership value $g(A)$ represents the degree of available evidence of the belief that a given element x belongs to a crisp subset A. The collection of these degrees of belief represents the *fuzziness* associated with several crisp alternatives. This type of uncertainty, which we call a monotone measure, is different from the uncertainty associated with the boundaries of a single set, which we call a fuzzy set.

Monotone measures are defined for a finite universal set by the following axioms

$$1 g(\varnothing) = 0 g(X) = 1 \tag{1.11}$$

$$2g(A) \leq g(B) \text{ for } A, B \in P(X), A \subseteq B \tag{1.12}$$

The first axiom represents the boundary conditions for the monotone measure, *g(A)*. It says that there is no evidence for the null set and there is a complete (i.e. unity) membership for the universe. The second axiom represents monotonicity by simply stating that if one set *A* is completely contained in another set *B* then the evidence supporting *B* is at least as great as the evidence supporting the subset *A*. A belief measure also represents a mapping from the crisp power set of a universe to the unit interval representing evidence, denoted by

$$Bel: P(X) \rightarrow [0,1] \tag{1.13}$$

A plausibility measure is also a mapping on the unit interval characterizing the total evidence, given by

$$Pl: P(X) \rightarrow [0,1] \tag{1.14}$$

Thus, plausibility can be expressed by Eq. (1.14) and belief can be expressed as

$$Bel(A) = 1 - Pl(\bar{A}) \tag{1.15}$$

For two disjoint sets A and $\bar{A}$, $Pl(A) + Pl(\bar{A}) \geq 1$.

Equation (1.9) states that for whatever evidence supports set A its plausibility measure is always at least as great as its belief measure.

$$Pl(A) \geq Bel(A) \tag{1.16}$$

We now define another function on the crisp sets ($A \in P(X)$) of a universe, denoted *m(A)*, which can be used to express and determine both belief and plausibility measures. This measure is also a mapping from the power set to the unit interval

$$m: P(X) \rightarrow [0,1] \tag{1.17}$$

This measure, called a *basic evidence assignment* (BEA), which is also termed a *basic probability assignment* (BPA) [5] has boundary conditions

$$m(\varnothing) = 0 \tag{1.18}$$

$$\sum{}_{A \in P(X)} m(A) = 1 \tag{1.19}$$

The measure *m(A)* is the degree of belief that a specific element, *x*, of the universe *X* belongs to the set *A*, *but not to any specific subset of A*. In this way, *m(A)* differs from both beliefs and plausibility. There is a distinct difference between a BEA and a probability density function (pdf). The former is defined on the sets of the power set of a universe (i.e. on $A \in P(X)$), whereas the latter is defined on the singletons of the universe (i.e. on $x \in P(X)$). The BEA is used to determine a belief measure by

$$Bel(A) = \Sigma_{B \subseteq A} m(B) \tag{1.20}$$

In Eq. (1.19), *m(A)* is the degree of evidence in set *A alone*, whereas *Bel(A)* is the total evidence in set *A* and all subsets *(B)* of *A*. The measure *m(A)* is used to determine a plausibility measure by

$$Pl(A) = \Sigma_{B \cap A \neq \varnothing} m(B) \tag{1.21}$$

Equation (1.21) shows that the plausibility of an event A is the total evidence in set A plus the evidence in all sets of the universe that intersect with A (including those sets that are also subsets of A). Hence, the plausibility measure in set A contains all the evidence contained in a belief measure (*Bel(A)*) plus the evidence in sets that intersect with set A. The two measures, *belief* and *plausibility*, are nonadditive. This can be interpreted as the sum of all the belief measures is not required to be 1 and similarly for the sum of the plausibility measures. It is possible to obtain the basic probability assignment from the *belief* measure with the following inverse function

$$m(A) = \Sigma_{B|B \subseteq A} (-1)^{|A-B|} Bel(B) \tag{1.22}$$

where $|A - B|$ is the difference of the cardinality of the two sets.

In addition to deriving these measures from the BEA (m), these two measures can be derived from each other. For example, *plausibility* can be derived from *belief* using Eq. (1.23), where $\bar{A}$ is the classical complement of A. This definition of plausibility in terms of belief comes from the fact that all basic assignments must sum to 1. The precise probability of an event lies within the lower and upper bounds of *belief* and *plausibility*, respectively.

$$Bel(A) = P(A) = Pl(A) \tag{1.23}$$

The probability is uniquely determined if *Bel(A)* = *Pl(A)*. In this case, it corresponds to classical probability, and all the probabilities, *P(A)*, are uniquely determined for all subsets A of the universal set X. Otherwise, *Bel(A)* and *Pl(A)* may be viewed as lower and upper bounds on probabilities, respectively, where the actual probability is contained in the interval described by the bounds.

1.4.2 The Dempster's Rule of Combination

The purpose of aggregation of information is to meaningfully summarize and simplify a corpus of data whether the data are coming from a single source or multiple sources. Combination rules are the special types of aggregation methods for data obtained from *multiple* sources. These multiple sources provide different assessments for the same frame of discernment, and the DST is based on the assumption that these sources are *independent*. In the situation where *all* sources are considered reliable, a conjunctive operation is appropriate (A and B and C...). In the case where there is one reliable source among many, we can justify the use of a disjunctive combination operation (A or B or C...). There are multiple possible ways in which evidence can be combined in the DST. The original combination rule of multiple BEAs, known as the Dempster rule, is a generalization of Bayes' rule. This rule strongly emphasizes the agreement between multiple sources and ignores *all* the conflicting evidence through a normalization factor. This can be considered a strict AND operation.

Dempster's rule combines multiple belief functions through their basic probability assignments (m). These belief functions are defined on the same frame of discernment but are based on *independent* arguments or bodies of evidence. Suppose the evidence for certain fuzzy measures comes from more than one source, say two experts, then the evidence obtained from two independent sources and expressed by two BEAs (m_1,m_2) on some power set *P(X)* can be combined to obtain a joint BEA, denoted by m_{12}, using Dempster's rule of combined evidence [6]. The equations for combined evidence are as follows

$$m_{12}(A) = \frac{\sum_{B \cap C = A} m_1(B) m_2(C)}{1 - K} \tag{1.24}$$

$\forall A \neq \varnothing$, and $m_{12}(\varnothing) = 0$

where

$$K = \Sigma_{B \cap C = \varnothing} m_1(B) \times m_2(C) \tag{1.25}$$

K represents basic probability mass associated with conflict. This is determined by summing the products of BEAs of all sets where the intersection is null. The denominator in Dempster's rule, $1 - K$, is a normalization factor. This has the effect of completely ignoring conflict and attributing any probability mass associated with conflict to the null set.

1.5 Fuzzy C-means (FCM) Clustering

Clustering techniques are mostly unsupervised methods that can be used to organize data into groups based on similarities among the individual data items. Clustering of numerical data forms the basis of many classifications and systems modeling algorithms. The purpose of clustering is to identify natural groupings of data from a large dataset to produce a concise representation of a system's behavior. *Clustering* refers to identifying the number of subclasses of c clusters in a data universe X comprised of n data samples, and partitioning X into c clusters ($2 \leq c < n$). Note that $c = 1$ denotes rejection of the hypothesis that there are clusters in the data, whereas $c = n$ constitutes the trivial case where each sample is in a cluster by itself. There are two kinds of c-partitions of data: hard (or crisp) and soft (or fuzzy). For numerical data one assumes that the members of each cluster bear more mathematical similarity to each other than to members of other clusters. Two important issues to consider in this regard are how to measure the similarity between pairs of observations and how to evaluate the partitions once they are formed.

One of the simplest similarity measures is distance between pairs of feature vectors in the feature space. If one can determine a suitable distance measure and compute the distance between all pairs of observations then one may expect that the distance between points in the same cluster will be considerably less than the distance between points in different clusters. Several circumstances, however, mitigate the general utility of this approach, such as the combination of values of incompatible features, as would be the case, for example, when different features have significantly different scales. The clustering method defines *optimum partitions* through a global criterion function that measures the extent to which candidate partitions optimize a weighted sum of squared errors between data points and cluster centers in feature space. Hard C-means is employed to classify data in a crisp sense. By this we mean that each data point will be assigned to one, and only one, data cluster. The problem of hard C-mean lies in assigning the point on the line of symmetry to a class. To which class should this point belong? Whichever class the algorithm assigns this point to, there will be a good argument that it should be a member of the other class.

Alternatively, the argument may revolve around the fact that the choice of two classes is a poor one for this problem. Three classes might be the best choice, but the physics underlying the data might be binary and two classes may be the only option. The solution to the above problem is FCM. This technique was originally introduced by Jim Bezdek [7] as an improvement on earlier clustering methods. It provides a method that shows how to group data points that populate some multidimensional space into a specific number of different clusters.

This method uses concepts in n-dimensional Euclidean space to determine the geometric *closeness* of data points by assigning them to various clusters or classes and then determining the distance between the clusters. FCM is a data clustering technique in which a dataset is grouped into n clusters with every data point in the dataset belonging to every cluster to a certain degree. For example, a certain data point that lies close to the center of a cluster will have a high degree of belonging or membership to that cluster and another data point that lies far away from the center of a cluster will have a low degree of belonging or membership to that cluster. Thus, each data point belongs to a cluster to some degree that is specified by a membership grade. To introduce this method, define a sample set of n data samples that we wish to classify: $X = \{\boldsymbol{x}_1, \boldsymbol{x}_2, \boldsymbol{x}_3, \ldots, \boldsymbol{x}_n\}$. Each data sample, $\boldsymbol{x}_i$, is defined by m features (i.e. $\boldsymbol{x}_i = \{x_{i1}, x_{i2}, x_{i3}, \ldots, x_{im}\}$ where each $\boldsymbol{x}_i$ in the universe X is an m-dimensional vector of m elements or m features). Since all the m features can have different units, in general, we have to normalize each of the features to a unified scale before classification. In a geometric sense, each $\boldsymbol{x}_i$ is a *point* in m-dimensional feature space, and the universe of the data sample, X, is a *point set* with n elements in the sample space.

1.5.1 FCM Algorithm

The FCM algorithm works by assigning membership to each data point corresponding to each cluster center on the basis of distance between the cluster and the data point. The more the data are near to the cluster center the more is their membership toward the particular cluster center. Clearly, the summation of membership of each data point should be equal to one. The algorithm is based on minimization of the following objective function

$$J_m = \Sigma_{i=1}^{N} \Sigma_{j=1}^{c} \mu_{ij}^{m} \left\| x_i - c_j^{\,2} \right\|^2, 1 < m \leq \infty \tag{1.26}$$

where m (the fuzziness exponent) is any real number greater than 1, N is the number of data points, c is the number of clusters, μ_{ij} is the degree of membership of x_i in the cluster j, x_i is the ith of d-dimensional measured data, c_j is the d-dimension center of the cluster, and $||*||$ is any norm expressing the similarity between any measured data and the center.

Fuzzy partitioning ($|U|$), is carried out through an iterative optimization of the objective function shown above, with the update of membership μ_{ij} and the cluster centers c_j by

$$\mu_{ij} = \frac{1}{\sum_{k=1}^{c} \left(\frac{\left\| x_i - c_j \right\|}{\left\| x_i - c_k \right\|} \right)^{\frac{2}{m-1}}} \tag{1.27}$$

$$= \mu_{ij} = \frac{1}{\left(\frac{\left\| x_i - c_j \right\|}{\left\| x_i - c_1 \right\|} \right)^{\frac{2}{m-1}} + \left(\frac{\left\| x_i - c_j \right\|}{\left\| x_i - c_2 \right\|} \right)^{\frac{2}{m-1}} + \ldots\ldots\ldots + \left(\frac{\left\| x_i - c_j \right\|}{\left\| x_i - c_k \right\|} \right)^{\frac{2}{m-1}}}$$

where$||x_i - c_j||$ is the distance from point i to current cluster center j, $||x_i - c_k||$ is the distance from point i to other cluster centers k.

$$c_j = \frac{\sum_{i=1}^{N} \mu_{ij}^{m} .x_i}{\sum_{i=1}^{N} \mu_{ij}^{m}} \tag{1.28}$$

The iteration will stop when $\max_{ij}\left\{\left|\mu_{ij}^{(K+1)} - \mu_{ij}^{(K)}\right|\right\} < \varepsilon$, where ε is a termination criterion between 0 and 1, whereas K are the iteration steps. This procedure converges to a local minimum or a saddle point J_m.

1.5.2 Bellman–Zadeh Approach

In the conventional approach to decision-making, the principal components of a decision process are (i) a set of alternatives, (ii) a set of constraints on the choice between alternatives, and (iii) a performance function which associates with each alternative the gain (or loss) resulting from the choice of that alternative.

An important facet of this model is its equilibrium with respect to goals and constraints – a balance which eliminates the difference between them and makes it possible to relate in a relatively simple way the concept of a decision to those of goals and constraints of a decision process. According to Bellman and Zadeh, an important aspect of this model is its symmetry with respect to the goals and the constraints – a symmetry that erases the difference between them and makes it possible to relate in a relatively simple way the concept of a decision to those of goals and constraints. Here we have a fuzzy goal G and fuzzy constraints $C_1, ..., C_n$ in a space of alternatives X. Then G and C combine to form a decision denoted as D, which is a fuzzy set resulting from the intersection of G and C. The normalization of the goals and the constraints is an important initial step for the Bellman–Zadeh (BZ) approach, which provides a common denominator for the fuzzy goals and the fuzzy constraints, thereby making it possible to treat them alike. It is done as a grade of the individual constraint divided by taking a sum of the same constraint across all the alternatives or sites. The same is done for the fuzzy goals. This concept explains why it is perfectly justified to regard the fuzzy concepts of the fuzzy goal and the constraints rather than the performance function as one of the major components for decision-making in a fuzzy environment. The normalization process makes it possible to treat the fuzzy goals and the constraints as identical in the formulation of a decision. After performing the normalization step, the decision can be obtained using Eq. (1.29) and the corresponding membership grades is given by Eq. (1.30), where G and C combine to form a decision D, which is a fuzzy set resulting from the intersection of G and C.

$$D = G \bigcap C \tag{1.29}$$

and correspondingly their membership grades are as follows

$$\mu D = \mu G \wedge \mu C \tag{1.30}$$

In order to generalize, suppose that we have n goals $G_1, ..., G_n$ and m constraints $C_1, ..., C_m$, the resultant decision is the intersection of the defined goals represented as $G_1, ..., G_n$ and defined constraints $C_1, ..., C_m$

$$D = G_1 \bigcap G_2 \bigcap G_n \bigcap C_1 \bigcap C_2 \bigcap C_m \tag{1.31}$$

and correspondingly the membership grades are represented as

$$\mu D = \mu G \wedge G2, \ldots, \wedge \mu Gn \wedge \mu C1 \wedge \mu C2 \wedge \mu Cm \tag{1.32}$$

In Eq. (1.31), the definition of decision as the intersection of the goals and the constraints reflects the interpretation of *and* states and the decision is viewed as a fuzzy set, which is a confluence of goals and constraints.

In summary, the perception of decision is given by

Decision = Confluence of Goals and Constraints

More generally, let D be a fuzzy decision represented by a membership function μ_D. Let K be the set of points in X on which μ_D attains its maximum, if it exists. The nonfuzzy but, in general, subnormal subset D^M of D is defined by

$$\begin{aligned} \mu_D{}^m(x) &= \text{Max}\, \mu_{D(x)} \text{ for } x \in K, \\ &= 0 \text{ otherwise} \end{aligned} \tag{1.33}$$

will be said to be the optimal decision and any x in the support of D^M will be referred to as a maximizing decision. To summarize, a maximizing decision is simply any alternative in X which maximizes $\mu_D(x)$. In many cases, the goals and constraints are fuzzy sets in different spaces. Let f be a mapping from $X = \{x\}$ to $Y = \{y\}$, with x representing an input (cause) and y, $y = f(x)$, representing output (effect).

Suppose that the goals are defined as fuzzy sets $G_1, \ldots, G_n$ in Y while the constraints $C_1, \ldots, C_m$ are defined as fuzzy sets in X. For a fuzzy set G, in Y, membership function in G, is given by the equality as shown in Eq. (1.35).

$$\mu_{Gi}(x) = \mu_{Gi}(f(x)), i = 1, \ldots, n \tag{1.34}$$

The decision D can be expressed as the intersection of $G_1, \ldots, G_n$ and $C_1, \ldots, C_m$.

Here, we can express $\mu_D(x)$ more explicitly as

$$\mu D(x) = \mu G1(f(x)) \wedge \mu Gn(f(x)) \wedge \mu C1(x) \wedge .. \wedge Cm(x), \tag{1.35}$$

where $f: X \rightarrow Y$.

In this way, the case where the goals and the constraints are defined as fuzzy sets in different spaces can be reduced to the case where they are defined in the same space.

Acknowledgment

The author is immensely grateful to their mentor and guide Dr. Ashok Deshpande for motivating them to write this chapter. His contribution in writing this chapter is commendable and worthy.

References

1 Zadeh, L.A. (2008). Is there a need for fuzzy logic? *Information Sciences* 178(13): 2751–2779.

2 Dempster, A. (1967). Upper and lower probabilities induced by a multivalued mapping. *Annals of Mathematical Statistics* 38: 325–339.

3 Shafer, G. (1976). *A Mathematical Theory of Evidence*. Princeton University Press.
4 Helton, J.C. (1997). Uncertainty and sensitivity analysis in the presence of stochastic and subjective uncertainty. *Journal of Statistical Computation and Simulation* 57(1–4): 3–76.
5 Boudraa, A.O., Bentabet, A., Salzenstein, F., Guillon, L. (2004). Dempster–Shafer's basic probability assignment based on fuzzy membership functions. *Electronic Letters on Computer Vision and Image Analysis* 4(1): 1–9.
6 Al-Ani, A., Deriche, M. (2002). A new technique for combining multiple classifiers using the Dempster–Shafer theory of evidence. *Journal of Artificial Intelligence Research* 17: 333–361.
7 Bezdek, J. (1981). *Pattern Recognition with Fuzzy Objective Function Algorithms*. Plenum. doi: 10.1007/978-1-4757-0450-1.

Part II

Fuzzy Logic for Environmental Risk Assessment

2

Fuzzy-based Integrated Risk Assessment of Methylmercury in Lake Phewa, Nepal

Gyan Chhipi-Shrestha[1], Manjot Kaur[1], Devna Singh Thapa[2], Manuel Rodriguez[3], Shichang Kang[4], Chhatra Mani Sharma[5], Kasun Hewage[1], and Rehan Sadiq[1]

[1] *School of Engineering, University of British Columbia (Okanagan), Kelowna, BC, Canada*
[2] *Kathmandu University, Kathmandu, Nepal*
[3] *Laval University, École Supérieure d'Aménagement du Territoire, Canada*
[4] *State Key Laboratory of Cryospheric Science, Northwest Institute of Eco-Environment and Resources, Chinese Academy of Sciences, Lanzhou, China*
[5] *Tribhuvan University, Central Department of Environmental Science, Kathmandu, Nepal*

2.1 Introduction

Mercury (Hg) is a persistent substance and a global pollutant that results from natural and anthropogenic activities. Hg can be released into the air through weathering of rock containing Hg ore, or through anthropogenic activities including the burning of fossil fuels and incineration of medical and municipal wastes. Once released into the air, it can travel long distances and impact distant sites [1]. Surface water can be contaminated by run-off containing Hg, originating from anthropogenic and/or natural sources or deposition from the atmosphere. The USEPA estimated that anthropogenic sources contribute around 4,000 tonnes of Hg per year into the atmosphere (i.e. 50 to 75% of the total annual input) [2].

Hg can occur in three valence states (Hg^0, Hg^{1+}, and Hg^{2+}) and can be found in metallic form (Hg^0) and in various inorganic and organic complexes [1]. The conversion of inorganic Hg to methylmercury (MeHg) is mediated by microorganisms (e.g. sulfate-reducing bacteria in anaerobic environments) [1, 3]. Once MeHg is formed, Hg bioaccumulates in biota and biomagnifies in the food chain, highly increasing its concentration [3, 4]. In phytoplankton, MeHg is thought to accumulate mainly by passive uptake from water through diffusion across the cell membrane [5]. In zooplankton, MeHg is accumulated through a net process of dietary intake of phytoplankton, water uptake, fecal elimination, and growth dilution [5, 6]. Moreover, large fishes consume microorganisms, phytoplankton, zooplankton, and smaller fishes resulting in very high concentrations of MeHg in some fish consumed by humans [1, 7]. In addition, Hg(II) can actively convert into MeHg in paddy fields [8]. The MeHg that is produced on-site and transported through contaminated irrigation water can be translocated from soil to paddy plants, bioaccumulating in rice grains and further threatening human health [8, 9].

MeHg is highly toxic to several other living structures [1, 10]. Human and animal brains are susceptible to the toxic effects of MeHg [1]. In vitro and animal studies have reported on the immunotoxicity of MeHg leading to damaged immune systems. In addition, MeHg can harm the cardiovascular system [1]. The MeHg exposure can result in reproductive toxicity due to chromosomal anomaly and sister-chromatid exchange [10].

An integrated risk assessment (IRA) framework has been proposed by the WHO as an integrated approach for combining human health risk assessment and ecological risk

Fuzzy Systems Modeling in Environmental and Health Risk Assessment, First Edition. Edited by Boris Faybishenko, Rehan Sadiq, and Ashok Deshpande.

assessment. IRA is defined as "a science-based approach that combines the processes of risk estimation for humans, biota, and natural resources in one assessment" [11]. The HEROIC[1] consortium further defined IRA as "the mutual exploitation of ecological risk assessment for human health risk assessment and vice versa in order to coherently and more efficiently characterize an overall risk to humans and the environment for better informing the risk analysis process." Human health risk assessment and ecological risk assessment for chemicals have generally developed independently for historical and practical reasons and are usually separated today, using broadly separate data, models, and assumptions to characterize health or environmental risks [11, 12]. IRA aims to improve the quality and efficiency of risk assessments through the mutual exchange of information between two existing risk assessment disciplines and provide more coherent efforts to the decision-making process [11]. A central feature of IRA is the sharing and combination of independent sources of toxicological and ecotoxicological data for coherent and efficient risk assessment [11, 13].

Lake Phewa is a Ramsar site (a wetland site designated internationally important by the Ramsar Convention) in Nepal, and it is a major tourist attraction of the city of Pokhara with an influx of approximately 250 000 tourists per year [14]. The local people and visitors consume fish harvested from Lake Phewa [15]. Recreational activities such as boating on the lake are popular in the country. In addition, the lake is an important source of water for irrigation and hydroelectricity generation. Due to its economic, social, and aesthetic values, the lake is to be preserved. Various studies from collaborative research teams such as Thapa et al. [16], Sharma et al. [15, 17], Rosseland et al. [18], and Rupakheti et al. [19] revealed the presence of Hg, including MeHg, in the lake. However, ecological risk assessment and complete human health risk assessment of MeHg in the lake have not yet been conducted. Moreover, uncertainty is an unavoidable and inevitable component of any environmental system [20]. The present research investigated the integrated human health and ecological risk assessment of MeHg in Lake Phewa using a fuzzy approach to incorporate uncertainty in decision-making. IRA is in the developmental stage and this study is one of the initial attempts to apply the IRA framework in case studies beyond the IRA consortium.

2.2 Methodology

2.2.1 Sampling, Laboratory Analysis, and Consumer Survey

Lake Phewa lies at 28.21N latitude, 83.95E longitude, at an elevation of 742 m ASL. The lake has a total surface area of 435 ha and a maximum depth of 22.5 m [19]. Although approximately 28 species of fishes are found in the lake, four fish species – tilapia (*Oreochromis niloticus*), spiny eel (*Mastacembelus armatus*), African catfish (*Clarias gariepinus*), and sahar (*Tor putitora*) – are regarded as commercially important fish for the lake [15]. Among those, three species were investigated by Sharma et al. [15] to estimate MeHg concentration in fish tissues, and by Thapa et al. [16] to assess human exposure to MeHg by fish consumption using crisp data. These three species were commercially important during the present study and tilapia was the dominant one.

For the collected fish species, total length and weight of fillets just behind the dorsal fin (axial muscles) were measured after the skin was removed for the analysis of Hg (both total Hg (THg) and MeHg species). Stomach contents were collected and preserved in 70% ethanol

---

1 HEROIC (Health and Environmental Risks: Organisation, Integration and Cross-fertilisation of Scientific Knowledge)

to characterize the dominant diets. The muscle samples were kept in a freezer until they were brought to the laboratory. The samples were analyzed in the laboratory of the Department of Biology, Hong Kong University of Science and Technology (HKUST) for the analysis of THg and MeHg. The freeze-dried muscles were ground into fine powders. For the analysis of THg, 4 N HCl was used to clean all containers. About 0.2 g dried tissues were weighed. The weighed tissues were digested at 190 °C with *aqua regia* (i.e. 2 mL HNO_3:6 mL HCl in a microwave digestion system). An aliquot of the digested samples was taken and diluted as per need. Stable yellow color was developed by adding bromine monochloride (0.5 % v/v) as required. Then, the samples were pre-reduced by adding $NH_2OH.HCl$. Finally, THg was measured using the single gold trap amalgamation technique by Cold Vapour Atomic Fluorescence Spectroscopy. Similarly, For the analysis of MeHg, approximately 40 mg of tissues were digested with 25 % KOH in methanol at 60 °C for 3 h. 20–50 μL of the extract was buffered with sodium acetate at pH 4.9. The extract solution was ethylated by sodium tetraethyl borate in a 40 mL Teflon line borate glass bottle. MeHg was automatically quantified using MeHg analyzer with gas chromatographic separation and pyrolysis following atomic fluorescence detection [15].

A face-to-face questionnaire survey was conducted with 170 persons consuming fish harvested from Lake Phewa to determine their fish consumption rate, occupation, and age group. The purposive sampling technique was used to include adult persons consuming fish from Lake Phewa. The respondents were classified into six occupations: Fishermen, Local people, Hotel owners, Government staff (Fisheries Research Center), Army/Police, and Others (visitors). The saturation point principle was used to fix sample size (i.e. no additional respondent in a group was included when new information could not be obtained). Detailed information on the questionnaire survey can be obtained from Thapa et al. [16].

2.2.2 Integrated Risk Assessment

The IRA framework was proposed by the WHO in 2001 [11]. The framework consists of three steps: problem formulation, evaluation, and risk characterization, as shown in Figure 2.1 [11, 13].

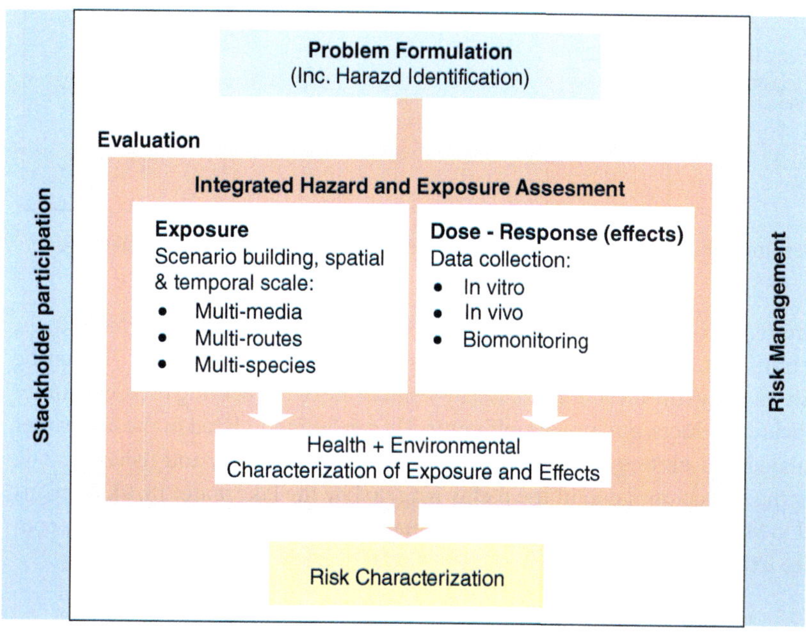

Figure 2.1 A framework for integrated human health and ecological risk assessment [11, 13].

2.2.2.1 Problem Formulation

This is the first step in which the nature and extent of integration was defined and overall goals, objectives, scope, activities, and resource needs were delineated for the IRA process. Risk managers, risk assessors, and other stakeholders were involved in the process. Hg present in lake water is converted into a highly neurotoxic organic form, MeHg, mainly by bacteria [1]. MeHg is bioaccumulated in biota and biomagnified across the food chain [4]. MeHg is hazardous to human health and ecosystem species such as fishes. Humans and fishes have common routes of exposure and similar effects [1]. In this research, the target organisms investigated are fishes and humans, specifically fishermen, local people, hotel owners, staff, army/police, and others (visitors). An ecosystem approach was used for IRA. The conceptual model of the sources and pathways of MeHg exposure to fishes and humans in Lake Phewa is shown in Figure 2.2. The potential chemical reactions involved in the formation and transport of MeHg were obtained from the National Research Council [1].

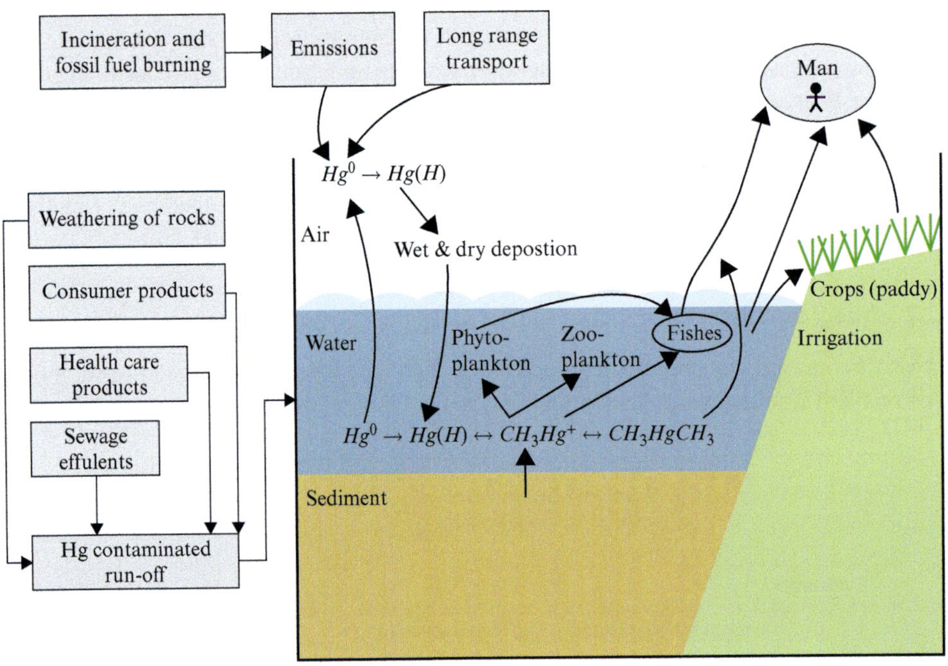

Figure 2.2 Conceptual model of routes of exposure of MeHg to fishes and humans in Lake Phewa.

Hg may be emitted into the air by the incineration of medical wastes, burning fossil fuels, and long-range transport. The chemical may reach Lake Phewa by air deposition. Hg may be added to the lake water by run-off contaminated by weathering of Hg-bearing rock, consumer products, healthcare products, and sewage effluents. Hg can be methylated in the lake water and bioaccumulated and biomagnified in phytoplankton, zooplankton, and fishes [1, 3–7, 15], as well as being bioaccumulated in the paddy irrigated by the lake water [8, 9]. Humans may be exposed to MeHg through ingesting fish, dermal contact with lake water, and by consuming local rice irrigated by the water.

2.2.2.2 Evaluation

This step consists of characterization of exposure and dose-response (effects) as follows:

A. Characterization of exposure

This involves estimating the doses of MeHg to which human population and environmental compartments are exposed. The sources of Hg were identified based on the literature and were discussed in the problem formulation step (Figure 2.3). MeHg can be transported to humans and fishes by various pathways. The common route of exposure for humans and fishes is contaminated water and foods. The transport and fate of Hg(II) and MeHg from water to fishes and their internal exposure can be modeled and predicted. However, the actual concentration (body burden) of MeHg in fish was measured by sampling and laboratory analysis, as described in Section 2.2.1.

The uncertainty in data was incorporated in the analysis using fuzzy set theory for IRA. The fuzzy set theory is applicable when parameters have uncertainty and data are limited to define any statistical distribution for advanced uncertainty analysis [21]. In this research the important parameters such as body burden of each fish species, bioaccumulation factors (BAFs) of rice and soil, contaminated body surface area, body weight, exposure frequency per year, and fish consumption rates of different population groups have uncertain and limited data. The collected data for different parameters were expressed in triangular fuzzy numbers (TFNs), i.e. (a, b, c), where a, b, and c denote the smallest, most probable, and largest values, respectively. TFNs are commonly used as they are simple for application [22]. The MeHg of fish tissue was measured using field sampling, whereas MeHg concentration in water was lacking. For the estimation of MeHg in water, BAFs of fishes obtained from Poste et al. [23] and Scudder Eikenberry et al. [24] were used. These literature-based BAFs were highly variable and were expressed in TFNs. Using the BAFs, the MeHg concentration in lake water was estimated using Eq. (2.1).

$$\widetilde{C_w} = \widetilde{C_f}/(\widetilde{BAF})_f \tag{2.1}$$

where $\widetilde{C_w}$, $(\widetilde{BAF})_f$, and $\widetilde{C_f}$ are TFNs for MeHg concentration in water (mg/L), BAF for fish (L/kg), and MeHg concentration in fish tissue (mg/kg), respectively. The tilde (~) on the top of a notation indicates a fuzzy number.

The lake water containing Hg was used for irrigation, which may have contaminated the agricultural crops. The major crop of Pokhara is paddy, which needs a large amount of inundated water for growth. Therefore, rice was also considered an additional media of MeHg exposure to humans. The concentration of MeHg in rice grains was calculated as follows [8]

$$\widetilde{C_r} = \widetilde{C_s}/(\widetilde{BAF})_f \tag{2.2}$$

where $\widetilde{C_r}$, $(\widetilde{BAF})_r$, and $\widetilde{C_s}$ are TFNs for MeHg concentration in rice grains, BAF for rice, and MeHg concentration in soil, respectively.

Equations (2.3) and (2.4) can be used for estimating exposure to MeHg and are given by [25]. The equations for estimating exposure to MeHg are given in Eqs. (2.3) and (2.4) [25].

$$\widetilde{CDI}_{ingestion"i"} = \frac{\widetilde{C_i} \otimes \widetilde{IR_i} \otimes \widetilde{EF} \otimes \widetilde{ED}}{\widetilde{BW} \otimes \widetilde{AT}} \tag{2.3}$$

$\widetilde{CDI}_{ingestion"i"}$ = Chronic daily intake through oral ingestion of exposure media "i" ($n = 2$; fish and rice) (mg/kg/day) and $\otimes$ denotes fuzzy multiplication

$\widetilde{C_i}$ = Concentration of MeHg in media "i" ($n = 2$; fish and rice, mg/kg)

$\widetilde{IR_i}$ = Ingestion rate of media "i" ($n = 2$)

$\widetilde{EF}$ = Exposure frequency (days/year)

$\widetilde{ED}$ = Exposure duration (years)

$\widetilde{BW}$ = Body weight (kg)

$\widetilde{AT}$ = Average time = $\widetilde{ED} \otimes 365$ (days)

$$\widetilde{CDI}_{dermal} = \frac{\widetilde{C_w} \otimes \widetilde{SA} \otimes \widetilde{Kp} \otimes \widetilde{ET} \otimes \widetilde{EF} \otimes \widetilde{ED} \otimes \widetilde{CF}}{\widetilde{BW} \otimes \widetilde{AT}} \tag{2.4}$$

where CDI_{dermal} = Chronic daily intake through dermal absorption exposure (mg/kg/day)

$\widetilde{C_w}$ = Concentration of MeHg in water (mg/L)

$\widetilde{SA}$ = Skin surface area available for contact for an adult during fishing (cm^2)

$\widetilde{Kp}$ = Chemical specific dermal permeability constant in water measured at 25 °C (cm/h)

$\widetilde{CF}$ = Conversion factor of 10^{-3} (L/cm^3)

$\widetilde{ET}$ = Exposure time (h/event)

The data for input parameters are given in Table 2.1.

Table 2.1 Input parameters for exposure assessment.

Parameters	TFN	Units	Source
EF (fish and rice)	(292, 329, 365)	d/yr	Assumed L, M, and U of TFN as 80%, 90% and 100% of total days in a year
ED	(20, 20, 20) for all except (0.083, 0.083, 0.083) for "others/visitors"	yr	Assumed 20 yr of service life in occupations and 1 m for others/visitors (constant)
BW	(70, 75, 80)	kg	Adapted from 80 kg recommended by USEPA [24]
SA	(0.942, 1.24, 1.538)	cm^2	Adapted from mean surface area of 1.24 cm^2 and 95th percentile of 1.538 cm^2 of arms, hands, legs, and feet [24]
Kp	(0.001, 0.001, 0.001)	cm/h	Assumed constant [24]
ET	(8, 8, 8)	h/event	Daily working hour for fishermen [24]
EF (dermal)	(52, 80, 108)	d/yr	Adapted from working days of mean of 80 d/yr and 95th percentile of 108 d/y for fishermen [24]

B. Characterization of effects

This step consists of dose-response assessment. It estimates the relationship between the level of exposure to a substance (dose) and the incidence and severity of an effect. Human and animal studies have reported the toxic effects of MeHg in the brain; however, fetal brains are more sensitive to MeHg than adult brains [1]. Due to an abundance of well-documented evidence, the developmental neurotoxic effects of MeHg in children exposed in utero have been used as a suitable endpoint for establishing reference dose (RfD) [1, 26]. RfD is widely used for modeling exposure and response for noncancerous diseases in humans. However, MeHg has been classified as a "possible" human carcinogen by the International Agency for

Research on Cancer (IARC) and the US Environmental Protection Agency (EPA) due to a lack of strong evidence of the association between Hg exposure and cancer rates [1]. As a result, cancer risk estimation for MeHg is not common.

The oral RfD for MeHg was obtained from the US EPA's Integrated Risk Information System (IRIS) database, which is 1.0E-4 mg/kg/d [26]. For ecosystem species or fishes, predicted no effect concentration (PNEC) is widely used for modeling exposure and response. Unlike RfD, the studies estimating the value of PNEC for ecosystem species are limited. The TFN of PNEC for fishes was estimated as (2.40E-5, 9.46E-5, 1.89E-4) μg/L based on the no-observed-adverse-effect level (NOAEL) derived as 50% of low-observed-adverse-effect level (LOAEL) [27] of 0.3 μg/g MeHg in fish [4, 28] and BAF of fish [24].

2.2.2.3 Risk Characterization

This step estimates the incidence and severity of the adverse effects likely to occur in a human population or environmental compartment due to actual or predicted exposure to a substance. For fishes, the risk to health (endpoint) (i.e. ecological risk of MeHg exposure) was estimated using the risk quotient as given in Eq. (2.5) [29].

$$\widetilde{RQ} = \frac{\left(\widetilde{EEC}\right)}{\left(\widetilde{PNEC}\right)} \tag{2.5}$$

where:

$\widetilde{RQ}$ = Risk quotient for fishes

$\widetilde{EEC}$ = Estimated or directly measured environmental concentration of MeHg

$\widetilde{PNEC}$ = PNEC or benchmark concentration for fishes

The rank of $\widetilde{RQ}$ (i.e. RQ) was calculated to obtain its crisp (defuzzified) value using Eq. (2.6) [30].

$$\text{Rank of } \widetilde{RQ}(a,b,c) = \frac{(a+2b+c)}{4} \tag{2.6}$$

RQ was interpreted as

$RQ < 1$ indicates low or extremely low risk or probability of effects

$RQ \geq 1$ may indicate potential risk or effects

For humans, the risk to health (endpoint) of MeHg exposure was estimated by combining three exposure routes: fish ingestion, dermal contact, and rice consumption. The overall health risk was estimated as [25]

$$\widetilde{HI} = \sum_{i=1}^{3} \frac{\left(\widetilde{CDI}\right)i}{RfD} \tag{2.7}$$

where $\widetilde{CDI_i}$ = CDI for all exposure routes and media ($n = 3$)

$\widetilde{HI}$ = Hazard index

RfD = Chemical specific RfD (mg/kg/d)

Finally, the rank of $\widetilde{HR}$ was obtained using Eq. (2.6) and interpreted similar to RQ.

Although the overall risk is estimated by defuzzifying the fuzzy risk value, the uncertainty in the decision-making is indicated by the range of overall risk from the lowermost to uppermost values of the TFN in each occupation and fishes.

2.3 Results and Discussion

The results obtained are presented and discussed below.

2.3.1 Exposure Assessment

The body burden of MeHg measured in different fish species is given in Table 2.2. The table also shows the size and weight of the fishes sampled. The highest median body burden of 0.042 mg/kg occurred in *Mastacembelus armatus*, whereas the lowest median body burden was found in *Oreochromis niloticus* with a value of 0.013 mg/kg. Overall, the median body burden of MeHg in the fishes was 0.025 mg/kg with fifth percentile (P5) of 0.010 mg/kg and 95th percentile (P95) of 0.063 mg/kg, which were used for risk assessment.

Table 2.2 Fish size and body burden of MeHg.

				MeHg concentration (mg/kg)		
Fish	***n***	**Length (cm)**	**Weight (g)**	**P5**	**P50 (median)**	**P95**
O. niloticus	21	25.7 ± 2.9	334.0 ± 111.8	0.009	0.013	0.028
M. armatus	20	31.0 ± 4.3	65.5 ± 20.8	0.029	0.042	0.070
Clarias gariepinus	19	25.1 ± 3.6	128.8 ± 55.8	0.012	0.025	0.047
Total	60	27.3 ± 4.4	176.1 ± 136.1	0.010	0.025	0.063

P5: 5th percentile; P50: 50th percentile; P95: 95th percentile.

The TFN of BAF of fishes was constructed based on the data obtained from the extensive study of Riva-Murray et al. [31]. The L, M, and U are assumed as the 5th percentile, median, and 95th percentile for TFN considering 90% of the distribution, which is (7.9E + 05, 1.6E + 06, 6.2E + 06) L/kg as given in Table 2.2. The estimated concentration of MeHg in water is given in Table 2.3. Similarly, Pokhara did not have any known Hg pollution sources such as Hg mining, effluent of chlor-alkali industry, and other mining (gold, zinc, and lead), and therefore background concentration of MeHg was considered in the agricultural soil of the study area. The global background concentration of MeHg for the unpolluted soil was obtained from the comprehensive review conducted on MeHg in rice by Rothenberg et al. [8]. For this study, the background MeHg concentration in the agricultural soil was estimated by increasing the global background concentration by 15% to incorporate the possible increment due to irrigation water containing Hg species (Table 2.3). Also, the mean BAF for rice given by Zhang et al. [32] was varied by ±15% to construct a TFN of BAF of rice, which was (3.77, 4.44, 5.11) (unitless). The estimated concentration of MeHg in rice grains in the study area is given in Table 2.3.

The fish consumption rates in different occupations consuming fishes of Lake Phewa are given in Table 2.4. The fish consumption rates were highly variable across the occupations. Among the surveyed occupations, the lowest rate of fish consumption was found in the Staff of Fisheries Research Center with a mean rate of 68.2 ± 0.2 g/d, whereas the highest rate of fish consumption was found in the hotel owners with a mean value of 190.6 ± 212.3 g/d. The mean and 95th percentile of the consumption rates in all these occupations are higher than the USEPA-recommended fish intake rate, except for the 95th percentile rate for army/police. The USEPA provided the fish consumption rates of mean 16 g/d and 95th percentile of 96 g/d per capita and that of mean 54.4 g/d with 95th percentile of 160 g/d for consumers only, considering a body weight of 80 kg [33].

Table 2.3 MeHg concentration in different media.

Media	Values in TFNs	Unit	Source
Fish (C_{fish})	(0.010, 0.025, 0.063)	mg/kg	Sampling and measurement
Water (C_{water})	(4.0E-06, 1.3E-05, 1.6E-05)	μg/L	Estimated using C_{fish} and BAF_{fish}
Soil (C_{soil})	(0.23, 0.67, 1.55)	ng/g	[31]
Rice (C_{rice})	(0.87, 2.97, 7.92)	ng/g	Estimated using C_{soil} and BAF_{rice}

C: concentration; BAF: bioaccumulation factor.

Table 2.4 Daily fish consumption rate (g/d).

Occupation	*n*	$\bar{x} \pm \sigma$	P5	P50	P95
Fishermen	45	132.7 ± 123.3	35.7	95.3	342.9
Locals	32	94.8 ± 89.9	35.7	71.4	175.0
Hotel owners	35	190.6 ± 212.3	33.1	107.1	589.3
Staff	26	68.2 ± 0.2	22.3	71.4	163.7
Army/police	20	73.2 ± 24.5	35.7	71.4	107.1
Others (visitors)	12	130.7 ± 139.7	32.6	83.4	358.9

For MeHg exposure from rice as a staple food, a routine rice consumption rate of 355 g/p/d for urban areas of Nepal was obtained from CBS [34]. The TFN of rice consumption rate of (0.151, 0.177, 0.204) kg/d was constructed assuming 50% of the total rice consumption was sourced from local harvest with ± 15% variation in the estimate.

Even without local Hg pollution sources, the estimated concentration of MeHg in fishes and water of Lake Phewa was likely to be present mainly as a result of the global atmospheric supply of Hg and its active methylation [24, 35, 36]. Such concentration may also reach a level higher than the acceptable range [35, 36]. The BAF for fishes, i.e. TFN (Triangular Fuzzy Number) of (7.9E + 05, 1.6E + 06, 6.2E + 06) L/kg considered in this study lies in the range of BAFs observed by the US Geological Survey, which is 2.3E + 04 to 3.9E + 07 L/kg with the mean of 2.1E + 6 for unmined sites in the period of 1998 to 2005 in the United States [36]. Similarly, the estimated concentration of MeHg in the paddy soil of Pokhara can be a result of the global atmospheric supply of Hg [35–38] and its conversion into MeHg in paddy fields [8, 37, 38], as well as by MeHg carried through the irrigation water [38].

2.3.2 Risk Characterization

The human health risks of MeHg from exposure to different media are given in Table 2.5. As shown in Table 2.5, the defuzzified individual health risk due to fish ingestion was very high for hotel owners. The health risk of MeHg by rice consumption was very low as the MeHg concentration in rice grains was primarily based on the background concentration of MeHg in unpolluted soil. The risk estimates were equal in all occupations due to the consideration of an equal rice consumption rate. The health risk from dermal exposure was present only in fishermen as they had dermal contact with lake water during fishing activities; the health risk in that case was negligible.

Table 2.5 Human health risk from exposure to different media.

	Fuzzy risk estimate for different media			Defuzzified risk value		
Occupation	**Fish**	**Rice**	**Dermal**	**Fish**	**Rice**	**Dermal**
Fishermen	(0.04, 0.29, 3.08)	(0.01, 0.06, 0.23)	(5.3E-13, 3.7E-12, 8.2E-12)	0.92	0.09	4.0E-12
Locals	(0.04, 0.21, 1.57)	(0.01, 0.06, 0.23)	—	0.51	0.09	—
Hotel owners	(0.03, 0.32, 5.30)	(0.01, 0.06, 0.23)	—	1.49	0.09	—
Govt. staff	(0.02, 0.21, 1.47)	(0.01, 0.06, 0.23)	—	0.48	0.09	—
Army/police	(0.04, 0.21, 0.96)	(0.01, 0.06, 0.23)	—	0.36	0.09	—
Other/ visitors	(0.03, 0.25, 3.23)	(0.01, 0.06, 0.23)	—	0.94	0.09	—

The fuzzy risk values estimated for each media in fact show an uncertainty in the estimation. Comparatively, the range of risk from "L" to "U" in the TFN was larger in fish ingestion with the range in two orders of magnitude, whereas the risk estimates for rice ingestion and dermal contact had an uncertainty range of one order of magnitude, although membership functions are different in the TFNs. The high uncertainty in the risk estimate in fish ingestion was due to a large variation of one order of magnitude in fish consumption rate, as well as high variability in MeHg concentration in fish tissue, human body weight, and exposure frequency.

The percent contribution by different routes of exposure in total risk is shown in Figure 2.3. The fish consumption had a dominant contribution in health risk, ranging from approximately 79% in army/police to 94% in hotel owners. Although the magnitudes of risk are equal for rice consumption, the percent contribution in total risk is different, ranging from 6% in hotel owners to 21% in army/police. Due to a very low value of health risk from dermal contact, the contribution in total risk is insignificant.

The overall health risks of MeHg exposure in different occupations (humans) and fish species (ecosystem) are shown in Figure 2.4 in fuzzy numbers and in Figure 2.5 in crisp

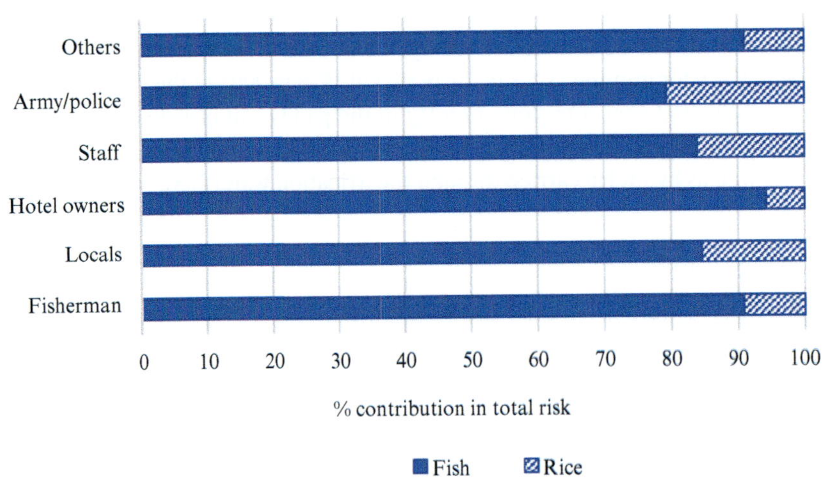

Figure 2.3 Percent contribution by different routes of exposure in total risk.

values. As given in Figure 2.5 for humans, the overall health risk of MeHg exposure was acceptable for all occupations except for hotel owners, fishermen, and others (visitors). The health risks to fishermen and others (visitors) were slightly higher than the acceptable level. Hotel owners, fishermen, and others had potential health risk or effects of MeHg exposure, primarily due to high consumption rates of fish. The average fish consumption rate of hotel owners, fishermen, and others (visitors) were 6.7, 6.0, and 5.2 times higher, respectively, than the average per capita fish consumption rate and 2.0, 1.8, and 1.5 times higher, respectively, than the consumers-only fish intake rate recommended by the USEPA [33]. In fact, the questionnaires were answered by respondents based on their memory rather than any actual weighing of fishes, which may have affected the results. In regards to hotel owners, who had the highest health risk, most of the respondents stated that they consumed the

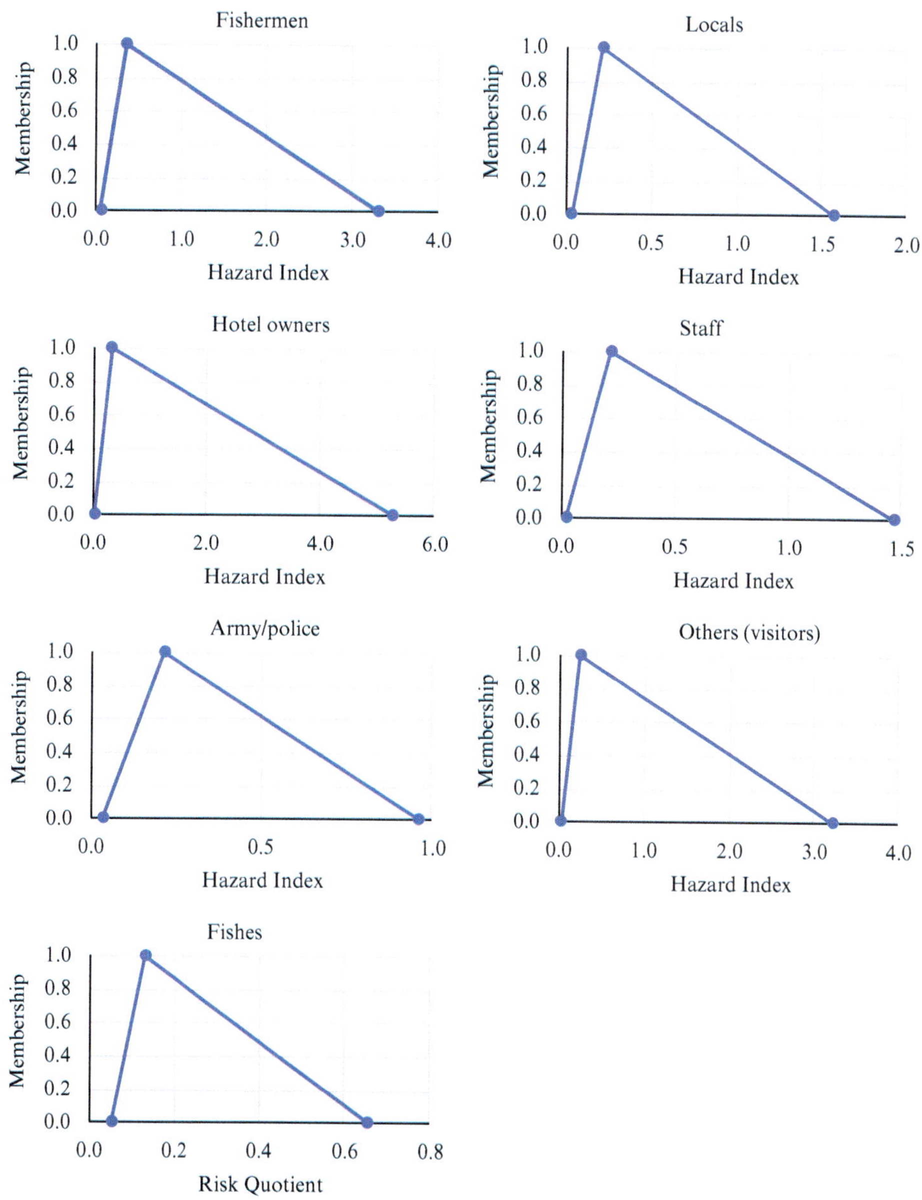

Figure 2.4 Health risks to humans and fishes in fuzzy numbers.

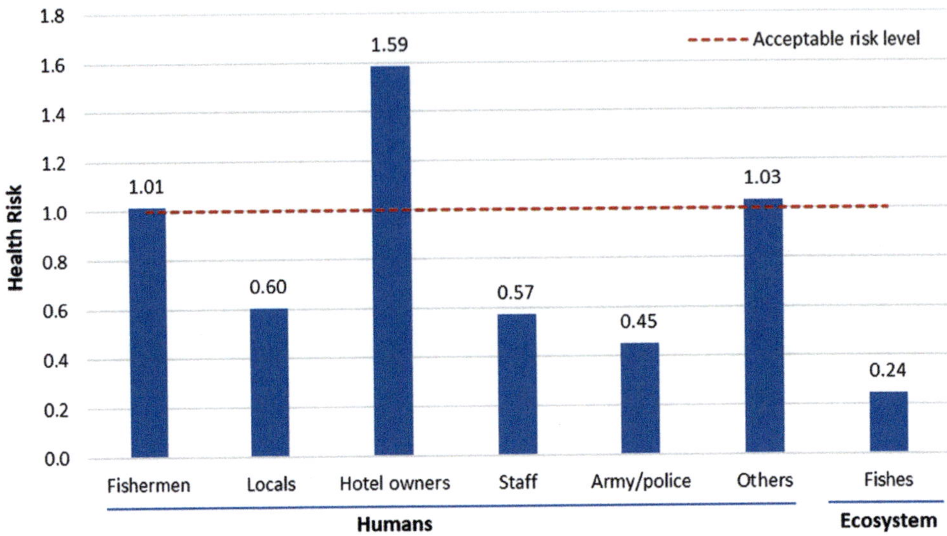

Figure 2.5 Overall health risk to humans by occupation and fishes of ecosystem.

leftover cooked fishes at the end of the day, which may have largely increased their fish consumption rate over other consumers.

The ecological risk of MeHg to fishes in terms of TFN was (0.05, 0.13, 0.66) (Figure 2.4) and its defuzzified value was 0.24 (Figure 2.5). The ecological risk was within the acceptable level, indicating low risk of probability of effects to fishes. The result is supported from the perspective of body burden. The entire range of MeHg body burden of fishes from 0.008 to 0.084 μg/g in Lake Phewa is lower even than the wildlife critical value of 0.1 μg/g of MeHg in fish for piscivorous mammals, indicating low ecotoxicity to fishes [39]. The TFN of ecological risk estimate indicates a range of risk values from 0.05 to 0.66, although the membership varies in between, indicating an uncertainty of one order of magnitude. This range of uncertainty is lower than that in the human health risk estimate.

The present research has limited IRA to humans and fish species of an ecosystem. Due to the lack of actual measurement data, MeHg concentration in lake water and rice grains was estimated with the help of the literature, which may add uncertainty to the results. Similarly, the fish consumption rate was estimated based on the questionnaire survey rather than actual weighing of fish to be eaten. Comparatively, the use of measured MeHg concentration in fish tissue for estimating MeHg in lake water can have less uncertainty compared to the entirely literature-based estimation of MeHg in rice grains. However, the use of a fuzzy approach can incorporate such parameters of uncertainty in analysis [21]. Fish is commonly used as a target organism in ecological risk assessment, and ecotoxicity data of fishes for certain chemicals are available. However, unlike the human toxicity database for human health risk assessment, the database on ecotoxicity of chemicals for ecological risk assessment is limited. In this context, IRA with mutual exchange of information between human health risk assessors and ecological risk assessors becomes more efficient and coherent. Also, IRA can save time and money by preventing a doubling of tasks in exposure assessment, for example chemical concentration measurements in lake water and the chemical transport from water to fishes are common in both human health risk assessment and ecological risk assessment. Furthermore, this research can be extended by including other target species, such as phytoplankton, zooplankton, and fish-eating birds.

2.4 Conclusion

Fuzzy-based IRA of MeHg in Lake Phewa was applied combining human health and ecological risk assessment to incorporate uncertainty in decision-making. The results show that fish consumption contributed approximately 90% or higher of the total human health risk and the remaining risk was attributed to rice consumption. Hotel owners, fishermen, and others (visitors) in particular had potentially significant health risks from MeHg in the lake. The higher health risk was primarily due to very high fish consumption rates in the three occupations. Among them, hotel owners had an especially high health risk. Moreover, the ecological risk to fishes was low and within the acceptable level. However, with a lack of actual measurement data, MeHg concentration in lake water and rice grains was estimated based on the literature, which may add uncertainty to the results. The application of fuzzy techniques in IRA allows risk assessors to approximate uncertainty in decision-making, building more confidence in decisions. This study can be extended by including other target species, such as phytoplankton, zooplankton, and fish-eating birds.

References

1 National Research Council. (2000). *Toxicological Effects of Methylmercury*. Washington: National Academies Press.

2 USEPA. (1997). *Mercury Study Report to Congress: Volume I: Executive Summary*. Washington: United States Environmental Protection Agency, Office of Air Quality Planning and Standards and Office of Research and Development, Office of Air Quality Planning and Standards and Office of Research and Development, Office of Air Quality Planning and Standards.

3 Kim, E., Mason, R.P., and Bergeron, C.M. (2008). A modeling study on methylmercury bioaccumulation and its controlling factors. *Ecological Modelling* 218 (3–4): 267–289.

4 Scheuhammer, A., Braune, B., Man, H., Frouin, H., Krey, A., Letcher, R., Loseto, L., Noël, M., Ostertag, S., Ross, P., and Wayland, M. (2015). Recent progress on our understanding of the biological effects of mercury in fish and wildlife in the Canadian Arctic. *Science of the Total Environment* 509–510: 91–103.

5 Schartup, A.T., Qureshi, A., Dassuncao, C., Thackray, C.P., Harding, G., and Sunderland, E.M. (2018). A model for methylmercury uptake and trophic transfer by marine plankton. *Environmental Science and Technology* 52 (2): 654–662.

6 Kainz, M., Lucotte, M., and Parrish, C.C. (2002). Methyl mercury in zooplankton: the role of size, habitat, and food quality. *Canadian Journal of Fisheries and Aquatic Sciences* 59 (10): 1606–1615.

7 Watras, C.J. and Bloom, N.S. (1992). Mercury and methylmercury in individual zooplankton: implications for bioaccumulation. *Limnology and Oceanography* 37 (6): 1313–1318.

8 Rothenberg, S.E., Windham-Myers, L., and Creswell, J.E. (2014). Rice methylmercury exposure and mitigation: a comprehensive review. *Environmental Research* 14 (4): 407–423.

9 Zhu, W., Song, Y., Adediran, G.A., Jiang, T., Reis, A.T., Pereira, E., Skyllberg, U., and Björn, E. (2018). Mercury transformations in resuspended contaminated sediment controlled by redox conditions, chemical speciation and sources of organic matter. *Geochimica et Cosmochimica Acta* 220: 158–179.

10 Hong, Y.-S., Kim, Y.-M., and Lee, K.-E. (2012). Methylmercury exposure and health effects. *Journal of Preventive Medicine & Public Health* 45 (6): 353–363.

11 World Health Organization. (2001). *Integrated Risk Assessment*. Geneva: WHO/UNEP/ILO.

12 Bridges, J. (2003). Human health and environmental risk assessment: the need for a more harmonised and integrated approach. *Chemosphere* 52 (9): 1347–1351.

13 Wilks, M.F., Roth, N., Aicher, L., Faust, M., Papadaki, P., Marchis, A., Calliera, M., Ginebreda, A., Andres, S., Kühne, R., and Schüürmann, G. (2015). White paper on the promotion of an integrated risk assessment concept in European regulatory frameworks for chemicals. *Science of the Total Environment* 521–522 (1): 211–218.

14 Tripathee, L., Kang, S., Rupakheti, D., Zhang, Q., Huang, J., and Sillanpää, M. (2016). Water-soluble ionic composition of aerosols at urban location in the foothills of Himalaya, Pokhara Valley, Nepal. *Atmosphere* 7 (8): 1–13.

15 Sharma, C.M., Basnet, S., Kang, S., Rosseland, B.O., Zhang, Q., Pan, K., Borgstrøm, R., Li, Q., Wang, W.X., Huang, J., Teien, H.C., and Sharma, S. (2013). Mercury concentrations in commercial fish species of Lake Phewa, Nepal. *Bulletin of Environmental Contamination and Toxicology* 91 (3): 272–277.

16 Thapa, D.S., Sharma, C.M., Kang, S., and Sillanpää, M. (2014). The risk of mercury exposure to the people consuming fish from Lake Phewa, Nepal. *International Journal of Environmental Research and Public Health* 11: 6771–6779.

17 Sharma, C.M., Kang, S., Sillanpää, M., Li, Q., Zhang, Q., Huang, J., Tripathee, L., Sharma, S., and Paudyal, R. (2015). Mercury and selected trace elements from a remote (Gosainkunda) and an urban (Phewa) lake waters of Nepal. *Water, Air, and Soil Pollution* 226 (2). DOI:10.1007/s11270-014-2276-3.

18 Rosseland, B.O., Teien, H.C., Basnet, S., Borgstrøm, R., and Sharma, C.M. (2017). Trace elements and organochlorine pollutants in selected fish species from Lake Phewa, Nepal. *Toxicological and Environmental Chemistry* 99 (3): 390–401.

19 Rupakheti, D., Tripathee, L., Kang, S., Sharma, C.M., Paudyal, R., and Sillanpää, M. (2017). Assessment of water quality and health risks for toxic trace elements in urban Phewa and remote Gosainkunda lakes, Nepal. *Human and Ecological Risk Assessment* 23 (5): 959–973.

20 Sadiq, R. and Tesfamariam, S. (2009). Environmental decision-making under uncertainty using intuitionistic fuzzy analytic hierarchy process (IF-AHP). *Stochastic Environmental Research and Risk Assessment* 23 (75): 75–91.

21 Chhipi-Shrestha, G., Mori, J., Hewage, K., and Sadiq, R. (2016). Clostridium difficile infection incidence prediction in hospitals (CDIIPH): a predictive model based on decision tree and fuzzy techniques. *Stochastic Environmental Research and Risk Assessment* 31: 1–14.

22 Chhipi-Shrestha, G., Hewage, K., and Sadiq, R. (2017). Selecting sustainability indicators for small to medium sized urban water systems using fuzzy-ELECTRE method. *Water Environment Research* 89 (3): 238–249.

23 Poste, A.E., Muir, D.C.G., Guildford, S.J., and Hecky, R.E. (2015). Bioaccumulation and biomagnification of mercury in African lakes: the importance of trophic status. *Science of the Total Environment* 506–507: 126–136.

24 Scudder Eikenberry, B., Riva-Murray, K., Knightes, C., Journey, C., Chasar, L., Brigham, M., and Bradley, P. (2015). Optimizing fish sampling for fish–mercury bioaccumulation factors. *Chemosphere* 135: 467–473.

25 Legay, C., Rodriguez, M.J., Sadiq, R., Sérodes, J.B., Levallois, P., and Proulx, F. (2011). Spatial variations of human health risk associated with exposure to chlorination by-products occurring in drinking water. *Journal of Environmental Management* 92 (3): 892–901.

26 USEPA. (2002). Chemical assessment summary: Methylmercury (MeHg) CASRN 22967-92-6. Washington.

27 Du, M., Wei, D., Tan, Z., Lin, A., and Du, Y. (2015). Predicted no-effect concentrations for mercury species and ecological risk assessment for mercury pollution in aquatic environment. *Journal of Environmental Sciences (China)* 28: 74–80.

28 Dillon, T., Beckvar, N., and Kern, J. (2010). Residue-based mercury dose-response in fish: an analysis using lethality equivalent test endpoints. *Environmental Toxicology and Chemistry* 29 (11): 2559–2565.

29 USEPA. (1998). *Guidelines for Ecological Risk Assessment*. Washington, DC: US Environmental Protection Agency.

30 Liou, T. and Wang, M.J. (1992). Ranking fuzzy numbers with integral value. *Fuzzy Sets and Systems* 50: 247–255.

31 Riva-Murray, K., Bradley, P.M., Scudder Eikenberry, B.C., Knightes, C.D., Journey, C.A., Brigham, M.E., and Button, D.T. (2013). Optimizing stream water mercury sampling for calculation of fish bioaccumulation factors. *Environmental Science and Technology* 47 (11): 5904–5912.

32 Zhang, H., Feng, X., Larssen, T., Shang, L., and Li, P. (2010). Bioaccumulation of methylmercury versus inorganic mercury in rice (*Oryza sativa* L.) grain. *Environmental Science and Technology* 44 (12): 4499–4504.

33 USEPA. (2011). *Exposure Factors Handbook: 2011 Edition*. EPA/600/R. Washington: US Environmental Protection Agency.

34 CBS. (2016). *Annual Household Survey 2014/2015: Major Findings*. Kathmandu, Nepal: Central Bureau of Statistics.

35 USEPA. (2001). *Mercury Maps: A Quantitative Spatial Link Between Air Deposition and Fish Tissue*. Washington: US Environmental Protection Agency, Office of Water.

36 Scudder, B.C., Chasar, L.C., Wentz, D.A., Bauch, N.J., Brigham, M.E., Moran, P.W., and Krabbenhoft, D.P. (2009). *Mercury in Fish, Bed Sediment, and Water from Streams Across the United States, 1998–2005*. Reston, VA: US Geological Survey.

37 Tang, S., Fan, H., Mao, T., and Huang, Z. (2017). Methylmercury accumulation in rice plants (*Oryza sativa* L.) in low Hg areas and the potential health: risk exposing in extreme micro-environments. *Advanced Materials Science* 2 (3): 1–6.

38 Zhao, L., Anderson, C.W.N., Qiu, G., Meng, B., Wang, D., and Feng, X. (2016). Mercury methylation in paddy soil: source and distribution of mercury species at a Hg mining area, Guizhou Province, China. *Biogeosciences* 13 (8): 2429–2440.

39 Yeardley, R.B., Lazorchak, J.M., and Paulsen, S.G. (1998). Elemental fish tissue contamination in northeastern US lakes: evaluation of an approach to regional assessment. *Environmental Toxicology and Chemistry* 17 (9): 1875–1884.

3

A Fuzzy Approach to Analyze Data Uncertainty in the Life Cycle Assessment of a Drinking Water System

A Case Study of the City of Penticton (CA)

Thais Ayres Rebello, Gyan Chhipi-Shrestha, Venkata U.K. Vadapalli, Emmi Matern, Rehan Sadiq, and Kasun Hewage

School of Engineering, University of British Columbia (Okanagan), Kelowna, BC, Canada

3.1 Introduction

Water is a critical element in the urban environment, representing 90% of the entire materials flow in an urban environment. Furthermore, water systems are the source of almost all urban water consumed, and only a small portion directly derives from alternative sources, such as rainwater or private underground captures [1]. The urban water cycle (UWC) is the spatiotemporal interaction between hydrological processes and water, considering the supply, treatment, distribution, consumption, use, collection, followed by wastewater collection, treatment, and disposal, reuse, and provision. In other words, the UWC is defined as the conjunction of natural and manmade water streams [2]. A typical manmade UWC consists of water extraction, treatment, distribution, use, collection, and wastewater treatment and disposal [3], which is a linear model that does not fulfill current needs.

An UWC is composed of four important inputs: water, contaminants, energy, and chemicals. Supply sources such as surface, groundwater, and precipitation provide most of the water in the urban cycle. Contaminants, on the other hand, enter the water cycle from the surface or groundwater flows, wastewater flows from water use (domestic, agricultural, and industrial), wastewater treatment flows, rainwater flows, and atmospheric pollutants conveyed by precipitation [2]. The third input is energy, a vital resource due to cost representativeness in urban systems and environmental impacts related to greenhouse gases and natural resource use. Energy consumption occurs in the treatment systems, water supply, and thermal water heating [2]. Additionally, energy can be produced during wastewater treatment, through biomass or biogas production [4]. Lastly, chemicals enter the UWC through the water and wastewater treatment systems, and they are often used as coagulants and disinfectants, as well as industrial discharges [2].

In this context, the analysis of the environmental impacts of urban water systems becomes more significant, and different strategies can be used to analyze the final environmental

Fuzzy Systems Modeling in Environmental and Health Risk Assessment, First Edition. Edited by Boris Faybishenko, Rehan Sadiq, and Ashok Deshpande.

impacts. For instance, diverse environmental indicators can be used [5], and the material flux analysis can be applied [6]. However, these methodologies can allocate different impacts from one life cycle stage to another. Hence, a comprehensive method that evaluates the entire life cycle, such as the life cycle assessment (LCA), is required [7].

While LCA is an important tool to assess environmental impacts throughout the entire life cycle of a product or system [8], the LCA process has many significant uncertainties, such as scenario, parameter, and model uncertainties [9]. Additionally, the application of uncertainty analysis to LCA frameworks is not common in current studies; only 20% of studies present this analysis [10].

Regarding urban water studies, few documents are available in the literature concerning the distribution system [11–16]. The lack of previous studies further increases the uncertainties related to this urban water system stage, especially regarding available data and considerations made throughout the analysis.

Some of the important uncertainties current urban water systems need to deal with are (i) data not available, (ii) available data not certain (uncertainty), and (iii) many values are available for a parameter (variability) [9]. The lack of an appropriate value refers to a situation in which researchers have to use data from another technology or estimate the value. Additionally, sometimes more than one value is available, especially for the analysis of technology that presents variations over time, such as urban water facilities. Several methodologies are available to study uncertainties, such as parameter and scenario variation, also referred to as sensitivity analysis; sampling methods, analytical methods, and nontraditional methods such as the fuzzy set methods, Bayesian methods, nonparametric statistics, robust statistics, neural networks, artificial intelligence methods, and Monte Carlo analyses [9]. However, some of these methods can greatly increase the complexity of the analysis, and large amounts of data and resources are required. For instance, Monte Carlo simulations require sufficient data to estimate a reliable standard deviation of a dataset and dedicated software at disposition, which may not always be available.

Fuzzy methodology allows the use of intervals and membership to different values within the interval, which can translate vague information into numerical factors while performing LCA [17]. Thus, fuzzy methodology can insert not only the data uncertainty but also the variability of stakeholders' opinions in the final aggregation.

Few papers are available considering urban water system uncertainties. Jeong et al. [18] quantify data uncertainty as the squared standard deviation based on the type of input and data quality inputs. The authors calculated the impact values 5000 times, using Monte Carlo methodology. Similarly, Pillot et al. [19] calculated a statistical distribution of life cycle impact assessment with a Monte Carlo simulation, using 10 000 iterations and using a 1–5% convergence criteria. Finally, García-Sánchez et al. [20] used data means and standard deviations, bottom and top outliers, and coefficients of variation to implement a Monte Carlo simulation with the SimaPro 9.0 software.

This chapter aims to analyze the uncertainties in the LCA of the water treatment and water distribution system of the city of Penticton. To achieve this, a triangular fuzzy number was used, considering three scenarios for the environmental impacts: optimistic, most probable number, and pessimistic. This study analyzed the production, construction, and water treatment and distribution systems' use. Afterward, the analysis combined all environmental indicators using the weighed product method (WPM) into a single score. The variability of the WPM single-score indicator was compared with the inbuilt results of SimaPro. All results were found using ReCiPe midpoint and ReCiPe endpoint. Additionally, the analysis considered each life cycle phase, as well as the total contributions to understanding how uncertainty affects the final midpoint and endpoint categories.

Finally, this chapter showcases a methodology that can be applied in future studies when sufficient data or computational resources are not available. The present contribution showcases the possibility of impact aggregation considering different weights from those assigned in the software to adapt site-specific features in the data and to include various environmental impact priorities in making informed decisions by stakeholders.

3.2 Methodology

The methodology of this chapter is divided into three parts: (i) LCA of the most probable number and identification of important processes, (ii) LCA of the optimistic and pessimistic scenarios, and (iii) data analysis, in which the fuzzy indicators are evaluated.

3.2.1 Life Cycle Assessment

To conduct the LCA assessment, ISO 14040:2006 [21] and ISO 14044:2006 [22] were used as the baseline method. These norms describe the LCA process, which consists of four main stages: (i) scope and goal definition, (ii) life cycle inventory, (iii) life cycle impact assessment, and (iv) interpretation.

3.2.1.1 Scope and Goal Definition

This work aimed to evaluate the uncertainties of the urban water treatment and distribution system of the city of Penticton. The analysis considered the construction and use of the water treatment system. For the distribution system pipe production, installation (construction), and use were included for analysis. Hence, the demolition and maintenance phases were excluded for both water treatment and distribution systems. This is due to the lack of data for these phases.

The functional unit of the study was the annual treatment and distribution of drinking water in 2016 in Penticton. The system boundaries considered energy and chemical uses in the production, construction, and operation of water treatment and distribution systems. Water extraction and water use were not analyzed in the present study (Figure 3.1).

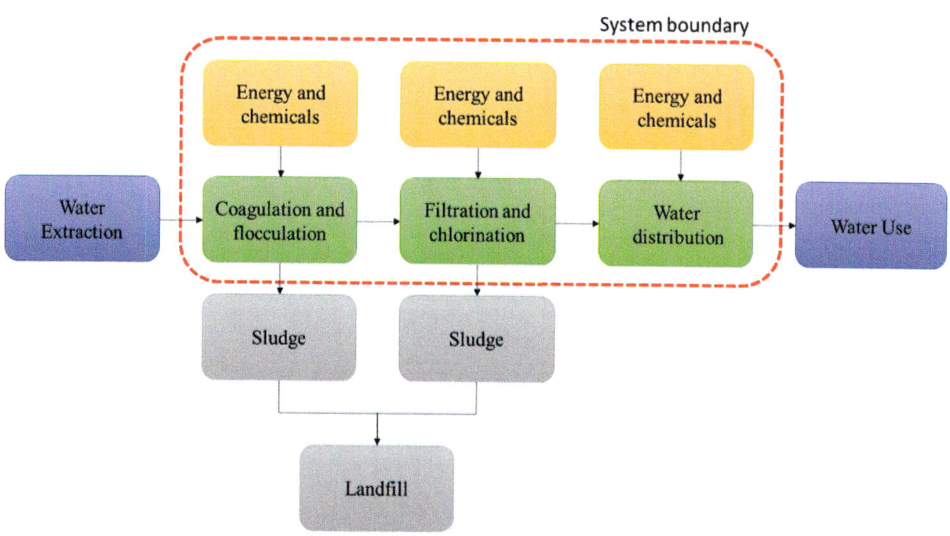

Figure 3.1 System boundaries used in the uncertainty analysis of urban water treatment and distribution in Penticton.

3.2.1.2 Life Cycle Inventory

This study considered the following environmental aspects: chemical use, energy consumption, construction of the facilities, and transportation of the construction materials. All of the background information used in ecoinvent 3.0 was either selected from a British Columbia database or a Canadian database, when available. If no such databases were available, a global database was selected. Table 3.1 shows all data used for the most probable scenarios.

Table 3.1 Life cycle inventory for the production of pipelines, pipeline construction, and water system construction and use.

Production of pipelines			
Process	Value	Unit	Source
Extrusion, plastic pipes	33.53	tonne	—
Pump	12.5	p	—
Water treatment and distribution systems construction			
Pipe installation	3940	p	—
Pump	0.06	p	—
Transport	74 513.28	tkm	—
Water treatment facility	0.142	p	—
Water treatment and distribution systems use			
Chlorine	7100	kg	—
Aluminum oxide	21726	kg	[23]
Hydrochloric acid	98 974	kg	[23]
Polyacrylamide	14 910	kg	—
Electricity	221	MWh	—
Electricity	1169	MWh	—

Note: In source column, non-described are foreground data from the specific facility.

To analyze the uncertainties, a variation in the most critical criteria for evaluation was made. Table 3.2 shows all variability input in the data.

3.2.1.3 Life Cycle Impact Assessment

Different environmental impact categories were analyzed for the midpoint (ReCiPe midpoint 1.13 (H) World (H)), endpoint, and the single score levels (ReCiPe endpoint 1.13 (H)

Table 3.2 Data variability used to construct the low and high environmental impacts scenarios.

Important processes in the assessment	Variation (%)	Reference
Electricity[3]	27	[24]
Hydrochloric acid[3]	15	Assumed
Aluminum oxide[3]	15	Assumed
Extrusion, plastic pipes[1]	1	[25]
Other processes to test uncertainty		
Transportation[2]	10	Assumed

[1] Pipeline manufacturing;
[2] transportation in the construction phase; and
[3] water treatment stage.

World (H/H)), as seen in Table 3.3. As remarked by Dekker et al. [26], the ReCiPe methodology harmonizes the midpoint and endpoint cause–effect calculations. The calculations and descriptions for each environmental impact category can be found elsewhere, including Goedkoop et al. [27].

The environmental impacts were calculated using the variations given in Table 3.2 to generate the optimistic, most probable, and pessimistic scores.

Table 3.3 Environmental impact categories from ReCiPe endpoint 1.13 (H) World (H/H) analyzed for the midpoint, endpoint, and single-score evaluations.

Impact category	Unit	Impact category	Unit	Damage category
Midpoint impact		**Endpoint impact**		
Climate change	kg CO_2 eq	Climate change Human health	DALY	Total
Ozone depletion	kg CFC-11 eq	Ozone depletion	DALY	Human Health
Terrestrial acidification	kg SO_2 eq	Human toxicity	DALY	Ecosystems
Freshwater eutrophication	kg P eq	Photochemical oxidant formation	DALY	Resources
Marine eutrophication	kg N eq	Particulate matter formation	DALY	—
Human toxicity	kg 1,4-DB eq	Ionizing radiation	DALY	—
Photochemical oxidant formation	kg NMVOC	Climate change Ecosystems	species · yr	—
Particulate matter formation	kg PM_{10} eq	Terrestrial acidification	species · yr	—
Terrestrial ecotoxicity	kg 1,4-DB eq	Freshwater eutrophication	species · yr	—
Freshwater ecotoxicity	kg 1,4-DB eq	Terrestrial ecotoxicity	species · yr	—
Marine ecotoxicity	kg 1,4-DB eq	Freshwater ecotoxicity	species · yr	—
Ionizing radiation	kBq U-235 eq	Marine ecotoxicity	species · yr	—
Agricultural land occupation	m^2a	Agricultural land occupation	species · yr	—
Urban land occupation	m^2a	Urban land occupation	species · yr	—
Natural land transformation	m^2	Natural land transformation	species · yr	—
Water depletion	m^3	Metal depletion	$	—
Metal depletion	kg Fe eq	Fossil depletion	$	—
Fossil depletion	kg oil eq	—	—	—

CO_2 eq: carbon dioxide equivalent per kilogram; CFC-11: chlorofluorocarbon-11; SO_2: sulfur dioxide; 1,4-DB: 1,4-dichlorobenzene; DALY: disability-adjusted life year; NMVOC: nonmethane volatile organic compounds; PM_{10}: particulate matter, particles with a diameter of 10 μm or less.

3.2.1.4 Interpretation

The analysis of the uncertainty in this work was performed using fuzzy logic (FL) combined with the WPM method. Additionally, the results were compared among themselves to understand the variability in each environmental impact category. Finally, the aggregated index, generated by the WPM, was analyzed and compared with the single score generated by the SimaPro software.

3.2.2 Fuzzy Approach

According to Yoon and Hwang [28], there are three ways to describe data: point, in which only one value is attributed; imprecise, for results approximation; and fuzzy, in which an interval with membership is attributed. Additionally, fuzzy numbers can be expressed in different ways. A triangular fuzzy number (TFN) is simple and commonly used [29]. In TFNs, a range of values constituted by low, most probable, and high values substitute the point values [30]. FL can be used in many different applications, including LCA for uncertainties analysis [28].

3.2.2.1 Fuzzification

TFNs are composed of an interval containing an optimistic, most probable, and pessimist number – Eq. (3.1) [31]. In the present contribution, the fuzzy number of each impact category was generated through multiple simulations, considering the variations presented in Section 3.2.2.1 (Table 3.2).

$$F_i = (O_i, A_i, P_i) \tag{3.1}$$

in which F_i is the fuzzy number for the impact category i, O_i is the optimistic value, using the inputs as calculated in Eq. (3.2), A_i is the average scenario, also referred to in the present contribution as the most probable scenario, and P_i refers to the pessimistic scenario, using the inputs calculated with Eq. (3.3).

$$I_j = M_j - V_j * M_j \tag{3.2}$$

$$I_j = M_j + V_j * M_j \tag{3.3}$$

where I_j is the input for the j flux, M_j is the medium value for the input (Table 3.1), and V_j is the flux variation (%). After generating each varying input, the ReCiPe methodology was applied to calculate the categories optimistic, most probable, and pessimist values, which correspond to the fuzzy number of each impact category.

After this approach, the WPM methodology was applied to generate an aggregated value for the midpoint categories. The weights used for the fuzzification of the WPM are described in Section 3.2.2.2. and the WPM method in Section 3.2.2.3. Finally, the WPM method was applied to each point in the fuzzy number (optimistic, most probable, and pessimist), generating a fuzzy WPM score, which also contained three points (optimistic, most probable, and pessimist). The arithmetic operations of fuzzy methods can be referred to elsewhere, including in Chhipi-Shrestha et al. [29].

3.2.2.2 Weights Used in the Analysis

In the fuzzy WPM analysis, different weights were used for the optimistic, most probable, and pessimistic levels (Table 3.4). A variation of ±20% was assumed in the eutrophication, toxicity, climate change, water use, and metal and fossil depletion categories, which are more important in analyzing urban water systems. For other less important impact categories, a variability of ±10% was assigned in the weights. All weights were then normalized, considering the most probable weight total (Table 3.5).

Table 3.4 Weights used for the midpoint categories.

Impact category	Fuzzy weights (optimistic; most probable*; pessimistic)
Climate change	(16.84, 21.06, 25.27)
Ozone depletion	(5.67, 6.31, 6.94)
Terrestrial acidification	(5.58, 6.2, 6.82)
Freshwater eutrophication	(2.24, 2.8, 3.36)
Marine eutrophication	(2.36, 2.96, 3.55)
Human toxicity	(3.17, 3.97, 4.76)
Photochemical oxidant formation	(4.30, 4.78, 5.25)
Particulate matter formation	(8.06, 8.96, 9.85)
Terrestrial ecotoxicity	(1.53, 1.92, 2.30)
Freshwater ecotoxicity	(1.53, 1.92, 2.30)
Marine ecotoxicity	(1.53, 1.92, 2.30)
Ionizing radiation	(4.50, 5.01, 5.51)
Agricultural land occupation	(7.14, 7.94, 8.73)
Urban land occupation	(7.14, 7.94, 8.73)
Natural land transformation	(7.14, 7.94, 8.73)
Water depletion	(6.88, 8.51, 10.21)
Metal depletion	(6.04, 7.55, 9.06)
Fossil depletion	(6.65, 8.32, 9.98)
Total	(98.316, 116.01, 133.70)

**Source:* Adapted from Sale et al. [32].

Table 3.5 Normalized weights used for the midpoint categories.

Impact category	Normalized weights
Climate change	(0.15, 0.18, 0.22)
Ozone depletion	(0.05, 0.05, 0.06)
Terrestrial acidification	(0.05, 0.05, 0.06)
Freshwater eutrophication	(0.02, 0.02, 0.03)
Marine eutrophication	(0.02, 0.03, 0.03)
Human toxicity	(0.03, 0.03, 0.04)
Photochemical oxidant formation	(0.04, 0.04, 0.05)
Particulate matter formation	(0.07, 0.08, 0.08)

(Continued)

Table 3.5 (Continued)

Impact category	Normalized weights
Terrestrial ecotoxicity	(0.01, 0.02, 0.02)
Freshwater ecotoxicity	(0.01, 0.02, 0.02)
Marine ecotoxicity	(0.01, 0.02, 0.02)
Ionizing radiation	(0.04, 0.04, 0.05)
Agricultural land occupation	(0.06, 0.07, 0.08)
Urban land occupation	(0.06, 0.07, 0.08)
Natural land transformation	(0.06, 0.07, 0.08)
Water depletion	(0.06, 0.07, 0.09)
Metal depletion	(0.05, 0.07, 0.08)
Fossil depletion	(0.06, 0.07, 0.09)

Source: Adapted from Sale et al. [32].

3.2.2.3 Life Cycle Impact Index

The weighted product method (WPM) is largely used to compare different alternatives for multicriteria analysis. In this chapter, the analysis applied the WPM to understand whether the final aggregated estimate would be different from the single score generated by the SimaPro software and if the variations identified would be comparable. This methodology was selected because it does not require the normalization of the data used, and only one alternative is analyzed primarily to evaluate the uncertainty involved [28]. Equation (3.4) shows the WPM applied.

$$\text{Impact Index}\left(\mathrm{I_i}\right) = \prod_{j=i}^{n} x_{ij}^{w_j}, \quad i=1 \tag{3.4}$$

In which I_i is a life cycle impact index, x_{ij} is the environmental impact calculated for each impact category, and w_j is the weight assigned for that impact category in the specific scenario evaluation (optimistic, most probable scenario, and pessimistic). To aggregate the total life cycle impact index, it was considered that all life cycle stages have the same weight ($w_j = 1$).

3.3 Results and Discussion

3.3.1 Pipe Production

For pipe production, the overall environmental impact variation for all categories (midpoint, endpoint, and single score) presented a variation of less than 1%. These results are important because the uncertainties analyzed did not greatly affect the environmental impact of this life cycle stage, also the variability for this life cycle stage is carried out through the different levels of environmental impacts (midpoint, endpoint, and single score calculations). Table 3.6 shows the optimistic, most probable, and pessimistic environmental impacts calculated for each midpoint environmental impact category tested. Categories from the endpoint and single score analysis presented a similar variation.

Table 3.6 Life cycle impacts of pipe production.

Impact category	Unit	TFN
Climate change	kg CO_2 eq	(15 529.55, 15 685.27, 15 840.98)
Ozone depletion	kg CFC-11 eq	(0.00094, 0.00095, 0.00096)
Terrestrial acidification	kg SO_2 eq	(77 027, 77 790, 78.55)
Freshwater eutrophication	kg P eq	(8.17, 8.25, 8.33)
Marine eutrophication	kg N eq	(3.55, 3.59, 3.62)
Human toxicity	kg 1,4-DB eq	(6173.12, 6229.07, 6285.02)
Photochemical oxidant formation	kg NMVOC	(41.71, 42.13, 42.54)
Particulate matter formation	kg PM10 eq	(44.20, 44.64, 45.08)
Terrestrial ecotoxicity	kg 1,4-DB eq	(0.92, 0.92, 0.93)
Freshwater ecotoxicity	kg 1,4-DB eq	(200.65, 202.54, 204.43)
Marine ecotoxicity	kg 1,4-DB eq	(188.06, 189.81, 191.57)
Ionizing radiation	kBq U-235 eq	(2399.55, 2423.68, 2447.80)
Agricultural land occupation	m^2a	(5492.64, 5548.02, 5603.39)
Urban land occupation	m^2a	(183.62, 185.45, 187.27)
Natural land transformation	m^2	(2.46, 2.49, 2.51)
Water depletion	m^3	(369.53, 373.27, 376.99)
Metal depletion	kg Fe eq	(778.09, 783.71, 789.32)
Fossil depletion	kg oil eq	(3928.95, 3968.35, 4007.75)

While the variation amongst the optimistic/pessimistic and the most probable scenario was small, some categories showed important variations for the absolute numbers. For instance, in the climate change category, the environmental impact is (15 529, 15 685, 15 840) kg CO_2 eq, which was mainly due to chemical use uncertainty and transportation uncertainties. Another important category that showed a similar fuzzy number variation was the human toxicity in which the environmental impact is (6173, 6229, 6285) kg 1,4 DB (dichlorobenzene) in the midpoint category.

3.3.2 Water Treatment Plant Construction and Pipeline Installation

Considering the water treatment system construction and pipe installation, most environmental impact categories did not present a high environmental impact variation, and all categories for the midpoint, endpoint, and single score (Table 3.9) presented a variation smaller than 1%. Table 3.7 shows the optimistic, most probable, and pessimistic environmental impacts calculated for each midpoint environmental impact category tested.

Furthermore, as in the pipeline analysis, although the percent variations were not important, the absolute fuzzy numbers had important variances. For instance, the final score for global warming potential was (832 062; 832 401; 832 739) kg CO_2 eq in the midpoint category, (1.1648; 1.1653; 1.1658) DALY (disability-adjusted life years) for global warming potential in the endpoint category for human health, and (0.0065; 0.0066; 0.00661) species · year for ecosystems' global warming potential.

Table 3.7 Life cycle impacts of drinking water treatment and distribution network construction.

Impact category	Unit	TFN
Climate change	kg CO_2 eq	(832 062.40, 832 401, 832 739.60)
Ozone depletion	kg CFC-11 eq	(0.071, 0.071265, 0.071)
Terrestrial acidification	kg SO_2 eq	(3542.73, 3544.56, 3546.39)
Freshwater eutrophication	kg P eq	(282.57, 282.59, 282.62)
Marine eutrophication	kg N eq	(208.11, 208.21, 208.31)
Human toxicity	kg 1,4-DB eq	(475 962.20, 476 077.9, 476 193.70)
Photochemical oxidant formation	kg NMVOC	(3780.02, 3782.85, 3785.69)
Particulate matter formation	kg PM10 eq	(2258.53, 2259.54, 2260.55)
Terrestrial ecotoxicity	kg 1,4-DB eq	(109.63, 109.84, 110.04)
Freshwater ecotoxicity	kg 1,4-DB eq	(16 154.08, 16 155.24, 16 156.4)
Marine ecotoxicity	kg 1,4-DB eq	(16 006.76, 16 009.15, 16 011.54)
Ionizing radiation	kBq U-235 eq	(40 168.38, 40 197.02, 40 225.67)
Agricultural land occupation	m^2a	(19 910.16, 19 915.14, 19 920.12)
Urban land occupation	m^2a	(13 3537.9, 13 3571.5, 13 3605.1)
Natural land transformation	m^2	(1109.12, 1109.26, 1109.40)
Water depletion	m^3	(16 419.35, 16 420.45, 16 421.55)
Metal depletion	kg Fe eq	(261 527.7, 261 537.4, 261 547.1)
Fossil depletion	kg oil eq	(203 918.1, 204 045.4, 204 172.7)

3.3.3 Water Treatment and Distribution System Operation

Water use and pipeline operation were the phases in the life cycle of drinking water treatment and distribution which presented the highest sensitivities to the changes applied. The final variations ranged from 3% in marine eutrophication to 15% in metal depletion for the midpoint categories. Endpoint categories presented a similar range for the variations analyzed, from 7% in fossil depletion to 15% in metal depletion. Finally, single score categories varied from 8 to 11%, for resources and human health, respectively. Table 3.8 shows the optimistic, most probable, and pessimistic midpoint environmental impacts calculated for each environmental impact category tested. The variations found through the different methodologies (midpoint, endpoint, and single score) were due to high sensitivity to the aluminum oxide and hydrochloric acid variations used in the operational phase. Finally, these results highlight the importance of the fuzzy methodology for systems that do not have an important number of data, showing that the input variations impact results over 10% in some impact categories.

Furthermore, the analysis indicates that the global warming impact was (118 408; 131 217; 144 026) kg CO_2 eq for the midpoint category and (0.17; 0.18; 0.20) DALY in the endpoint category, considering human health and (0.00094; 0.00104; 0.001142) species · year for the climate change ecosystems. These results show an important variability, due to the variations in the inputs. In this life cycle stage, the main reason for the variations was the uncertainty in chemical use (aluminum oxide and hydrochloric acid).

Table 3.8 Life cycle impacts of drinking water treatment and pipeline operation.

Impact category	Unit	TFN
Climate change	kg CO_2 eq	(118 408.5, 131 217.3, 144 026)
Ozone depletion	kg CFC-11 eq	(0.091, 0.105, 0.119)
Terrestrial acidification	kg SO_2 eq	(725.67, 807.33, 889.00)
Freshwater eutrophication	kg P eq	(57.24, 65.35, 73.45)
Marine eutrophication	kg N eq	(153.67, 158.33, 162.98)
Human toxicity	kg 1,4-DB eq	(117 034.6, 135 420.2, 153 805.8)
Photochemical oxidant formation	kg NMVOC	(389.95, 439.40, 488.85)
Particulate matter formation	kg PM10 eq	(316.61, 359.08, 401.55)
Terrestrial ecotoxicity	kg 1,4-DB eq	(84.44, 98.72, 112.99)
Freshwater ecotoxicity	kg 1,4-DB eq	(2356.63, 2725.13, 3093.63)
Marine ecotoxicity	kg 1,4-DB eq	(2906.82, 3372.51, 3838.21)
Ionizing radiation	kBq U-235 eq	(10 589.14, 11 968.22, 13 347.29)
Agricultural land occupation	m^2a	(11 368.11, 13 123.01, 14 877.90)
Urban land occupation	m^2a	(1659.82, 1910.94, 2162.06)
Natural land transformation	m^2	(103.94, 121.18, 138.43)
Water depletion	m^3	(6240.70, 7199.98, 8159.26)
Metal depletion	kg Fe eq	(16 487.24, 19 326.44, 22 165.63)
Fossil depletion	kg oil eq	(40 940.34, 44 065.42, 47 190.49)

3.3.4 Life Cycle Impact Index of Drinking Water Systems

Considering the midpoint results for the different life cycle stages, the final aggregated number (i.e. life cycle impact index) presented an important variability in the midpoint categories. The production stage, which was the least affected, presented a 60% decrease and a 154% increase, and the fuzzy WPM final aggregation score was (89.35; 225.88; 572.64), which shows an important variability, due to input uncertainties.

Additionally, the construction phase showed a 79% decrease and a 378% increase in the overall WPM analysis and the final score was (3233, 15 454, 73 868). The use phase presented a final score variation of 76% in the optimistic and 325% in the pessimistic scenario (Table 3.9). Finally, the differences were increased in the final aggregation with the WPM method, due to the calculation method.

Table 3.9 Life impact index based on midpoint categories.

Life cycle stage	TFN
Production	(89.35, 225.88, 572.64)
Construction	(3233.75, 15 454.61, 73 868.79)
Use	(728.31, 3029.83, 12 866.75)

While the difference in the final WPM and single score numbers was expected, due to the differences in the calculation methodologies, the SimaPro single score calculated presented a higher resilience to the variations applied, with a 10% variability for the life cycle stage, for the present contribution.

3.3.5 Software-based Aggregated Life Cycle Impact Analysis

The uncertainty variances applied in this study have influenced the results less than 2% for most midpoint impact categories, and up to 8.35% in ozone depletion. Other categories with high percentage variation were freshwater eutrophication (±2.31%), human toxicity (±3.00%), and terrestrial ecotoxicity (6.92%). The fuzzy numbers for those categories were (0.16; 0.17; 0.19) kg CFC-11 eq, for ozone depletion, (194.99; 209.48; 223.97) kg 1,4-DB eq, (599 169; 617 727; 63 6284) kg 1,4-DB eq for human toxicity.

The endpoint categories presented the same range, and ozone depletion varied 9.23% when compared to the most probable scenario. The fuzzy number for the ozone depletion in the endpoint category was (0.00047; 0.00052; 0.00056) DALY, while the terrestrial ecotoxicity number was (2.94E-05; 3.16E-05; 3.38E-05) species · year.

Single score categories given by the software presented a lower variation than what was observed in the previous analysis (1–2%) (Table 3.10). The tendencies of the fuzzy analysis showed that the variations were diluted in the total LCA, but they continued to present important ranges for midpoint and endpoint categories.

Table 3.10 Impact scores for different life cycle stages.

Damage category	Total life cycle impact (TFN)
Total	(134.49, 136.43, 138.36)
Human health	(53.97, 54.92, 55.86)
Ecosystems	(5.78, 5.86, 5.93)
Resources	(74.74, 75.65, 76.57)
Pipe production (TFN)	
Total	(1.78, 1.79, 1.81)
Human health	(0.83, 0.83, 0.84)
Ecosystems	(0.09, 0.09, 0.09)
Resources	(0.86, 0.87, 0.88)
Water treatment and distribution construction (TFN)	
Total	(115.14, 115.19, 115.24)
Human health	(45.88, 45.89, 45.91)
Ecosystems	(5.12, 5.12, 5.12)
Resources	(64.15, 64.17, 64.20)
Water treatment and distribution use (TFN)	
Total	(17.57, 19.44, 21.32)
Human health	(7.27, 8.19, 9.11)
Ecosystems	(0.58, 0.65, 0.72)
Resources	(9.73, 10.61, 11.49)

3.3.6 Total Life Cycle

The final score aggregation for the midpoint categories was (3988.36; 20 228.81; 103 211.30), which represents a 90% decrease and 410% increase when comparing optimistic to the most probable number and the pessimistic and the most probable number, respectively. For the endpoint categories, the final aggregating score was (0.036756; 0.049464; 0.066917), which shows a 26% decrease and a 35% increase in the final aggregation index.

When considering the total LCA, the WPM presented a higher sensitivity to the uncertainty analysis than the SimaPro methodology. These disparities are mainly due to aluminum oxide and hydrochloric acid variations. The results emphasize the importance of the fuzzy methodology in the total LCA measurements as well as the stage comparisons, and the methodology can be applied to studies that have small availability of data, but important variability in data points.

3.3.7 Overall Discussion

Few studies have analyzed the uncertainty of water systems, especially due to the lack of data in the field. Jeong et al. [18], for instance, quantified the data uncertainty using the squared standard deviation based on the type of input. Additionally, the authors applied data quality indicators, developed by ecoinvent: reliability, completeness, temporal correlation, geographic correlation, further technological correlation, and sample size [33]. The authors found that, for their scenarios analyzed, the uncertainties studied were significant, even at the midpoint analysis level. The variations encountered in their work varied from 16% in global warming to 232% in noncarcinogenic toxicity. The differences between the uncertainty results from Jeong et al. [18] and the present contribution are due to different methodologies used for the life cycle impact assessment and uncertainty evaluation.

Similarly, Pillot et al. [19] used the ReCiPe methodology and ran a Monte Carlo simulation up to 10 000 times. The authors found that the variability of the results could be as high as 40% in the single-score ReCiPe categories (resources, human health, and ecosystem). In this case, the authors applied a different methodology for uncertainty analysis, using higher variation in their inputs. Finally, García-Sánchez et al. [20] also applied a Monte Carlo simulation with 50 000 iterations. The authors quantified the means, bottom and top 2.5% outliers, and variation coefficients. The uncertainties related in their paper showed a 104.66% variation for noncarcinogenic toxicity and 83% in freshwater eutrophication.

While all authors have previously used the Monte Carlo approach, it requires a large dataset to calculate the average and the variances of the points. When there are no such resources available, the fuzzy analysis can be performed to simplify the uncertainty calculations. Additionally, this approach can also be performed in research groups with scarce resources that cannot access software with the Monte Carlo evaluation.

The uncertainty analysis using a fuzzy approach presented some interesting trends. First of all, similar variations were identified in the midpoint and endpoint categories, when considering the individual impact categories. The final aggregation using the WPM, however, showed an increase in the variability when all impact categories are considered in a single indicator. This behavior was not seen for the SimaPro results, in which the single-score simulations presented a higher resilience when compared to the previous endpoint and midpoint analysis. Figure 3.2 shows the tendencies in the uncertainties as analyzed.

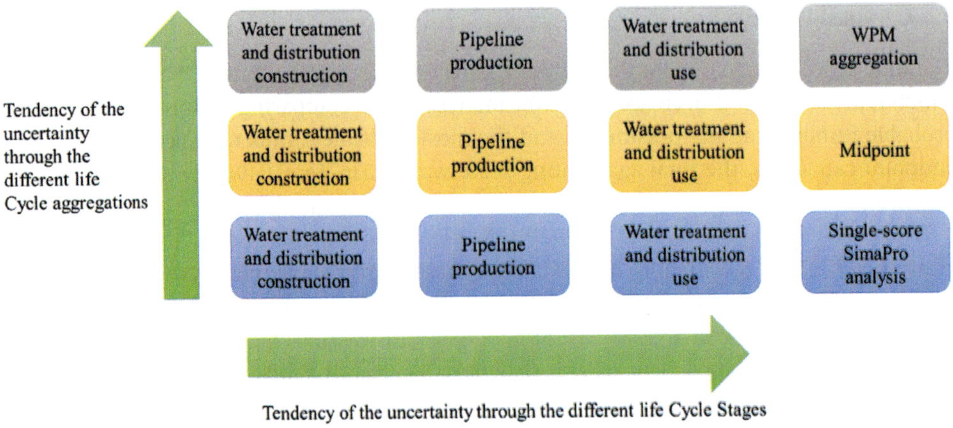

Figure 3.2 Final tendencies assessed in the uncertainty analysis.

3.3.8 Study Limitations

While this work presents an important demonstration of how to use FL on datasets to analyze the uncertainty of results, there are some limitations to the work, which are listed below:

- Use of a background database, which may not be representative of the reality of the study;
- The analysis focused on the main inputs concerning environmental impact, thus future works should apply the same framework to all data inputs; and
- While all data uncertainties were analyzed together, the work could not provide insights on which data input made the greatest contribution, which can be obtained in future works through an ANOVA analysis.

3.4 Conclusion

This work aimed to analyze the importance of different uncertainty variations to drinking water treatment and distribution systems to understand how data can affect the outcomes of an environmental impact analysis. A fuzzy WPM-based life cycle impact was used to understand if the variations would greatly impact the final ranking alternatives. The methodology presented in this chapter can be applied in studies with high data variability and uncertainty but that do not present enough data points to calculate a representative average and standard deviations using the Monte Carlo simulation method or by research groups or enterprises that do not have access to Monte Carlo simulation software.

Additionally, this study showed that the percentage differences between the individual environmental impact categories were not significant. However, important variations were perceived in the final aggregation through the WPM methodology, varying from 154 to 378% in production and construction, respectively. Furthermore, the total LCA for the midpoint categories also presented small percentage variations. When the results were aggregated to a life cycle impact index, however, the percentual variation presented an 80% decrease and a 410% increase for the midpoint aggregation.

Finally, this work shows that, while individual environmental impact categories may show a small percentual variation through the use of FL, the final aggregation (impact index) of those categories presents an important percentual variation. Another important point is that the optimistic, most probable, and pessimistic scenarios had important variations in the absolute numbers of the categories. Furthermore, individual life cycle stages tend to present smaller variations than the overall impact of combining all life cycle stages, which emphasizes the importance of testing different scenarios and different datasets.

References

1 Chini, C.M., Konar, M., and Stillwell, A.S. (2017). Direct and indirect urban water footprints of the United States. *Water Resources Research* 53 (1): 316–327. https://agupubs.onlinelibrary.wiley.com/doi/full/10.1002/2016WR019473.

2 Peña-Guzmán, C.A., Melgarejo, J., Prats, D., Torres, A., and Martínez, S. (2017). Urban water cycle simulation/management models: a review. *Water* 9 (4): 1–29. doi: 10.3390/w9040285.

3 Lee, M., Keller, A.A., Chiang, P.C., Den, W., Wang, H., Hou, C.H., Yan, J. (2017). Water-energy nexus for urban water systems: a comparative review on energy intensity and environmental impacts in relation to global water risks. *Applied Energy* 205: 589–601. doi: 10.1016/j.apenergy.2017.08.002.

4 Arashiro, L.T., Montero, N., Ferrer, I., Acién, F.G., Gómez, C., and Garfí, M. (2018). Life cycle assessment of high rate algal ponds for wastewater treatment and resource recovery. *Science of the Total Environment* 622: 1118–1130. doi: 10.1016/j.scitotenv.2017.12.051.

5 Maurya, S.P., Singh, P.K., Ohri, A., and Singh, R. (2020). Identification of indicators for sustainable urban water development planning. *Ecological Indicators* 108: 105691.

6 Chèvre, N., Guignard, C., Rossi, L., Pfeifer, H.R., Bader, H.P., and Scheidegger, R. (2011). Substance flow analysis as a tool for urban water management. *Water Science and Technology* 63 (7): 1341–1348.

7 Baitz, M., Albrecht, S., Brauner, E., Broadbent, C., Castellan, G., Conrath, P., . Tikana, L. (2013). LCA's theory and practice: like ebony and ivory living in perfect harmony? *The International Journal of Life Cycle Assessment* 18: 5–13. doi: 10.1007/s11367-012-0476-x.

8 Vieira, D.R., Calmon, J.L., and Coelho, F.Z. (2016). Life cycle assessment (LCA) applied to the manufacturing of common and ecological concrete: a review. *Construction and Building Materials* 124: 656–666.

9 Heijungs, R. and Huijbregts, M.A. (2004). A review of approaches to treat uncertainty in LCA. *International Congress of Environmental Modelling and Software*. Osnabruck, Germany. https://scholarsarchive.byu.edu/iemssconference/2004/all/197 (accessed 4 October 2022).

10 Bamber, N., Turner, I., Arulnathan, V., Li, Y., Ershadi, S.Z., Smart, A., and Pelletier, N. (2020). Comparing sources and analysis of uncertainty in consequential and attributional life cycle assessment: review of current practice and recommendations. *The International Journal of Life Cycle Assessment* 25 (1): 168–180.

11 Amores, M.J., Meneses, M., Pasqualino, J., Antón, A., and Castells, F. (2013). Environmental assessment of urban water cycle on Mediterranean conditions by LCA approach. *Journal of Cleaner Production* 43: 84–92.

12 Hofman-Caris, R., Bertelkamp, C., de Waal, L., van den Brand, T., Hofman, J., van der Aa, R., and van der Hoek, J.P. (2019). Rainwater harvesting for drinking water production: a sustainable and cost-effective solution in the Netherlands? *Water* 11 (3): 511.

13 Lemos, D., Dias, A.C., Gabarrell, X., and Arroja, L. (2013). Environmental assessment of an urban water system. *Journal of Cleaner Production* 54: 157–165.

14 Sørup, H.J., Brudler, S., Godskesen, B., Dong, Y., Lerer, S.M., Rygaard, M., and Arnbjerg-Nielsen, K. (2020). Urban water management: can UN SDG 6 be met within the Planetary Boundaries? *Environmental Science & Policy* 106: 36–39.

15 Venkatesh, G., Chan, A., and Brattebø, H. (2014). Understanding the water-energy-carbon nexus in urban water utilities: comparison of four city case studies and the relevant influencing factors. *Energy* 75: 153–166.

16 Zhang, Q., Liu, S., Wang, T., Dai, X., Baninla, Y., Nakatani, J., and Moriguchi, Y. (2019). Urbanization impacts on greenhouse gas (GHG) emissions of the water infrastructure in China: trade-offs among sustainable development goals (SDGs). *Journal of Cleaner Production* 232: 474–486.

17 Igos, E., Benetto, E., Meyer, R., Baustert, P., and Othoniel, B. (2019). How to treat uncertainties in life cycle assessment studies? *The International Journal of Life Cycle Assessment* 24 (4): 794–807.

18 Jeong, H., Minne, E., and Crittenden, J.C. (2015). Life cycle assessment of the city of Atlanta, Georgia's centralized water system. *The International Journal of Life Cycle Assessment* 20 (6): 880–891.

19 Pillot, J., Catel, L., Renaud, E., Augeard, B., and Roux, P. (2016). Up to what point is loss reduction environmentally friendly?: the LCA of loss reduction scenarios in drinking water networks. *Water Research* 104: 231–241.

20 García-Sánchez, M. and Güereca, L.P. (2019). Environmental and social life cycle assessment of urban water systems: the case of Mexico City. *Science of the Total Environment* 693: 133464.

21 ISO. International Standard 14040:2006. (2006). Life cycle assessment: principles and framework. ISO 14040, 3: 28.

22 ISO. International Standard 14044:2006. (2006). Environmental management: life cycle assessment: requirements and guidelines. ISO 14044 2006 (7): 652–668.

23 Postacchini, L., Ciarapica, F.E., and Bevilacqua, M. (2018). Environmental assessment of a landfill leachate treatment plant: impacts and research for more sustainable chemical alternatives. *Journal of Cleaner Production* 183: 1021–1033.

24 Sarbu, I. (2016). A study of energy optimisation of urban water distribution systems using potential elements. *Water* 8 (12): 593.

25 Hiesgen, G., Saul, K., and Rauwendaal, C. (2016, October). Temperature induced dimensional variation in extrusion. *AIP Conference Proceedings* 1779 (1): 030005.

26 Dekker, E., Zijp, M.C., van de Kamp, M.E., Temme, E.H., and van Zelm, R. (2019). A taste of the new ReCiPe for life cycle assessment: consequences of the updated impact assessment method on food product LCAs. *The International Journal of Life Cycle Assessment* 25 2315–2324.

27 Goedkoop, M., Heijungs, R., Huijbregts, M., De Schryver, A., Struijs, J., and Van Zelm, R. (2009). ReCiPe 2008: a life cycle impact assessment method which comprises harmonised category indicators at the midpoint and the endpoint level. https://www.researchgate.net/publication/230770853_Recipe_2008 (accessed 4 October 2022), 1–126.

28 Yoon, K.P. and Hwang, C.L. (1995). *Multiple Attribute Decision Making: An Introduction*. Sage.

29 Chhipi-Shrestha, G., Mori, J., Hewage, K., and Sadiq, R. (2016). *Clostridium difficile* infection incidence prediction in hospitals (CDIIPH): a predictive model based on decision tree and fuzzy techniques. *Stochastic Environmental Research and Risk Assessment* 1–14. doi: 10.1007/s00477-016-1227-5.

30 Li, R.J. (1999). Fuzzy method in group decision making. *Computers & Mathematics with Applications* 38 (1): 91–101.

31 Chang, S.C. (1999). Fuzzy production inventory for fuzzy product quantity with triangular fuzzy number. *Fuzzy Sets and Systems* 107 (1): 37–57.

32 Sala, S., Cerutti, A.K., and Pant, R. (2018). *Development of a Weighting Approach for the Environmental Footprint*. Luxembourg: Publications Office of the European Union.

33 Frischknecht, R., Jungbluth, N., Althaus, H.-J., Doka, G., Dones, R., Heck, T., Hellweg, S., Hischier, R., Nemecek, T., Rebitzer, G., and Spielmann, M. (2004). The ecoinvent database: overview and methodological framework. *The International Journal of Life Cycle Assessment* 10: 3–9.

4

Environmental Quality Assessment Using Fuzzy Logic

Jyoti Yadav[1] *and Kedar Rijal*[2]

[1] *Department of Computer Science, Savitribai Phule Pune University, Pune, India*
[2] *Tribhuvan University, Department of Environmental Science, Kathmandu, Nepal*

4.1 Introduction

Information on the status of environmental quality is necessary to formulate sound public policies and the effective implementation of environmental quality programs. One of the effective ways of communicating such information in general, and air and water quality in particular, to policymakers and ultimately to end users is with indices. The efforts on the development of WQIs, expressed in an interval scale in numeric terms, were initiated in the 1990s in USA. An AQI or WQI is, in essence, a function used to simplify large numbers of data into a more useful form which might convey an image of overall air/water quality to a variety of users. The limitation of the selected indices is their lack of power in dealing with uncertainty and subjectivity present in the expert's perception in describing air/water quality in linguistic terms, such as: *good*, *fair*, *poor*, and so on for the defined purpose.

There exists *imprecision/fuzziness* in the perception of domain experts about rating environmental quality parameters in various linguistic terms as *epistemic uncertainty* – an unavoidable feature of most humanistic systems. Uncertainty is also present due to the inherent variability of randomness in environmental quality parametric data and is known as *statistical* or *aleatory uncertainty*. We believe that fuzzy logic (FL) concepts could be considered as a useful formalism for modeling, both *aleatory* and *epistemic uncertainty* [1, 2].

4.2 Conventional Air Quality Index (CAQI)

The US Environmental Protection Agency (EPA) formulated a uniform AQI for the use of state and local agencies. AQIs are numbers used by government agencies to characterize the quality of air at a given location. As the AQI increases, an increasingly large percentage of the population is likely to experience severe adverse health effects. The computation of an AQI requires an air pollutant concentration from a monitoring station or a model. The function used to convert from air pollutant concentration to AQI varies by pollutant, and varies from country to country. AQI values are divided into ranges, and each range is assigned a descriptor and a color code. The pollutant with the highest concentration is considered the *responsible pollutant*. The pollutant index can be computed using the following formula

$$\text{Index} = \frac{\text{Pollutant Concentration}}{\text{Pollutant Standard Level}} * 100 \tag{4.1}$$

Fuzzy Systems Modeling in Environmental and Health Risk Assessment, First Edition. Edited by Boris Faybishenko, Rehan Sadiq, and Ashok Deshpande.

4.3 Fuzzy Air Quality Index (FAQI)

In this approach [3–8], different weighting factors were assigned to each pollutant according to its priority, and trapezoidal membership functions were defined for the classification. A set of inference rules were used in the final index computation. In the first step, the pollutants were processed by the inference system and normalized, thus resulting in the FAQI. Each of the rules had only one antecedent in order for the weight assignment to be facilitated. The structure of the fuzzy rules applied is as follows: If PM_{10} is *very good* then FAQI is *good*. The classical fuzzy inference system (FIS) Mamdani was used. PM_{10} and NO_X were assigned maximum weights, as these are the two pollutants emitted by vehicles.

4.4 Fuzzy Description of Air Quality Classification

Air pollution experts can describe air quality directly in linguistic terms (NO_2 is *very good* and PM_{10} is *poor*, and so on) once the information about the pollutants is made available to them. It is our belief that a mathematical framework to be used in decision research should be close to human thinking processes, and FL is an appropriate kind of mathematical framework to deal with perception-based modeling. The following sections discuss in brief the novel Zadeh–Deshpande (ZD) method to directly describe air quality linguistically with a degree of linguistic certainty attached to each description.

The identification of relevant parameters for the defined purpose, number of sampling stations, time and frequency of observations, etc., are crucial and invariably based on the expert's knowledgebase. Collecting pollution parametric data from sampling locations is the first step. On the basis of probability distribution fitting, mean and variance values can be used in further analysis. Often, collected data from the sampling locations is inadequate for carrying out statistical analysis (in fact it may not even be necessary). In such a situation, the defined statistical technique, a bootstrap method, can be employed to ensure data reliability.

4.4.1 Zadeh–Deshpande (ZD) Formalism

ZD formalism is similar to that of the fuzzy thinking of the human brain which invariably does not work with numerical values. Furthermore, a linguistic degree of certainty (DC) is attached to each linguistic term to ensure the confidence level. The approach could also be labelled as fuzzy stat modeling.

4.4.1.1 Convex Normalized Fuzzy Number (CNFN)

Randomness in air quality data or, for that matter, any parametric data can be simulated using Gaussian distribution with the probability density function given by

$$f(x) = \frac{1}{\sigma\sqrt{2\pi}} e^{-\frac{1}{2}\left(\frac{x-\mu}{\sigma}\right)^2} \tag{4.2}$$

where x is a random variable taking values on real line, μ is the mean, and σ is the standard deviation. The continuous random variable can be discretized using distribution

$$\varnothing(x) = P[x \in X] \tag{4.3}$$

such that the probability of any x_i in the parametric domain, denoted by $p(x_i)$, is

$$p(x_i) = \varnothing\left(x_i + \frac{dx}{2}\right) - \varnothing\left(x_i - \frac{dx}{2}\right) \tag{4.4}$$

where dx is a small interval about x_i for $i = 1, 2, 3, 4, \ldots, n$.

Thus, if x_i is some point on the parametric domain x for which $p(x_i)$ is maximum then define function $\mu_A(x)$ as

$$\mu_A(x) = p(x) / p(x_i) \tag{4.5}$$

Equation (4.5) transforms a random variable into a CNFN A with membership grade $\mu_A(x)$, thereby characterizing the dynamic behavior of the governing parameter. The construction of a fuzzy number or fuzzy sets for modeling the perception of the experts in classifying each parametric domain linguistically involves:

- Selection of linguistic terms such as *very good*, *good*, *fair*, etc., which allows referencing of all possible parametric values to be described.
- Classification of the parametric domain and assigning linguistic terms to each class linearly by experts reflecting the imprecision in their perception.
- The set of values for which all the experts assign the same linguistic term are given $\mu = 1.0$ while none of the expert assigning that term are given $\mu = 0.0$. The breakeven points of membership grades 0.0 and 1.0 are connected by continuous monotonic function, which presupposes that the degree of consensus among the experts goes on increasing as the parametric values approach the core of the fuzzy number for the specified linguistic term.

Based on the data obtained using the bootstrap method, a probability distribution fitting to air quality parametric data can be carried out. As we intend to compare probability distribution with the expert fuzzy sets drawn for the selected parameters, it is necessary to transform probability distribution into possibility distribution of the parametric data, using the concept of CNFN. Probability distribution can be transformed into a CNFN with membership grade function $\mu_A(x)$ thereby characterizing the dynamic behavior of the pollutants into a possibility distribution. Uncertainty in an expert's perception, termed epistemic uncertainty, is modeled by constructing fuzzy sets/fuzzy numbers.

4.4.1.2 Degree of Match (DM)

A computational scheme of DM can be used with a view to estimate matching between an expert's perception and the antecedent part of the rule, in order to describe air quality fuzzily with a certain DC. The fuzzy number for field data (A) on parameters and the fuzzy numbers (A') characterizing linguistic terms are matched together to arrive at a measure called the degree of match, or DM, defined by

$$DM_{ff}(A, A') = \frac{\int \mu_{A \cap A'}(x)dx}{\int \mu_A(x)dx}, x \in X \tag{4.6}$$

in which X denotes the universe, and $\mu_{A \cap A'}(x)$ is the membership grade for $A \cap A$. Furthermore, if A and A' are the discrete possibility distributions the measure is defined as

$$DM_{ff}(A, A') = \frac{\sum \mu_{A \cap A'}(x)dx}{\sum \mu_A(x)dx}, x \in X \tag{4.7}$$

4.4.2 Type-1 Fuzzy Inference System (FIS)

Fuzzy inference is the process of formulating the mapping from a given input to an output using FL. The mapping then provides a basis from which decisions can be made. The process of fuzzy inference involves fuzzification, defuzzification, implication, and aggregation. The basic structure of an FIS consists of three conceptual components:

- A rule base, which contains a selection of fuzzy rules.
- A database, which defines the membership functions used in the fuzzy rules.
- A reasoning mechanism, which performs the inference procedure upon the rules and given facts to derive a reasonable output or conclusion.

Figure 4.1 represents the FIS used for describing air quality linguistically. A set of fuzzy rules are constructed for classifying air quality as: *very good*, *good*, *fair*, *poor*, and *very poor* in order to aggregate the set of attributes. Each rule has antecedent propositions connected together using the AND operator, resulting in some consequences. These linguistic descriptions are invariably imprecise/vague/fuzzy, taking into account the inadequate information on the health implications of each parameter on the users and the aggregated effect of all the parameters on human health. A fuzzy rule based system (FRBS) can be developed for knowledge representation or a reasoning process. The fuzzy rules are defined by the air quality/water quality domain experts based on their expertise.

Following are the sample rules stored in the knowledge base:

1) If PM_{10} is <*fair*> and NO_X is <*poor*> and SO_X is <*good*> then air quality is *fair*.
2) If PM_{10} is <*fair*> and NO_X is <*good*> and SO_X is <*fair*> then air quality is *fair*.

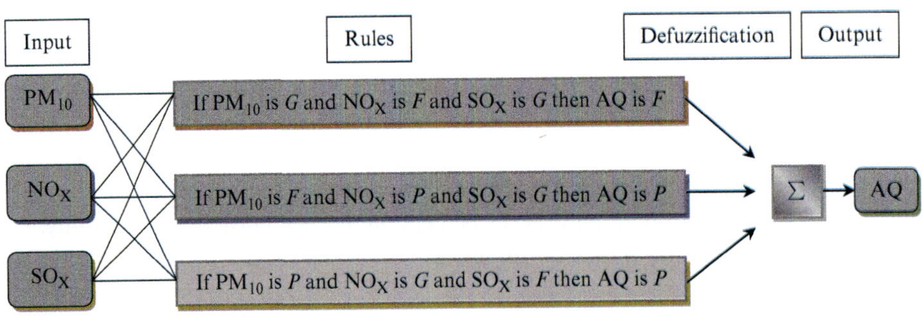

Figure 4.1 Fuzzy inference system describing air quality linguistically.

The degree of match of each classification rule indicates the certainty value of the classification. The greater the degree of match, the greater the possibility that air quality is classified in that class. The rules are processed using conjunction and disjunction operators. The optimal acceptance strategy is usually that for which the degree of assertion is the maximum.

4.4.3 Matching Between Two Fuzzy Values

The fuzzy number for pollutant data *(A)* on parameters and the fuzzy number characterizing linguistic terms are matched together to arrive at a measure called the DM (Figures 4.2 and 4.3), which is defined by

$$DM_{ff}(A, A') = \frac{\int \mu_{A \cap A'}(x)dx}{\int \mu_A(x)dx}, x \in X \tag{4.8}$$

in which X denotes the universe, and $\mu_{A \cap A'}(x)$ is the membership grade for $A \cap A'$. A (fact or the data) and A' (expert's perception) are the possibility distributions the measure is defined as

$$DM_{ff}(A, A') = \frac{\sum \mu_{A \cap A'}(x)dx}{\sum \mu_A(x)dx}, x \in X \tag{4.9}$$

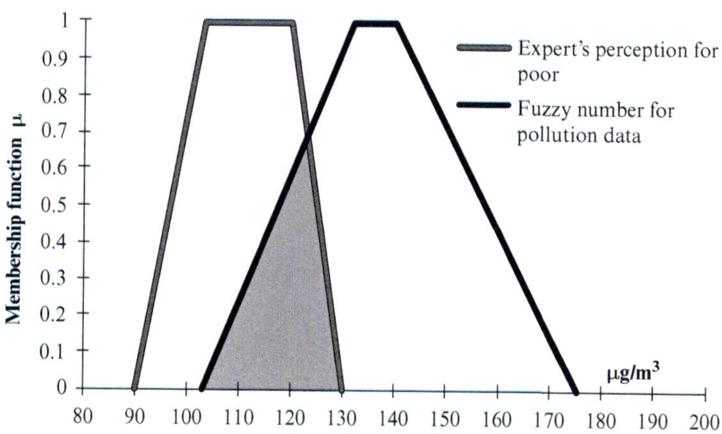

Figure 4.2 Degree of match of PM_{10} with linguistic term poor for PCMC location.

4.5 Fuzzy Rule Based System (FRBS)

The computational scheme of the DM can be used with a view to estimate matching between expert's perception and the antecedent part of the rule, in order to describe air quality fuzzily with certain DC. The degree of match of each classification rule indicates the certainty value of classification, in the present case air and water quality. The greater the DG, the greater the possibility that air and water quality is classified in that class. An FRBS was developed for knowledge representation or reasoning process. A set of fuzzy rules is constructed for classifying air quality as: *very good*, *good*, *fair*, *poor*, and *very poor* in order to aggregate the set of attributes. These linguistic descriptions are invariably imprecise/vague/fuzzy, keeping in view the inadequate information on the health implications of each parameter on the users and the aggregated effect of all the parameters on human health. Sample rules displayed in Table 4.1 are stored in the knowledge base. The rules are processed using conjunction and disjunction operators. The optimal acceptance strategy is usually that for which the degree of assertion is the maximum. Figure 4.4 presents the hierarchical structure for water quality classification and Figure 4.1 represents the FIS for air quality classification.

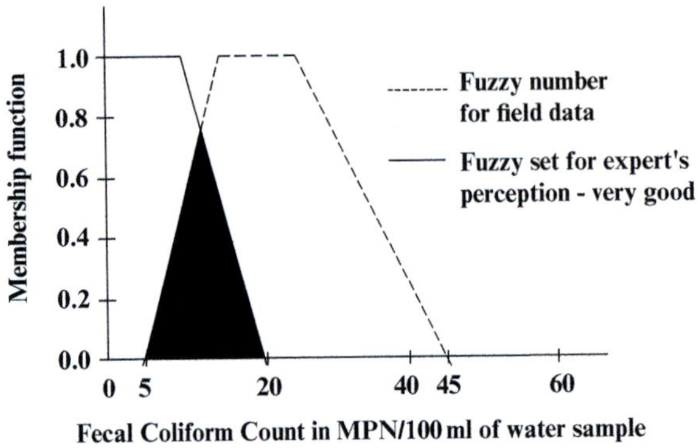

Figure 4.3 Fuzzy matching between A and A'.

4.6 Case Study

The application of fuzzy set theory has been a long-standing requirement when dealing with engineering systems in general and air and water quality classifications in particular.

4.6.1 Linguistic Description of Air Quality with a Degree of Linguistic Certainty

Air pollutants are added to the atmosphere from a variety of sources that change the composition of the atmosphere and affect the biotic environment. The concentration of air pollutants depends not only on the quantities that are emitted from air pollution sources but also on the ability of the atmosphere either to absorb or to disperse these emissions. The air pollution concentration varies spatially and temporally, causing air pollution patterns to change with different locations and times due to changes in meteorological and topographical conditions. The sources include vehicular traffic, emissions of hazardous gases from industries, and those from domestic and natural environments.

Air pollution is the fifth leading cause of death in India after high blood pressure, indoor air pollution, tobacco smoking, and poor nutrition, with about 620 000 premature deaths occurring from air pollution related diseases [9]. Especially in developing countries, a large portion of vehicles on city roads use diesel fuel and contribute greatly to the emissions of particulates, especially those that are less than 10 microns in size and are respirable. In many cities, transportation fuels contain lead and high amounts of sulfur and use older engine designs that emit more toxic pollutants than modern ones. Various other factors like the predominance of older vehicles, lack of maintenance, limited use of emission control technologies, poor traffic management, and poor road conditions help accentuate the level of automobile pollution. The number of two-/three-wheeler vehicles which use two-stroke engines and produce up to 10 times size hydrocarbons are on the increase.

The design of an air quality monitoring network requires a structured approach which involves: identification of pollutants, selection of locations, frequency and duration of sampling, sampling techniques, infrastructural facilities, labor, maintenance costs, and so on. The design also depends on the type of pollutants in the atmosphere through various common sources, called common urban air pollutants, such as suspended particulate matter (SPM), particulate matter (PM_{10}), sulfur dioxide (SO_2), oxides of nitrogen (NO_X), and carbon monoxide (CO), etc. Factors to be considered for the identification of air quality monitoring locations include: high traffic density, industrial growth, human population and its distribution, emission source, public complaints, if any, the land use pattern, etc. Realizing the increase in air pollution levels, it was considered relevant to initiate a study to classify the air quality in PCMC.

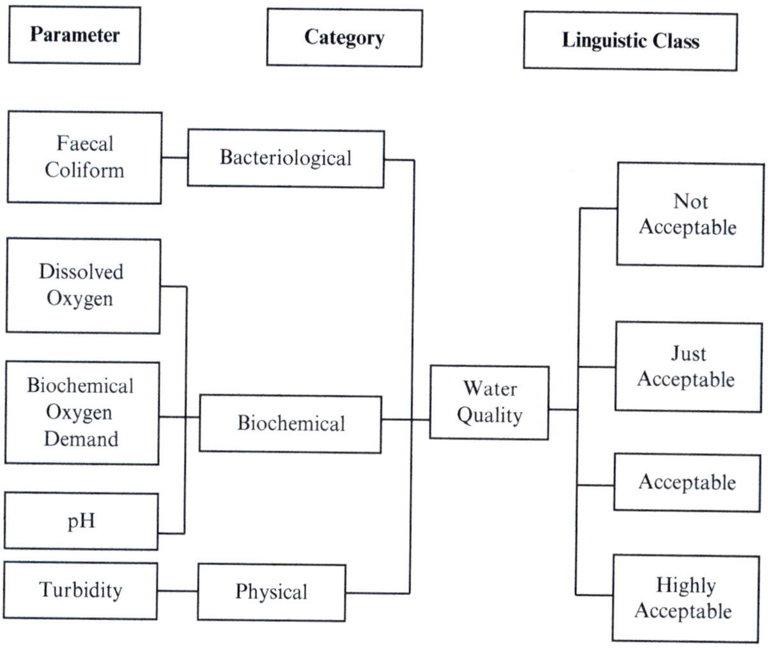

Figure 4.4 Typical hierarchical structure for water quality classification.

The case study relates to fuzzy air quality description with the available air quality data. The locations of air quality monitoring stations are (1) PCMC and (2) Bhosari in PCMC. Auto exhaust pollution is on the increase in PCMC. From an air quality standpoint, the winter months are more important due to the phenomenon of temperature inversion. In November 2010, air quality parametric data were considered for the assessment of ambient air quality in PCMC. Three pollutants monitored by the Maharashtra Pollution Control Board (MPCB), namely oxides of sulfur (SO_X), oxides of nitrogen (NO_X), and re-suspended particulate matter (PM_{10}) were considered for the assessment of air quality following the concept of an FIS with a degree of match as detailed in this chapter.

Table 4.1 Fuzzy rules by air and water quality domain experts.

Station	Air quality fuzzy rules	Water quality fuzzy rules
1.	If RSPM is *good* AND NO_X is *poor* AND SO_X is *good* THEN air quality is *good*	If BOD is *good* AND DO is *very good* AND pH is *very good* THEN biochemical status of water is *good*
2.	If RSPM is *fair* AND NO_X is *good* AND SO_X is *fair* THEN air quality is *fair*	If bacteriological status of water is *fair* AND biochemical status of water is *very good* AND physical status of water is *very good* THEN overall water quality at water intake well is *fair*

RSPM: respirable suspended particulate matter; BOD: biochemical oxygen demand; DO: dissolved oxygen.

4.6.2 Fuzzy Description of Pawana River Water Quality Near PCMC Intake Well

Developing countries have been witnessing pollution of water resources which has assumed a serious threat to humans due to an increase in the incidence of water-related diseases. The World Health Organization (WHO) in a 2002 report [10] states that around 21% of communicable diseases in India are water related, and out of these, diarrheal diseases alone killed over 700 000 Indians in 1999. The present study refers to the water quality of an intake well at a water treatment plant in PCMC. PCMC depends mainly on water from the Pawana River. Water quality surveillance from the source to consumer tap is a routine activity of major water suppliers. In India, the issue of sample size using statistical methods is invariably ignored. How many samples to be collected from a sampling location? In this study, we have implemented the concept described below.

The total number of coliform bacteria is one of the important parameters in describing water quality for drinking purposes. A monitoring study was intended to estimate the mean concentration of total coliform near the intake well of PCMC near the water treatment plant. A preliminary survey consisting of fourteen ($n = 14$) representative observations were: 1200, 1300, 1600, 1100, 400, 600, 1800, 300, 1000, 440, 350, 120, 460, and 1250 expressed in MPN (most probable number)/100 mL of sample. Biochemical oxygen demand (BOD) expressed as mg/L was also in the range of 6–18 mg/L, indicating pollution at the intake point on the Pawana River. Looking at these high values, the then municipal commissioner (MC), who was also an engineer, requested the first author on suggesting the sample size as he was keen to direct PCMC officials to carry out sampling based on basic statistics. The purpose of these investigations was to initiate action on the defaulters (a sugar mill owner, sewage from nearby colonies, villages to take necessary corrective measures). What sample size was needed to estimate the true mean within ± 200 units? Also, suggest a remedial measure to bring down drastically the bacterial load, say ≥ 500 MPN/100 mL of water sample.

Table 4.2 CAQI and fuzzy description of air quality with DC in PCMC location by four AQ experts.

		AQ description and DC in linguistic terms			
Location	**CAQI**	**Expert 1**	**Expert 2**	**Expert 3**	**Expert 4**
PCMC	151.13 (VP)	Very Poor (1.00-VG)	Very Poor (0.92-VG)	Poor (0.92-VG)	Poor (0.88-G)
Bhosari	100.38 (P)	Poor (0.53-F)	Poor (0.30-P)	Fair (0.30-P)	Fair (0.42-F)

Assume data will follow a normal distribution

$$\text{Number of Samples } N = \frac{\left(Z_{\frac{\alpha}{2}}, \sigma\right)^2}{E} \tag{4.10}$$

where σ is the standard deviation (524 MPN/100 mL of water sample), Z is $\left[\frac{x_i - \bar{x}}{\mu}\right] = 1.96$; for $\alpha/2 = 0.025$, and E (true mean within ± 200 units) = half-length $E = 200$. The computed number of samples will be 26.37, and so a total of 30 samples should be collected. Therefore, PCMC officials should collect additional 30 − 14 = 16 samples for the estimation of total coliform. A similar procedure was followed for deciding the number of samples for all the other water quality parameters, namely DO, BOD, pH, and turbidity, and was used in deciding the water quality near the PCMC intake well using FL based formalism.

4.7 Results and Discussion

The air quality description with the DM method describes air quality straightway in linguistic terms with a *linguistic* DC. The degree of uncertainty is modeled using type-1 FL. First of all, the uncertainty in the expert's perception is modeled and then the second uncertainty related to the DC is modeled by defining fuzzy sets for DC as *very good*, *good*, *fair*, *poor*, and *very poor*.

From Table 4.2 it can be seen that the air quality computed using the DC method is on a par with the air quality description using the CAQI. We can say that air quality at the PCMC location is *very poor* with a *very good* DC. Variability in experts' perception is also modeled using FL. The four air quality experts are almost close in their opinion about describing air quality at the PCMC monitoring location (see Table 4.3).

Table 4.3 Fuzzy description of water quality with DC.

Linguistic description	DC
Very good	0
Good	0.2 (*fair*)
Fair	0.75 (*good*)
Poor	0.1

Water quality at PCMC is *fair* with DC = 0.75, *good* with DC = 0.2, and *poor* with DC = 0.1

- It could be inferred from fuzzy modeling that the water quality (Figure 4.3) at the intake well of the water treatment plant is *fair* with DC 0.75 (*good*) and is tending toward *good* with DC 0.2 (*fair*). It is necessary to improve the Pawana River water quality as *very good* with a high DC. The strength of FL concepts lies in defining, in this case, river water quality at intake, straightway in linguistic terms with some DC attached to each linguistic terms.
- There is visible organic pollution due to domestic waste water from the nearby villages. A few industrial discharges are also located in about 2 km from the intake well. Therefore, there will be an increase in biochemical oxygen demand resulting in a high bacteriological load in the Pawana River.
- In case the pollution remains unchecked, the possibility of introducing pre chlorination at the intake resulting into possible trihalomethane (THM) formation cannot be overruled. PCMC, therefore, must undertake pollution abatement measures. The team of experts will look into the issue and locate the sources of pollution for further action.

4.8 Concluding Remarks

FL via computing with words is a humancentric logic which has been successfully used where the perceptions of domain experts are of primary concern. This is more relevant in decision-making wherein the decision-makers have no or scant numerical information of the governing parameters. It is well known that the expert knowledge base uses only linguistic terms. Why not compute with words and why always numbers?

It can be observed from the first case study that air quality computed using DC method is on a par with the air quality description using the CAQI. The air quality at the PCMC location can be classified as *very poor* with a *very good* DC. Variability in experts' perception is also modeled using FL. The four air quality experts are very close in their opinion in describing air quality at the PCMC location.

The experimental results for water quality at the intake well of the water treatment plant have been classified as *fair* with DC as *good* but is tending toward *good* with a *fair* DC.

The research study has conclusively demonstrated the utility of soft computing techniques with a focus on FL in water and air quality management. For the types of complex and imprecise problems that arise in environmental policy, the ability to model complex behaviors as a collection of simple if–then rules makes FL via computing with words an appropriate modeling tool. Would decision-makers and the public accept expressions of water or air quality goals in linguistic terms with computed degrees of certainty? Resistance is likely. In many regions, such as the United States and European Union (EU), both decision-makers and members of the public seem more comfortable with the current system – in which government agencies avoid confronting uncertainties by setting guidelines that are crisp and often fail to communicate uncertainty. Perhaps someday a more comprehensive approach that includes exposure surveys, toxicological data, and epidemiological studies coupled with fuzzy modeling will go a long way toward resolving some of the conflict, divisiveness, and controversy in the current regulatory paradigm.

Acknowledgment

Jyoti and Kedar would like to sincerely thank our guide Dr. Ashok Deshpande for his untiring efforts in helping us write this chapter.

References

1 Zadeh, L.A. (2002). Toward a perception-based theory of probabilistic reasoning with imprecise probabilities. *Journal of Statistical Planning and Inference* 105: 233–264.

2 Zadeh, L.A. (1978). Fuzzy sets as a basis for a theory of possibility. *Fuzzy Sets and Systems* 1(1): 3–28

3 Nagendra, S.M., Venugopal, K., and Jones, S.L. (2007). Assessment of air quality near traffic intersections in Bangalore city using air quality indices. *Transportation Research Part D* 12: 167–176.

4 Sowlat, H.M., Gharibi, H., Yunesian, M., Mahmoudiand, M.T., and Lotfi, S. (2011). A novel, fuzzy-based air quality index (FAQI) for air quality assessment. *Atmospheric Environment* 45 (12): 2050–2059.

5 Deshpande, A.W. and Raje, D.V. (2003). Fuzzy logic applications to environmental management systems: case studies, 364–368. *IEEE. Proceedings of the IEEE International Conference on Industrial Informatics, 2003 (INDIN 2003).* https://ieeexplore.ieee.org/stamp/stamp.jsp?tp=&arnumber=1300356 (accessed 22 October 2022). Banff, AB, Canada. IEEE.
6 Mckone, T.E. and Deshpande, A.W. (2005). Can fuzzy logic bring complex environmental problems into focus? *Environmental Science and Technology* 39 (2): 42A–47A.
7 Ersoy, K. and Cho, S. (1992). An algorithm to compute the degree of match in fuzzy systems. *Journal of Fuzzy Sets and Systems* 49: 285–299.
8 Deshpande, A.W., Raje, D.V., and Khanna, P. (1996). Fuzzy description of river water quality. *International Conference-EUFIT*, Germany.
9 *The Economic Times* 2013. *Air pollution fifth leading cause of death in India: study*. https://economictimes.indiatimes.com/air-pollution-fifth-leading-cause-of-death-in-india-study/articleshow/18494588.cms?from=mdr (accessed 22 October 2022).
10 World Health Organization. (2002). The World Health Report: 2002: Reducing risks, promoting healthy life. https://apps.who.int/iris/bitstream/handle/10665/42510/WHR_2002.pdf (accessed 4October 2022).

5

Assessing Spatiotemporal Water Quality Variations in Polluted Rivers with Uncertain Flow Variations

An Application of Triangular Type-2 Fuzzy Sets

Husnain Haider[1], *Pushpinder Singh*[2], *Majed Alinizzi*[1], *Saleem S. AlSaleem*[1], *and Rehan Sadiq*[3]

[1] *Qassim University, Buraydah, Qassim, Saudi Arabia*
[2] *Lovely Professional University, Department of Mathematics, Chaheru, Phagwara, Punjab, India*
[3] *School of Engineering, University of British Columbia (Okanagan), Kelowna, BC, Canada*

5.1 Introduction

Rivers in developing countries are facing multifarious water quality problems. In recent decades, many rivers in such countries have experienced extreme flow variations, in a year, due to seasonal rainfalls, the impacts of climate change, and increasing irrigation requirements. As nearby cities are growing to accommodate rapidly increasing populations, these rivers also face high pollution loads from city outflows [1, 2]. Consequently, anaerobic conditions prevail during low flows, and the rivers meet desired water quality standards for a designated use during high flow conditions [3]. Wastewater entering a river consists of different conservative (e.g. total dissolved solids, metals, etc.) and nonconservative (e.g. pathogens, organics, etc.) pollutants [4]. Water quality assessment for both the present and future (with increased pollution loads) scenarios is mandatory for sustainable management of natural resources [5]. A water quality index (WQI) aggregates the resultant levels of different pollutants, depending on their relative importance, and effectively summarizes the state of a river for engineers, policymakers, and the public.

For water quality modeling, the river is subdivided into different reaches depending on the locations of pollution sources entering a river, including outfalls and natural drains [3]. Outfalls collect wastewater from urban catchments through sewers, and surface drains primarily contain surface run-off. Each outfall may discharge wastewater with different characteristics depending on the land use activities in its catchment area. Further, pollution load (i.e. a product of pollutant concentration and wastewater flow) at an outfall is a function of the type and level of treatment. As the aim of the present study is to describe each river reach using a WQI, we know that the concentrations of nonconservative substances change temporally and spatially with changing river flow and velocity in the reach.

Two types of uncertainties exist in the development of a WQI for rivers with extreme flow variations. The first type results from variations in minimum flows during a season [3] (i.e. low flow analysis is done for a given assessment period), a limited number of wastewater and river samples [6], mistakes in sample collection and storage [7], and possible errors in laboratory analysis [6]. The second type of uncertainties caters for the variability in nonconservative pollutants' concentrations along the length (i.e. direction of the flow) in a given reach. These changes are significant in long and slow-moving reaches. In this chapter, a robust WQI simultaneously addresses both types of uncertainties using type-2 fuzzy sets.

Fuzzy Systems Modeling in Environmental and Health Risk Assessment, First Edition. Edited by Boris Faybishenko, Rehan Sadiq, and Ashok Deshpande.

Zadeh [8] first extended the conventional type-1 fuzzy set to a type-2 fuzzy set. In the type-2 fuzzy set, membership values are type-1 fuzzy sets on [0, 1] [3], and its membership function provides an additional degree of freedom to directly model uncertainties. Being three-dimensional, the type-2 fuzzy sets improve the interfaces originated from accumulated uncertainties due to imprecise knowledge, limited data, and fuzziness in the elusive or subjective information.

The real numbers in R are ordered linearly by $\geq$, while fuzzy numbers cannot be ordered in such a way. As possibility distributions characterize the fuzzy numbers, the chance of overlapping increases the difficulty to estimate how much one fuzzy number is higher or lower than the other. In this chapter, we propose a new approach for the ranking of triangular type-2 fuzzy sets to resolve the above-stated uncertainties that exist in the development of the WQI for a river with large flow variations. We consider a union of primary membership function (PMF) in terms of both the centroid of secondary membership function (SMF) and footprints of uncertainties.

The remainder of the chapter is organized as follows: Section 5.2 briefly outlines the basic concepts and definitions of type-2 fuzzy sets. The approach used to aggregate the pollutants' levels in river, using triangular type-2 fuzzy sets, is described in Section 5.3. Section 5.4 explains a multicriteria analysis problem based on the proposed approach. Finally, the application of the purposed approach to develop a WQI for the River Ravi is explained in Section 5.5.

5.2 Triangular Type-2 Fuzzy Sets

DEFINITION 1 [9]. Let X be the universe of discourse. A triangular type-2 fuzzy sets, denoted as $\tilde{A}=\{(a,c,b;\xi(c)),(a',c',b';b)\}$, is characterized by a linear function $\xi_{\tilde{A}}(u)$ as shown in Figure 5.1a. It is defined as

$$\xi_{\tilde{A}}(u)=\begin{cases} 0, 0\leq u<a, b\leq u<1 \\ \dfrac{\xi(c)(u-a)}{(c-a)}, a\leq u<c \\ \dfrac{\xi(c)(u-b)}{(c-b)}, c\leq u<b \end{cases}$$

where $a < c < b$ and $\xi(c)$ is the maximum value of the secondary membership $\xi(u)$, that is $\max(\xi(u)) = \xi(c)$. u is the domain of SMF, called the PMF, and it is characterized by a linear function $u_{\tilde{A}}(x)$ as shown in Figure 5.1b. It is defined as

$$u_{\tilde{A}}(x)=\begin{cases} 0, 0\leq x<a', b'\leq x<1 \\ \dfrac{b(x-a')}{(c'-a')}, a'\leq x\leq c' \\ \dfrac{b(x-b')}{(c'-b')}, c'\leq x\leq b' \end{cases}$$

where $0<a'<c'<b'<1$ and b is the maximum value of the primary membership u, that is $\max(u) = b$.

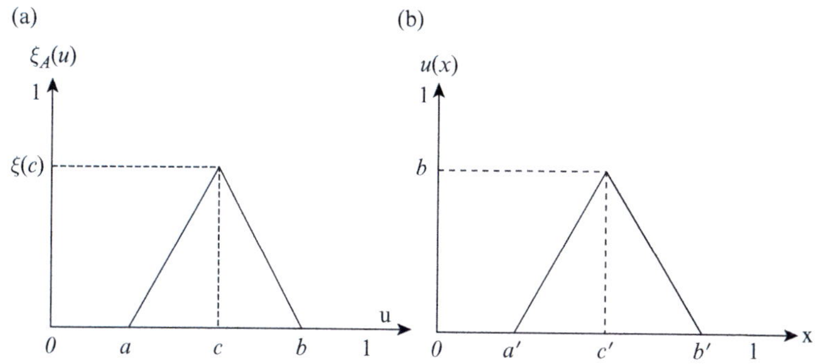

Figure 5.1 Membership functions: (a) SMF of $\tilde{A}$; (b) PMF of $\tilde{A}$.

The triangular type-2 fuzzy set $A_i = \{(a_i, c_i, b_i; \xi(c_i)), (a'_i, c'_i, b'_i; b_i)\}$, together with primary and SMF, is shown in Figure 5.2.

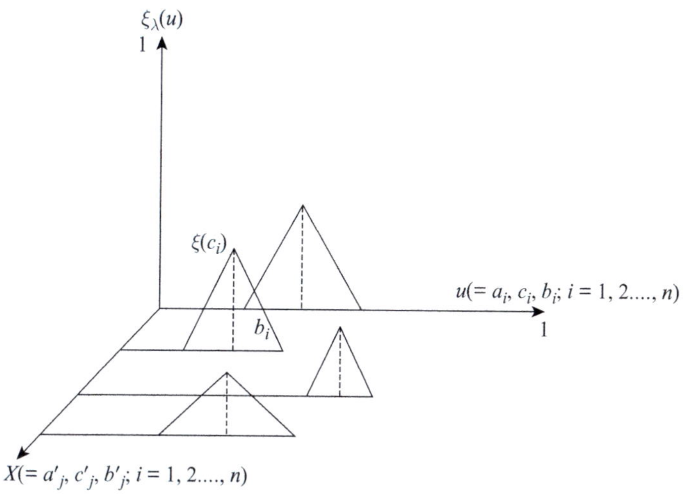

Figure 5.2 Triangular type-2 fuzzy sets.

5.3 The Proposed Aggregation Approach for Triangular Type-2 Fuzzy Sets

DEFINITION 2 [9]. For any triangular type-2 fuzzy set $\tilde{A}_i$, the expectation value of centroid of SMF is defined as

$$M(\tilde{A}_i) = \frac{\int_{a_i}^{b_i} u\xi_{\tilde{A}_i}(u)du}{\int_{a_i}^{b_i} \xi_{\tilde{A}_i}(u)du} \tag{5.1}$$

Now

$$\int_{a_i}^{b_i} u\xi_{\tilde{A}_i}(u)du = \int_{a_i}^{c_i} u\xi_{\tilde{A}_i}(u)du + \int_{c_i}^{b_i} u\xi_{\tilde{A}_i}(u)du$$

implies

$$\int_{a_i}^{c_i} \frac{\xi(c_i)(u^2 - a_i u)}{(c_i - a_i)} du + \int_{c_i}^{b_i} \frac{\xi(c_i)(u^2 - b_i u)}{(c_i - b_i)} du$$

and this implies

$$\frac{\xi(c_i)}{6(c_i - a_i)(c_i - b_i)} \{c_i^3(b_i - a_i) + c_i(b_i^3 - a_i^3) - a_i b_i (a_i^2 + b_i^2)\}$$

Hence

$$\int_{a_i}^{b_i} u\xi_{\tilde{A}_i}(u)du = \frac{\xi(c_i)}{6(c_i - a_i)(c_i - b_i)} \{c_i^3(b_i - a_i) + c_i(b_i^3 - a_i^3) - a_i b_i (a_i^2 + b_i^2)\} \tag{5.2}$$

Again

$$\int_{a_i}^{b_i} \xi_{\tilde{A}_i}(u)du = \int_{a_i}^{c_i} \xi_{\tilde{A}_i}(u)du + \int_{c_i}^{b_i} \xi_{\tilde{A}_i}(u)du$$

implies

$$\int_{a_i}^{c_i} \frac{\xi(c_i)(u - a_i)}{(c_i - a_i)} du + \int_{c_i}^{b_i} \frac{\xi(c_i)(u - b_i)}{(c_i - b_i)} du$$

and this implies

$$\frac{\xi(c_i)}{2(c_i - a_i)(c_i - b_i)} \{c_i^2(b_i - a_i) + c_i(b_i^2 - a_i^2) - a_i b_i (a_i - b_i)\}$$

Hence

$$\int_{a_i}^{b_i} u\xi_{\tilde{A}_i}(u)du = \frac{\xi(c_i)}{2(c_i - a_i)(c_i - b_i)} \{c_i^2(b_i - a_i) + c_i(b_i^2 - a_i^2) - a_i b_i (a_i - b_i)\} \tag{5.3}$$

On substituting Eqs. (5.2) and (5.3) in Eq. (5.1) we get

$$M(\tilde{A}_i) = \frac{1}{3} \frac{\{c_i^3(b_i - a_i) + c_i(b_i^3 - a_i^3) - a_i b_i (a_i^2 - b_i^2)\}}{\{c_i^2(b_i - a_i) + c_i(b_i^2 - a_i^2) - a_i b_i (a_i - b_i)\}} \tag{5.4}$$

DEFINITION 3 [9]. The uncertainty in the primary membership of type-2 fuzzy sets $\tilde{A}_i$ consists of a bounded region that we call the footprint of uncertainty (FOU).

Thus, the FOU was obtained by using the integral value of the PMF. Assume that $\varphi_{\tilde{A}_i}^L = \frac{b_i(x - a_i')}{(c_i' - a_i')}$ is the left PMF and $\varphi_{\tilde{A}_i}^R = \frac{b_i(x - b_i')}{(c_i' - b_i')}$ is the right PMF of two triangular type-2 fuzzy sets. Suppose that $\varsigma_{\tilde{A}_i}^L$ and $\varsigma_{\tilde{A}_i}^R$ are inverse functions of $\varphi_{\tilde{A}_i}^L$ and $\varphi_{\tilde{A}_i}^R$, respectively, as shown in Figure 5.3.

Therefore, the area under the PMF or FOU is given by the equation $\int_0^{b_i} \varsigma_{\tilde{A}_i}^L(y)dy + \int_0^{b_i} \varsigma_{\tilde{A}_i}^R(y)dy$.

On solving the above equation, we get

$$FOU(\tilde{A}_i) = \frac{b_i\left(a_i^{'} + 2c_i^{'} + b_i^{'}\right)}{2} \tag{5.5}$$

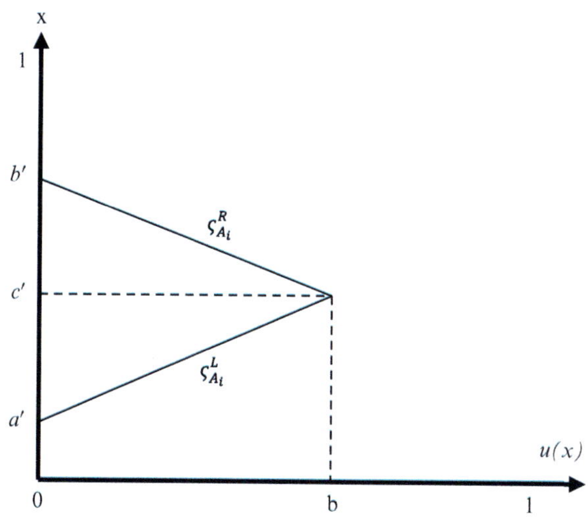

Figure 5.3 Inverse image of PMFs.

From Definitions 2 and 3 the ranking function for triangular type-2 fuzzy set $\tilde{A}_i$ is

$$\mathfrak{R}(\tilde{A}_i) = \frac{1}{2}\left\{M(\tilde{A}_i) + FOU(\tilde{A}_i)\right\}$$

From Eqs. (5.4) and (5.5) we get

$$\mathfrak{R}(\tilde{A}_i) = \frac{1}{2}\left\{\frac{1}{3}\frac{\left\{c_i^3\left(b_i - a_i\right) + c_i\left(b_i^3 - a_i^3\right) - a_i b_i(a_i^2 - b_i^2)\right\}}{\left\{c_i^2\left(b_i - a_i\right) + c_i\left(b_i^2 - a_i^2\right) + a_i b_i(a_i - b_i)\right\}} + \frac{b_i\left(a_i^{'} + 2b_i^{'} + c_i^{'}\right)}{2}\right\} \tag{5.6}$$

5.3.1 Algorithms

Consider n type-2 fuzzy sets $\tilde{A}_i$, i = 1, 2, …, n, each with a triangular membership function. For each pair of fuzzy sets, say $\tilde{A}_1$ and $\tilde{A}_2$, the pairwise comparison proceeds as

Let $\tilde{A}_1 = \left\{(a_1, c_1, b_1; \xi(c_1)), (a_1^{'}, c_1^{'}, b_1^{'}; b_1)\right\}$ and $\tilde{A}_2 = \left\{(a_2, c_2, b_2; \xi(c_2)), (a_2^{'}, c_2^{'}, b_2^{'}; b_2)\right\}$ be two type-2 fuzzy sets with triangular secondary and primary memberships.

$$\xi_{\tilde{A}_1}(u) = \begin{cases} 0, & 0 \le u \le a_1,\ b_1 \le u < 1, \\ \dfrac{\xi(c_1)(u - a_1)}{(c - a_1)} & a_1 \le u < c_1, \\ \dfrac{\xi(c_1)(u - b_1)}{(c_1 - b_1)} & c_1 \le u < b_1. \end{cases}$$

$$\xi_{\tilde{A}_2}(u)=\begin{cases} 0, & 0\le u\le a_2,\ b_2\le u<1, \\ \dfrac{\xi(c_2)(u-a_2)}{(c-a_2)} & a_2\le u<c_2, \\ \dfrac{\xi(c_2)(u-b_2)}{(c_1-b_2)} & c_2\le u<b_2, \end{cases}$$

$$u_{\tilde{A}_1}(x)=\begin{cases} 0, & 0\le x\le a_1',\ b_1'\le x<1, \\ \dfrac{b_1(x-a_1')}{(c_1'-a_1')} & a_1'\le x<c_1', \\ \dfrac{b_1(x-b_1')}{(c_1'-b_1')} & c_1'\le u<b_1', \end{cases}$$

$$u_{\tilde{A}_2}(x)=\begin{cases} 0, & 0\le x\le a_2',\ b_2'\le x<1, \\ \dfrac{b_2(x-a_2')}{(c_2'-a_2')} & a_2'\le x<c_2', \\ \dfrac{b_2(x-b_2')}{(c_2'-b_2')} & c_2'\le u<b_2', \end{cases}$$

Use the following steps to compare $\tilde{A}_1$ and $\tilde{A}_2$:

Step 1: Standardize each triangular type-2 fuzzy sets by transforming $\tilde{A}_1$, $\tilde{A}_2$ into $\tilde{A}_1^*$, $\tilde{A}_2^*$ where $\tilde{A}_1^*=\{(a_1,c_1,b_1;\xi(c),(a_1',c_1',b_1';b_1)\}$ and $\tilde{A}_2^*=\{(a_2,c_2,b_2;\xi(c),(a_2',c_2',b_2';b_2)\}$.

The secondary and PMF of $\tilde{A}_1^*$ and $\tilde{A}_2^*$ are given as

$$\xi_{\tilde{A}_1^*}(u)=\begin{cases} 0, & 0\le u\le a_1,\ b_1\le u<1, \\ \dfrac{\xi(c)(u-a_1)}{(c-a_1)} & a_1\le u<c_1, \\ \dfrac{\xi(c)(u-b_1)}{(c_1-b_1)} & c_1\le u<b_1. \end{cases}$$

$$\xi_{\tilde{A}_2^*}(u)=\begin{cases} 0, & 0\le u\le a_2,\ b_2\le u<1, \\ \dfrac{\xi(c)(u-a_2)}{(c-a_2)} & a_2\le u<c_2, \\ \dfrac{\xi(c)(u-b_2)}{(c_1-b_2)} & c_2\le u<b_2, \end{cases}$$

$$u_{\tilde{A}_1^*}(x)=\begin{cases} 0, & 0\le x\le a_1',\ b_1'\le x<1,\\ \dfrac{b\left(x-a_1'\right)}{\left(c_1'-a_1'\right)} & a_1'\le x<c_1',\\ \dfrac{b\left(x-b_1'\right)}{\left(c_1'-b_1'\right)} & c_1'\le u<b_1', \end{cases}$$

$$u_{\tilde{A}_2^*}(x)=\begin{cases} 0, & 0\le x\le a_2',\ b_2'\le x<1,\\ \dfrac{b\left(x-a_2'\right)}{\left(c_2'-a_2'\right)} & a_2'\le x<c_2',\\ \dfrac{b\left(x-b_2'\right)}{\left(c_2'-b_2'\right)} & c_2'\le u<b_2', \end{cases}$$

where $\xi(c)=\min(\xi(c_1),\xi(c_2))$ and $b=\min(b_1,b_2)$.

Step 2: Using Eqs. 5.4 and 5.5 calculate $M_1(\tilde{A}_1^*)$, $M_2(\tilde{A}_2^*)$ and $FOU(\tilde{A}_1^*)$, $FOU(\tilde{A}_2^*)$.

Step 3: Using Eq. (5.6) calculate $\mathfrak{N}(\tilde{A}_1^*)$ and $\mathfrak{N}(\tilde{A}_2^*)$. Then, $\tilde{A}_1$ and $\tilde{A}_2$ are compared as:

Case 1. $\tilde{A}_1 \succeq \tilde{A}_2$ if $\mathfrak{N}(\tilde{A}_1^*)\succeq\mathfrak{N}(\tilde{A}_2^*)$.

Case 2. $\tilde{A}_1 \preceq \tilde{A}_2$ if $\mathfrak{N}(\tilde{A}_1^*)\preceq\mathfrak{N}(\tilde{A}_2^*)$.

Case 3. $\tilde{A}_1 \sim \tilde{A}_2$ if $\mathfrak{N}(\tilde{A}_1^*)\sim\mathfrak{N}(\tilde{A}_2^*)$.

5.4 Multicriteria Analysis Problem Based on the Proposed Aggregation Approach

Let $A=\{A_1,A_2,\ldots,A_n,\}$ be a set of alternatives and let $C=\{C_1,C_2,\ldots,C_n,\}$ be a set of criteria. The values of an alternative on a given criteria $C_j(j=1,2,\ldots,n)$ are type-2 fuzzy sets S_{ij}, which specifies the degree that an alternative A_i satisfies or does not satisfy the criterion C_j allocated by decision-makers according to linguistically defined values of type-2 fuzzy sets. The value of the ranking weight w_j for a type-2 fuzzy set was obtained by Eq. (5.6). The normalized weights are estimated using

$$w_j=\frac{\Re(w_j)}{\sum_{j=1}^{n}\Re(w_j)} \tag{5.7}$$

Therefore, the weighted ranking value for an alternative $A=\{A_1,A_2,\ldots,A_n,\}$ is given by

$$\Re_{w_j}(\tilde{A}_i)=\sum_{j=1}^{n}w_j\Re(S_{ij}) \tag{5.8}$$

Thus, using Eqs. (5.7) and (5.8) the obtained weighted ranking value for an alternative is used to rank the alternatives. In the present research, alternatives correspond to river reaches and the criteria represent water quality parameters. Ranks are calculated as the WQIs for each reach.

To develop the WQI for a river with extreme flow variations, the following steps were performed:

1) Calculate the weights (w_j) for criterion $C = \{C_1, C_2, \ldots, C_n,\}$ by using Eqs. (5.6) and (5.7).
2) Find the water pollution index (WPI) by calculating the weighted pollution levels for all the reaches, $A = \{A_1, A_2, \ldots, A_n,\}$, by using Eqs. (5.6) and (5.8).
3) Find the WQI using the following relationship:

$$WQI = 1 - WPI \tag{5.9}$$

5.5 Application Example

5.5.1 Study Area

The proposed methodology is applied to the spatiotemporal water quality assessment of the River Ravi. The study area is shown in Figure 5.4. The Ravi River is one of the five largest rivers of Pakistan and presently is the most polluted river in the country. Flow in the river varies between 10 m^3/s and 10,000 m^3/s. Figure 5.4 shows that untreated wastewater enters the River Ravi through five outfalls and two drains. The QB Link canal joins the river at 85 km downstream of Siphon. Consequently, the river was subdivided into nine reaches between Siphon and Balloki Headworks. As of 2010, the river received around 40 m^3/s of untreated wastewater from the city of Lahore through five wastewater outfalls and two surface drains, shown in Figure 5.4. The projected wastewater flow will be almost six times

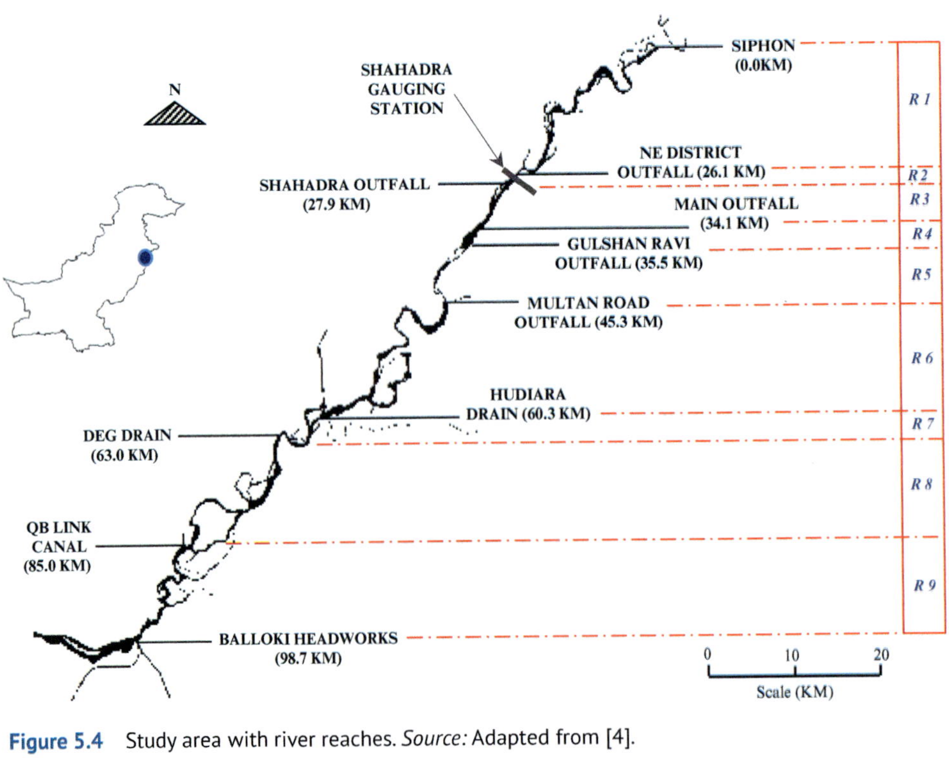

Figure 5.4 Study area with river reaches. *Source:* Adapted from [4].

higher than the minimum river flow (i.e. 56 m^3/s) by 2025 [10, 11]. Presently, almost all the designated uses of the river – including recreation, fishing, drinking water supply, and irrigation – have been restricted, particularly during low flow conditions. Assessing spatiotemporal water quality variations for a river with significant geographical importance, considering all the uncertainties identified in the introduction section, would be of great use for both decision-makers and the public.

5.5.2 Hydrodynamics Model and Low Flow Analysis

Gauge-discharge data and river cross-sections were obtained from the Survey and Discharge Divisions of the Punjab Irrigation Department. Power functions were developed to model the hydrodynamics (i.e. velocity, depth, and width) of the river. Two different power functions for river flow, (Q_r) < 283 m^3/s and Q_r > 283 m^3/s, accounts for the extreme flow variations between 10 and 10,000 m^3/s. Interested, readers can see Haider and Ali [3] for details on the development of the hydrodynamic model for the River Ravi.

First of all, an average seven consecutive days of flow was estimated for the period between 1995 and 2005 based on available data. Each year was divided into four seasons, i.e. winter (December to February), spring (March to May), summer (June to August), and fall (September to November). Five years with the lowest flow conditions during this period were selected for low flow analysis. The purpose of this study is to develop a WQI for a given season under the existing state of the river in the absence of any wastewater treatment. Therefore, minimum, average, and maximum flows were estimated from an average seven consecutive days of flow in Table 5.1.

Table 5.1 Summary of low flow analysis. Flow values are given in m^3/sec.

Year	Winter	Spring	Summer	Fall
1	21.0	16.9	173.9	73.6
2	23.9	22.2	22.1	69.5
3	5.5	5.4	429.4	6.6
4	5.1	5.4	69.2	13.2
5	5.6	6.5	423.4	12.6
MIN	5.1	5.4	22.1	6.6
AVG	12.2	11.3	223.6	35.1
MAX	23.9	22.2	429.4	73.6

5.5.3 Water Quality Model

In the present study, five important water quality parameters were selected for developing the WQI, including dissolved oxygen (DO), ultimate biochemical oxygen demand (UBOD), un-ionized ammonia (un-ionized NH_3), fecal coliforms (F. coli), and total dissolved solids (TDS). For water quality assessment, calibrated and verified models are used to predict concentrations of different parameters in the river. In the past, Haider and Ali developed, calibrated, and verified these models for the River Ravi [3, 12–14]. These models are described in Table 5.2. Monitoring surveys accompanied by laboratory analyses were conducted to determine the model inputs (i.e. average characteristics and flows of wastewater entering the River Ravi; see Table 5.3).

Table 5.2 Models used for assessment of Ravi River water quality. Adapted from [6].

Process	Equation/Model	Parameters	Eq.
Mass Balance for TDS only	$S_0 = \frac{Q_u S_u + Q_e S_e}{Q_u + Q_e}$	S_0 = TDS concentration in the river, mg/L Q_u = River flow u/s of the outfall, m^3/s S_u = TDS concentration in the river water upstream of the outfall, mg/L Q_e = Outfall flow, m^3/s S_e = TDS conc. in the wastewater, mg/L	(5.10)
CBOD	$L_c = L_{co} e^{-K_{cr}\frac{x}{U}}$	L_{cr} = CBOD, mg/L L_{co} = Initial CBOD in river at outfall, mg/L K_{cr} = River CBOD removal rate, day^{-1} x = distance in the river, m U = velocity in the river, m/day	(5.11)
Nitrification (NBODU)	$L_n = 4.57TKN$	L_n = NBODU, mg/L TKN = Total Kjeldahl nitrogen, mg/L	(5.12)
Nitrification (Org-N → NH_3-N)	$N_o = N_O oe^{-K_{oa}\frac{x}{U}}$	N_o = organic nitrogen (Org – N), mg/L $N_o o$ = initial concentration of Org-N, mg/L K_{oa} = nitrification rate coefficient for conversion of Org-N into ammonium	(5.13)
Nitrification (NH_3-N → NO_3-N)	$N_a = N_a oe^{-K_n\frac{x}{U}} + \frac{K_{oa} N_o o}{K_n - K_{oa}}(e^{-K_{oa}\frac{x}{U}} - e^{-K_n\frac{x}{U}}$	N_a = ammonia nitrogen, mg/L $N_a o$ = initial concentration of NH_4-N, mg/L K_n = nitrification rate coefficient for conversion of NH_4-N into NO_3-N, day^{-1}	(5.14)
Reaeration O'Connor-Dobbins (1956)	$K_a = 3.93\frac{U^{0.5}}{H^{1.5}}$	H = depth, m U = river velocity, m/s	(5.15)
DO (Modified Streeter-Phelps)	$D = D_0 e^{-K_a t} + \frac{K_c L_o}{K_a - K_{cr}}(e^{-K_{cr} t} - e^{-K_a t}) + \frac{K_n L_{no}}{K_a - K_n}(e^{-K_n t} - e^{-K_a t})$	K_c = CBOD deoxygenation rate, day^{-1}	(5.16)
Temperature corrections	$(K_a)_T = (K_a)_{20}(1.024)^{T-20}$ $(K_{cr})_T = (K_{cr})_{20}(1.047)^{T-20}$ $(K_n)_T = (K_n)_{20}(1.08)^{T-20}$	$(K_a)_T = K_a$ @ river temperature, day^{-1} $(K_a)_{20} = K_a$ @ 20°C, day^{-1} $(K_{cr})_T = K_r$ @ river temp, day^{-1} $(K_{cr})_{20} = K_r$ @ 20°C, day^{-1} $(K_n)_T = K_n$ @ river temp, day^{-1} $(K_n)_{20} = K_n$ @ 20°C, day^{-1}	(5.17a) (5.17b) (5.17c)
Ratio between CBODU and $CBOD_5$	$f = \frac{CBODU}{CBOD_5} = \frac{1}{1 - e^{-5k}}$	$CBODU$ = ultimate CBOD, mg/L $CBOD_5$ = 5 days CBOD, mg/L K = BOD bottle rate constant, day^{-1}	(5.18)

Table 5.3 Wastewater data used for Ravi River water quality assessment. Adapted from [6].

Wastewater outfalls/ surface drains	Flow (m^3/s)	BOD_5 (mg/L)	$CBOD_5$ (mg/L)	TKN (mg/L)	NH_3-N (mg/L)	TDS (mg/L)	*f*	CBODU (mg/L)
NE District Outfall	9.2	240	203	38.0	33.6	418	1.41	291
Shahadra Outfall	2.0	465	447	43.6	33.0	501	1.47	654
Main Outfall	5.7	198	189	45.6	34.2	429	1.47	264
Gulshan Ravi Outfall	3.9	246	212	34.6	29.1	545	1.40	297
Multan Road Outfall	3.0	234	234	44.0	37.5	515	1.36	316
Hudiara Drain	9.85	270	218	32.5	25.8	798	1.48	319
Deg Drain	169.9	7	5	2.9	2.6	80.5	2.02	11
QB Link Canal	509.2	6	4	0.9	0.6	85	1.40	6.3

5.5.4 Water Quality Assessment Using Fuzzy Type-2 Sets

The Ravi River study reach was sub-divided into nine reaches in Figure 5.4 (i.e. R_1, R_2, R_3, R_4, R_5, R_6, R_7, R_8, and R_9. Five water quality parameters were used: C_1: TDS; C_2: BOD_5; C_3: DO; C_4: NH_3_unionized; and C_5: F. coli). Using the models given in Table 5.2, water quality simulations for each season were performed to calculate the concentrations of all the parameters for minimum, average, and maximum of seven consecutive days of flows mentioned in Table 5.1. Temperature corrections were applied for all the seasons using Eq. (5.17). Accordingly, triangular fuzzy numbers were generated for all the parameters in a given reach. The water quality for each reach was evaluated using the linguistically defined PMFs of type-2 fuzzy sets given in Table 5.4. SMFs account for uncertainties due to water quality variations within a river reach depending on the length of the reach.

Weights of the criteria (i.e. water quality parameters) were allocated by four experts working in the field of water quality assessment. Table 5.5 presents the estimated values based on the water quality simulations and membership functions defined in Table 5.4.

Figure 5.5 illustrates the calculated values of WQI using Eq. (5.9) for all the seasons in each reach. It can be seen in Figure 5.5 that water quality remains *very high* throughout the year in

Table 5.4 Linguistically defined PMFs for water quality parameters.

Linguistically defined pollution levels	PMF	PMF	TDS (mg/L)	BOD_5 (mg/L)	DO (mg/L)	NH_3-unionized (mg/L)	F. coli (MPN/100 mL)
Very low (VL)	(0, 0.15, 0.3; 0.1)	(0, 0.125, 0.25; 0.1)	0–80	≤30	≥5	≤0.02	<1000
Low (L)	(0.2, 0.35, 0.5; 0.3)	(0.2, 0.325, 0.45; 0.2)	81–150	31–80	4–5	0.02–0.04	1000–20 000
Medium (M)	(0.35, 0.5, 0.65; 0.5)	(0.4, 0.525, 0.65; 0.3)	151–350	81–120	3–4	0.04–0.1	20 000–50 000
High (H)	(0.6, 0.7, 0.9; 0.7)	(0.6, 0.725, 0.85; 0.4)	351–500	121–300	2–3	0.1–0.2	50 000–150 000
Very high (VH)	(0.7, 0.9, 1; 0.9)	(0.8, 0.9, 1; 0.5)	>500	>300	<2	>0.2	>150 000

Table 5.5 Estimated values for reaches, seasons, and criteria based on water quality simulations and expert knowledge base.

River Reach	Season	C1: TDS								C2: BOD_5								
		SMF				PMF				SMF				PMF				
R1	S1	0	0.13	0.25	0.1	0	0.15	0.3	0.1	0	0.13	0.25	0.1	0	0.1	0.2	0.1	0
	S2	0	0.13	0.25	0.1	0	0.15	0.3	0.1	0	0.13	0.25	0.1	0	0.1	0.2	0.1	0
	S3	0	0.13	0.25	0.1	0	0.15	0.3	0.1	0	0.13	0.25	0.1	0	0.1	0.2	0.1	0
	S4	0	0.13	0.25	0.1	0	0.15	0.3	0.1	0	0.13	0.25	0.1	0	0.1	0.2	0.1	0
R2	S1	0	0.13	0.25	0.1	0.35	0.5	0.65	0.5	0	0.13	0.25	0.1	0.55	0.7	0.85	0.7	0.4
	S2	0	0.13	0.25	0.1	0.35	0.5	0.65	0.5	0	0.13	0.25	0.1	0.55	0.7	0.85	0.7	0.4
	S3	0	0.13	0.25	0.1	0.2	0.35	0.5	0.3	0	0.13	0.25	0.1	0.2	0.35	0.5	0.3	0.4
	S4	0	0.13	0.25	0.1	0.35	0.5	0.65	0.5	0	0.13	0.25	0.1	0.55	0.7	0.85	0.7	0.4
R3	S1	0	0.13	0.25	0.1	0.35	0.5	0.65	0.5	0.2	0.33	0.45	0.2	0.55	0.7	0.85	0.7	0.4
	S2	0	0.13	0.25	0.1	0.35	0.5	0.65	0.5	0.2	0.33	0.45	0.2	0.55	0.7	0.85	0.7	0.4
	S3	0	0.13	0.25	0.1	0.2	0.35	0.5	0.3	0.2	0.33	0.45	0.2	0.2	0.35	0.5	0.3	0.4
	S4	0	0.13	0.25	0.1	0.35	0.5	0.65	0.5	0.2	0.33	0.45	0.2	0.55	0.7	0.85	0.7	0.4
R4	S1	0	0.13	0.25	0.1	0.35	0.5	0.65	0.5	0	0.13	0.25	0.1	0.55	0.7	0.85	0.7	0.2
	S2	0	0.13	0.25	0.1	0.35	0.5	0.65	0.5	0	0.13	0.25	0.1	0.55	0.7	0.85	0.7	0.2
	S3	0	0.13	0.25	0.1	0.2	0.35	0.5	0.3	0	0.13	0.25	0.1	0.2	0.35	0.5	0.3	0.2
	S4	0	0.13	0.25	0.1	0.35	0.5	0.65	0.5	0	0.13	0.25	0.1	0.55	0.7	0.85	0.7	0.2
R5	S1	0	0.13	0.25	0.1	0.35	0.5	0.65	0.5	0.4	0.53	0.65	0.5	0.55	0.7	0.85	0.7	0.8
	S2	0	0.13	0.25	0.1	0.35	0.5	0.65	0.5	0.4	0.53	0.65	0.5	0.55	0.7	0.85	0.7	0.8
	S3	0	0.13	0.25	0.1	0.2	0.35	0.5	0.3	0.4	0.53	0.65	0.5	0.2	0.35	0.5	0.3	0.8
	S4	0	0.13	0.25	0.1	0.35	0.5	0.65	0.5	0.4	0.53	0.65	0.5	0.55	0.7	0.85	0.7	0.8
R6	S1	0	0.13	0.25	0.1	0.35	0.5	0.65	0.5	0.6	0.73	0.85	0.4	0.55	0.7	0.85	0.7	0.2
	S2	0	0.13	0.25	0.1	0.35	0.5	0.65	0.5	0.6	0.73	0.85	0.4	0.55	0.7	0.85	0.7	0.2
	S3	0	0.13	0.25	0.1	0.2	0.35	0.5	0.3	0.6	0.73	0.85	0.4	0.2	0.35	0.5	0.3	0.2
	S4	0	0.13	0.25	0.1	0.35	0.5	0.65	0.5	0.6	0.73	0.85	0.4	0.55	0.7	0.85	0.7	0.2
R7	S1	0	0.13	0.25	0.1	0.55	0.7	0.85	0.7	0.2	0.33	0.45	0.2	0.55	0.7	0.85	0.7	0.4
	S2	0	0.13	0.25	0.1	0.55	0.7	0.85	0.7	0.2	0.33	0.45	0.2	0.55	0.7	0.85	0.7	0.4
	S3	0	0.13	0.25	0.1	0.35	0.5	0.65	0.5	0.2	0.33	0.45	0.2	0.35	0.5	0.65	0.5	0.4
	S4	0	0.13	0.25	0.1	0.55	0.7	0.85	0.7	0.2	0.33	0.45	0.2	0.55	0.7	0.85	0.7	0.4
R8	S1	0	0.13	0.25	0.1	0.35	0.5	0.65	0.5	0.4	0.53	0.65	0.5	0.2	0.35	0.5	0.3	0.2
	S2	0	0.13	0.25	0.1	0.35	0.5	0.65	0.5	0.4	0.53	0.65	0.5	0.2	0.35	0.5	0.3	0.2
	S3	0	0.13	0.25	0.1	0.2	0.35	0.5	0.3	0.4	0.53	0.65	0.5	0.2	0.35	0.5	0.3	0.2
	S4	0	0.13	0.25	0.1	0.2	0.35	0.5	0.3	0.4	0.53	0.65	0.5	0.2	0.35	0.5	0.3	0.2
R9	S1	0	0.13	0.25	0.1	0.2	0.35	0.5	0.3	0.2	0.33	0.45	0.2	0.2	0.35	0.5	0.3	0.2
	S2	0	0.13	0.25	0.1	0.2	0.35	0.5	0.3	0.2	0.33	0.45	0.2	0.2	0.35	0.5	0.3	0.2
	S3	0	0.13	0.25	0.1	0.2	0.35	0.5	0.3	0.2	0.33	0.45	0.2	0	0.15	0.3	0.1	0.2
	S4	0	0.13	0.25	0.1	0.2	0.35	0.5	0.3	0.2	0.33	0.45	0.2	0	0.15	0.3	0.1	0.2
wj	S1	0	0.1	0.2	0.1	0	0.1	0.2	0.1	0.3	0.5	0.7	0.4	0.4	0.5	0.6	0.5	0.7
	S2	0.1	0.2	0.4	0.2	0.2	0.3	0.4	0.3	0.3	0.5	0.7	0.4	0.4	0.5	0.6	0.5	0.4
	S3	0	0.1	0.2	0.1	0	0.1	0.2	0.1	0.3	0.5	0.7	0.4	0.4	0.5	0.6	0.5	0.7
	S4	0	0	0	0	0	0	0	0	0.3	0.5	0.7	0.4	0.4	0.5	0.6	0.5	0.4

C3: DO								C4: NH_3_unionized								C5: F. Coli						
SMF				PMF				SMF				PMF				SMF				PMF		
0.13	0.25	0.1	0	0.15	0.3	0.1	0	0.13	0.25	0.1	0	0.15	0.3	0.1	0	0.13	0.25	0.1	0	0.15	0.3	0.1
0.13	0.25	0.1	0	0.15	0.3	0.1	0	0.13	0.25	0.1	0	0.15	0.3	0.1	0	0.13	0.25	0.1	0	0.15	0.3	0.1
0.13	0.25	0.1	0	0.15	0.3	0.1	0	0.13	0.25	0.1	0	0.15	0.3	0.1	0	0.13	0.25	0.1	0	0.15	0.3	0.1
0.13	0.25	0.1	0	0.15	0.3	0.1	0	0.13	0.25	0.1	0	0.15	0.3	0.1	0	0.13	0.25	0.1	0	0.15	0.3	0.1
0.53	0.65	0.5	0.55	0.7	0.85	0.7	0.2	0.33	0.45	0.2	0.55	0.7	0.85	0.7	0.4	0.53	0.65	0.5	0.55	0.7	0.85	0.7
0.53	0.65	0.5	0.55	0.7	0.85	0.7	0.2	0.33	0.45	0.2	0.55	0.7	0.85	0.7	0.4	0.53	0.65	0.5	0.55	0.7	0.85	0.7
0.53	0.65	0.5	0	0.15	0.3	0.1	0.2	0.33	0.45	0.2	0.2	0.35	0.5	0.3	0.4	0.53	0.65	0.5	0.2	0.35	0.5	0.3
0.53	0.65	0.5	0.55	0.7	0.85	0.7	0.2	0.33	0.45	0.2	0.55	0.7	0.85	0.7	0.4	0.53	0.65	0.5	0.55	0.7	0.85	0.7
0.53	0.65	0.5	0.55	0.7	0.85	0.7	0.2	0.33	0.45	0.2	0.55	0.7	0.85	0.7	0.2	0.33	0.45	0.2	0.55	0.7	0.85	0.7
0.53	0.65	0.5	0.55	0.7	0.85	0.7	0.2	0.33	0.45	0.2	0.55	0.7	0.85	0.7	0.2	0.33	0.45	0.2	0.35	0.5	0.65	0.5
0.53	0.65	0.5	0	0.15	0.3	0.1	0.2	0.33	0.45	0.2	0.2	0.35	0.5	0.3	0.2	0.33	0.45	0.2	0.2	0.35	0.5	0.3
0.53	0.65	0.5	0.55	0.7	0.85	0.7	0.2	0.33	0.45	0.2	0.55	0.7	0.85	0.7	0.2	0.33	0.45	0.2	0.2	0.35	0.5	0.3
0.33	0.45	0.2	0.7	0.85	1	0.9	0.2	0.33	0.45	0.2	0.55	0.7	0.85	0.7	0.4	0.53	0.65	0.5	0.55	0.7	0.85	0.7
0.33	0.45	0.2	0.7	0.85	1	0.9	0.2	0.33	0.45	0.2	0.55	0.7	0.85	0.7	0.4	0.53	0.65	0.5	0.55	0.7	0.85	0.7
0.33	0.45	0.2	0.2	0.35	0.5	0.3	0.2	0.33	0.45	0.2	0.35	0.5	0.65	0.5	0.4	0.53	0.65	0.5	0.2	0.35	0.5	0.3
0.33	0.45	0.2	0.55	0.7	0.85	0.7	0.2	0.33	0.45	0.2	0.55	0.7	0.85	0.7	0.4	0.53	0.65	0.5	0.35	0.5	0.65	0.5
0.9	1	0.5	0.7	0.85	1	0.9	0.4	0.53	0.65	0.5	0.55	0.7	0.85	0.7	0.6	0.73	0.85	0.4	0.35	0.5	0.65	0.5
0.9	1	0.5	0.7	0.85	1	0.9	0.4	0.53	0.65	0.5	0.55	0.7	0.85	0.7	0.6	0.73	0.85	0.4	0.35	0.5	0.65	0.5
0.9	1	0.5	0.35	0.5	0.65	0.5	0.4	0.53	0.65	0.5	0.35	0.5	0.65	0.5	0.6	0.73	0.85	0.4	0.2	0.35	0.5	0.3
0.9	1	0.5	0.7	0.85	1	0.7	0.4	0.53	0.65	0.5	0.55	0.7	0.85	0.7	0.6	0.73	0.85	0.4	0.2	0.35	0.5	0.3
0.33	0.45	0.2	0.7	0.85	1	0.9	0.6	0.73	0.85	0.4	0.55	0.7	0.85	0.7	0.4	0.53	0.65	0.5	0.55	0.7	0.85	0.7
0.33	0.45	0.2	0.7	0.85	1	0.9	0.6	0.73	0.85	0.4	0.55	0.7	0.85	0.7	0.4	0.53	0.65	0.5	0.35	0.5	0.65	0.5
0.33	0.45	0.2	0.35	0.5	0.65	0.5	0.6	0.73	0.85	0.4	0.35	0.5	0.65	0.5	0.4	0.53	0.65	0.5	0.2	0.35	0.5	0.3
0.33	0.45	0.2	0.7	0.85	1	0.7	0.6	0.73	0.85	0.4	0.55	0.7	0.85	0.7	0.4	0.53	0.65	0.5	0.35	0.5	0.65	0.5
0.53	0.65	0.5	0.7	0.85	1	0.9	0.2	0.33	0.45	0.2	0.55	0.7	0.85	0.7	0.4	0.53	0.65	0.5	0.7	0.85	1	0.9
0.53	0.65	0.5	0.7	0.85	1	0.9	0.2	0.33	0.45	0.2	0.55	0.7	0.85	0.7	0.4	0.53	0.65	0.5	0.7	0.85	1	0.9
0.53	0.65	0.5	0.35	0.5	0.65	0.5	0.2	0.33	0.45	0.2	0.35	0.5	0.65	0.5	0.4	0.53	0.65	0.5	0.35	0.5	0.65	0.5
0.53	0.65	0.5	0.7	0.85	1	0.7	0.2	0.33	0.45	0.2	0.55	0.7	0.85	0.7	0.4	0.53	0.65	0.5	0.55	0.7	0.85	0.7
0.33	0.45	0.2	0.55	0.7	0.85	0.7	0.2	0.33	0.45	0.2	0.35	0.5	0.65	0.5	0.2	0.33	0.45	0.2	0.35	0.5	0.65	0.5
0.33	0.45	0.2	0.55	0.7	0.85	0.7	0.2	0.33	0.45	0.2	0.35	0.5	0.65	0.5	0.2	0.33	0.45	0.2	0.35	0.5	0.65	0.5
0.33	0.45	0.2	0.35	0.5	0.65	0.5	0.2	0.33	0.45	0.2	0.35	0.5	0.65	0.5	0.2	0.33	0.45	0.2	0.2	0.35	0.5	0.3
0.33	0.45	0.2	0.55	0.7	0.85	0.7	0.2	0.33	0.45	0.2	0.35	0.5	0.65	0.5	0.2	0.33	0.45	0.2	0.2	0.35	0.5	0.3
0.33	0.45	0.2	0.35	0.5	0.65	0.5	0.2	0.33	0.45	0.2	0	0.15	0.3	0.1	0.2	0.33	0.45	0.2	0.2	0.35	0.5	0.3
0.33	0.45	0.2	0.35	0.5	0.65	0.5	0.2	0.33	0.45	0.2	0.2	0.35	0.5	0.3	0.2	0.33	0.45	0.2	0.2	0.35	0.5	0.3
0.33	0.45	0.2	0.2	0.35	0.5	0.3	0.2	0.33	0.45	0.2	0.2	0.35	0.5	0.3	0.2	0.33	0.45	0.2	0.2	0.35	0.5	0.3
0.33	0.45	0.2	0.35	0.5	0.65	0.5	0.2	0.33	0.45	0.2	0.2	0.35	0.5	0.3	0.2	0.33	0.45	0.2	0.2	0.35	0.5	0.3
0.8	0.9	0.9	0.7	0.8	0.9	0.9	0.4	0.6	0.8	0.5	0.5	0.6	0.7	0.7	0.1	0.2	0.4	0.2	0.2	0.3	0.4	0.3
0.6	0.8	0.5	0.5	0.6	0.7	0.7	0.4	0.6	0.8	0.5	0.5	0.6	0.7	0.7	0	0.1	0.2	0.1	0	0.1	0.2	0.1
0.8	0.9	0.9	0.7	0.8	0.9	0.9	0.4	0.6	0.8	0.5	0.5	0.6	0.7	0.7	0	0.1	0.2	0.1	0	0.1	0.2	0.1
0.6	0.8	0.5	0.5	0.6	0.7	0.7	0.3	0.5	0.7	0.4	0.4	0.5	0.6	0.5	0.1	0.2	0.4	0.2	0.2	0.3	0.4	0.3

R1 without receiving wastewater flows from the city of Lahore. Therefore, this 26 km reach is suitable for swimming, fishing, boating, and irrigation. Impact on groundwater quality is supposed to be negligible. The water quality in R2 remains *high* during the summer season and *low* in the rest of the year. The situation becomes critical from R3 to R7 as the river receives significant untreated wastewater flows from different outfalls (see Figure 5.4). Water quality improves in R8 and R9 from *very low* to *medium* after the confluence of Deg Drain and QB Link Canal. Water quality in R9 at Balloki Headworks, a popular recreational resort, seems safe for boating and irrigation. However, swimming should be prohibited in R9, except in the summer season. High levels of unionized ammonia are also reported in the literature [11, 12].

Table 5.6 presents the applicability of the WQI developed in this study, along with recommendations and required water quality management actions. The results illustrated in

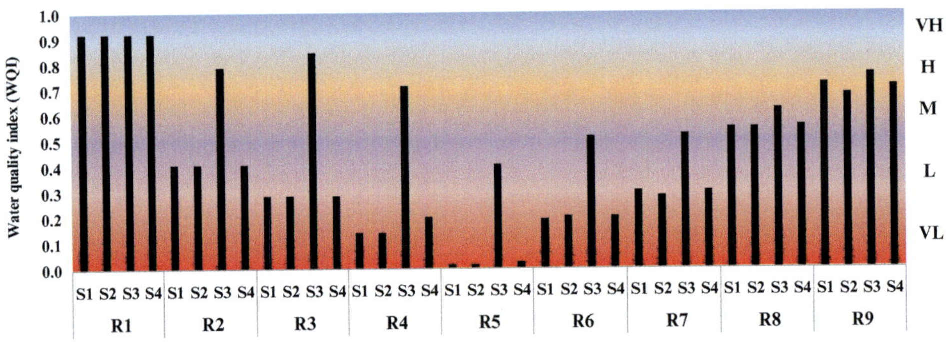

Figure 5.5 Water quality index for Ravi River, "R" represents the river reaches in Figure 5.4 while "S" denotes seasons as S1: winter, S2: spring, S3: summer, and S4: fall.

Table 5.6 Water quality index with recommended used and mitigation actions.

WQI	Suitable designated uses: Boating	Drinking	Irrigation	Swimming	Fishery	Proposed mitigation actions / guidelines
Very low (VL)	×	×	×	×	×	High degree of wastewater treatment required (i.e. activated sludge process followed by tertiary treatment for removal of metals)
Low (L)	✓	×	×	×	×	High degree of wastewater treatment required and the reaches should be prohibited for use through strict regulations
Medium (M)	✓	✓[2]	×	×	×	Wastewater treatment and planned monitoring for boating
High (H)	✓	✓[2]	✓[5]	×	×[3]	Planned monitoring with possible wastewater treatment to control NH_3-unionized and coliforms
Very high (VH)	✓	✓[1]	✓[4]	✓	✓	Planned water quality monitoring

[1]After conventional surface water treatment with removal of organics also check for metals.
[2]Conventional surface water treatment + removal of organics + possible removal of metals.
[3]Might not be suitable for most of fish species due to high unionized ammonia.
[4]Unrestricted irrigation.
[5]Restricted irrigation.

Figure 5.5 provide insights for spatiotemporal water quality variations. The guidelines developed in Table 5.6 facilitate the decision-makers (i.e. design engineers, municipality managers, environmental planners, researchers, and policymakers) for the development of a detailed water quality management plan for the River Ravi.

5.6 Conclusions and Recommendations

Most of the large rivers in developing countries face large seasonal flow variations and receive high pollution loads from the cities. In the absence of appropriate wastewater treatment, anaerobic conditions prevail in low flows. During floods, in monsoon season, the water quality in the river may be suitable for some of the beneficial uses.

A WQI is a useful measure to present the existing and future conditions of a river by aggregating the levels of different pollutants in a river, such as BOD_5, DO, NH_3-unionized, TDS, and F. coli. As these rivers face large flow variations, the resultant concentrations of different pollutants vary depending on the dilution available.

The WQI developed in the present study accommodates two main types of uncertainties for assessing the water quality in a river reach using fuzzy type-2 sets. The first type of uncertainties associated with river flow variations in a season, a limited number of wastewater and river samples, and errors in sample analyses. The second type caters for changes in the pollutants' concentrations along the length of the river reach, particularly for nonconservative substances.

Decision-makers can use the findings of this research for developing plans for the designated uses of the river and pollution control strategies. Although WQI was developed for the existing conditions when the river received untreated wastewater, the approach can also effectively predict future WQI after implanting water quality control actions. Additional water quality parameters can also be included in the development of the proposed WQI. The approach is applicable to rivers in developing countries and elsewhere with similar conditions.

References

1 Ma, T., Zhao, N., Ni, Y., Yi, J., Wilson, J.P., He, L., Du, Y., Pei, T., Zhou, C., Song, C., and Cheng, W. (2020). China's improving inland surface water quality since 2003. *Science Advances* 6 (1): 10.1126/sciadv.aau379.

2 Luo, P., Kang, S., Zhou, M., Lyu, J., Aisyah, S., Binaya, M., Regmi, R.K., and Nover, D. (2019). Water quality trend assessment in Jakarta: a rapidly growing Asian megacity. *PLoS One* 14 (7): e0219009.

3 Haider, H. and Ali, W. (2010). Development of dissolved oxygen model for a highly variable flow river: case study of Ravi River in Pakistan. *Environmental Modeling & Assessment* 15 (6): 583–599.

4 Chapra, S.C. (1997). *Surface Water Quality Modeling*. Singapore: McGraw-Hill International Editions.

5 Luo, P., Sun, Y., Wang, S., Wang, S., Lyu, J., Zhou, M., Nakagami, K., Takara, K., and Nover, D. (2020). Historical assessment and future sustainability challenges of Egyptian water resources management. *Journal of Cleaner Production* 263: 121154.

6 Rode, M. and Suhr, U. (2007). Uncertainties in selected river water quality data. *Hydrology and Earth System Sciences* 11: 863–874.

7 Harmel, R.D., Cooper, R.J., Slade, R.M., Haney, R.L., and Arnold, J.G. (2006). Cumulative uncertainty in measured streamflow and water quality data for small watersheds. *Transactions of the ASABE* 49 (3): 689–701.

8 Zadeh, L.A. (1965). Fuzzy sets. *Information and Control* 8: 338–353.

9 Singh, P., Kumar, M., and Kang, S.M. (2016). Ranking of triangular type-2 fuzzy sets and its application in multicriteria decision making problem. *International Journal of Pure and Applied Mathematics* 109 (3): 631–649.

10 Haider, H., Singh, P., Ali, W., Tesfamariam, S., and Sadiq, R. (2015). Sustainability evaluation of surface water quality management options in developing countries: multicriteria analysis using fuzzy UTASTAR method. *Water Resources Management* 29 (8): 2987–3013.

11 Haider, H. and Ali, W. (2013). Evaluation of water quality management alternatives to control dissolved oxygen and un-ionized ammonia for Ravi River in Pakistan. *Environmental Modeling & Assessment* 18 (4): 451–469.

12 Haider, H. and Ali, W. (2013). Calibration and verification of a dissolved oxygen management model for a highly polluted river with extreme flow variations in Pakistan. *Environmental Modeling & Assessment* 185 (5): 4231–4244.

13 Haider, H., Ali, W., and Haydar, S. (2013). Evaluation of various relationships of reaeration rate coefficient for modeling dissolved oxygen in a river with extreme flow variations. *Hydrological Processes* 27 (2013): 3949–3963.

14 Haider, H. and Ali, W. (2011). Fecal coliform management using a coupled hydrodynamics and water quality model for the River Ravi in Pakistan. *Pakistan Journal of Engineering and Applied Sciences* 9: 48–57.

6

Optimal Ranking of Air Quality Monitoring Stations and Thermal Power Plants in a Fuzzy Environment

Jyoti Yadav[1] *and Kalyani Salla*[2]

[1] *Department of Computer Science, Savitribai Phule Pune University, Pune, India*
[2] *Modern College of Arts, Commerce and Science, Pune, India*

Case Study I: Optimal Ranking of Air Quality Monitoring Stations (AQMS)

6.1 Introduction

The present status of poor air quality in cities in the developing world has been a talking point and calls for initiating and prioritizing mitigation measures. It can be inferred that because of deteriorating air quality in several cities in the world there is a need to initiate an action plan on air pollution abatement on risk-based optimal ranking of AQMS. In the Zadeh–Deshpande (ZD) approach, only air quality criteria pollutants are considered and in the final computation air quality is straightway described in linguistic terms such as *very high risk*, *low risk*, etc., with a degree of certainty (DC) in numeric terms attached to each description. However, there are some other constraints, such as wind velocity, population density, temperature, humidity, etc., which might affect air quality status that are not considered in the computations. The term *constraint* is used in this chapter to define the features that control the environment within particular limits. These constraints could be imprecise/fuzzy. For example, there could be several AQMS in a city with *high risk* but with varying DC in numeric terms. In such a scenario, the risk-based optimal ranking of AQMS for the identification of pollution mitigation measures assumes importance. This is basically a complex multiconstraint decision-making problem. Many a time, multiple conflicting constraints need to be evaluated in analyzing decisions in our daily lives or in professional settings. For example, in purchasing any commodity, its price, measure of quality, comfort, safety, etc., may be some of the main constraints that should be considered. In a similar vein, in air quality assessment, in addition to criteria pollutants which are primarily responsible

Fuzzy Systems Modeling in Environmental and Health Risk Assessment, First Edition. Edited by Boris Faybishenko, Rehan Sadiq, and Ashok Deshpande.

for air pollution, some other constraints based on micro meteorological data need to be considered. It could be argued that these additional constraints might be responsible for high/low air pollution levels. Invariably, all these constraints are imprecise/fuzzy. We usually weigh multiple constraints implicitly and are faced with the consequences of such decisions that are made based on intuition/perception. But in the current scenario it is important to properly structure the problem and explicitly evaluate multiple constraints. In summary, the decision-maker needs to take a decision in a fuzzy environment. The Bellman–Zadeh (BZ) method addressed this issue and presented a mathematical framework [1]. This case study, therefore, presents a combination of ZD and BZ formalism to arrive at a risk-based optimal ranking of AQMS with a case study.

6.2 Literature Survey

The study on the theory and methodology of fuzzy optimization has been active since the concept of fuzzy decision-making and the decision model under fuzzy environments were proposed by Bellman and Zadeh in the 1970s [1]. Peter Ekel et al. presented results of research and experience in the use of fuzzy set theory methodology for handling and overcoming various forms of uncertainty in the design and control of power systems and subsystems [2]. Approaches to solving fundamental problems of constructing fuzzy estimates of uncertain parameters, comparing alternatives in a fuzzy environment, and developing basic principles and methods of monocriteria and multicriteria making in a fuzzy environment are briefly described in [3]. The results of [3] are of a universal character and have found applications in: the design of distribution systems; optimization of reliability in distribution systems, state estimation, fast analysis, and operation of distribution systems; load systems (power and energy, different levels of territorial, temporal, and situational hierarchy of management) and in industrial power systems (red time control of active load). Various models and approaches to fuzzy linear programming [4, 5], fuzzy multi-objective programming [6], fuzzy integer programming [7], fuzzy dynamic programming [8], and fuzzy nonlinear programming [9] have been developed over the years by many researchers.

6.3 Methodology

This section gives a brief description of the techniques used and the methodology.

The authors have used conventional air quality index (AQI), ZD formalism for air quality classification, and the BZ method in optimal ranking on the basis of risk for thermal power plants (TPPs). The theoretical details of these methods are presented in Chapter 1.

Soft clustering is an unsupervised learning algorithm of partitioning data points into groups (or clusters) depending on their locality and connectivity within an m-dimensional feature space. In the clustering process, it should be more similar to the sets of elements in the cluster, but the similarities should be less to points outside the cluster. The essence of soft clustering is to divide the data into fuzzy partitions that overlap one another. For this reason, the belongingness of a data point in a cluster is described with a membership grade between 0 and 1. An unsupervised fuzzy C-means (FCM) algorithm is a technique based on the objective function. FCM ignores the importance of features in the clustering process. This affects its accurateness and exactness. This problem can be overcome by appropriately assigning weights to features according to their clustering importance. The weighting factor, usually denoted by m, controls the fuzziness of FCM clustering results, and the experimental results show that the m value interval is [2, 2.4]. The FCM algorithm works by assigning membership to each data point corresponding to each cluster center on the basis of distance between the cluster and the data point. The nearer the data are to the cluster center, the more is their membership toward the particular cluster center. Clearly, summation of membership of each data point should be equal to one.

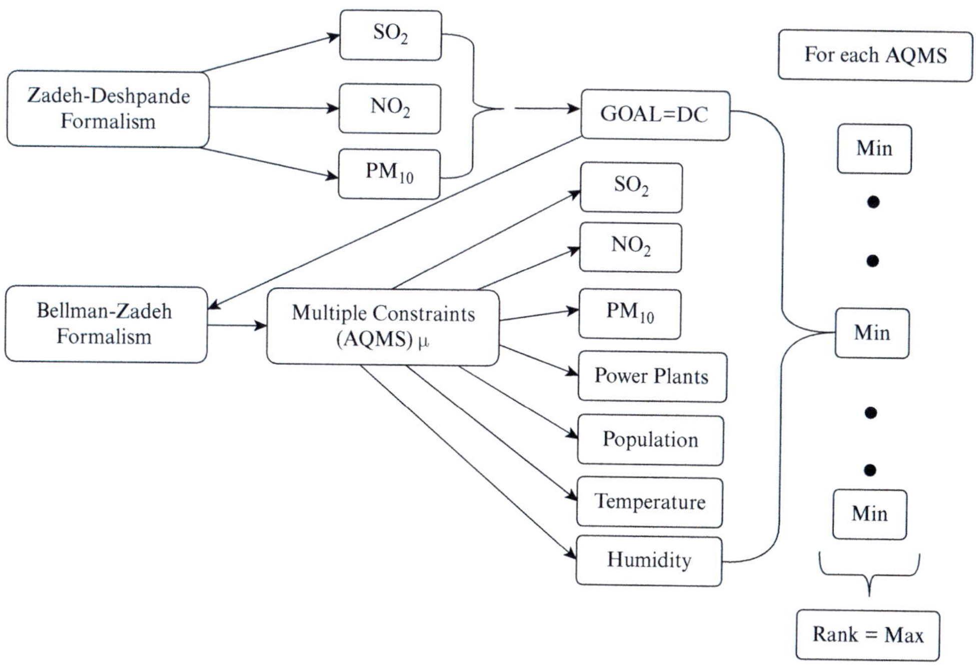

Figure 6.1 Overall framework for the combined ZD and BZ approach.

It is pertinent to mention that the BZ method has been used, in this case study in a different setting. Figure 6.1 presents an overall framework for the combined ZD and BZ approach. The air quality classification obtained for each city using the ZD formalism is described linguistically using the concept of DC. Like traditional AQI, the ZD method uses only three criteria pollutants to arrive at the air quality classification. This DC estimated in numeric terms using ZD formalism is used to define BZ formalism. BZ formalism along with the three criteria pollutants uses other constraints such as number of power plants in the city, population, temperature, and humidity to classify ambient air quality at the twelve AQMS.

6.4 Case Study I: Optimal Ranking of Air Quality Monitoring Stations (AQMS)

The limitations of the conventional air quality index (CAQI) calls for devising fuzzy logic (FL) based formalism, known as the ZD approach [4–6], wherein air quality/water quality is described in linguistic terms with a linguistic DC attached to each description. The degree of uncertainty is modeled using FL concepts. First of all, the uncertainty in the expert's perception is modeled and then the second uncertainty related to DC is modeled by defining fuzzy sets for DC. Based on ZD formalism Aurangabad, Chandrapur, Jalna, Lote, Mumbai, Nagpur,

Nanded, Nashik, Pune, Sangli, Solapur, and Thane are the locations of the twelve AQMS with *very unhealthy* air quality status with varying DC potential from an air quality viewpoint. Furthermore, the cities are also classified using FCM in five clusters with varying weighting factors based on the reference group proposed by the second author of this chapter. BZ formalism is then applied which ranks the cities as per their pollution levels taking into consideration additional parameters like wind velocity, temperature, etc., for air quality classification. The phenomenon increase in vehicular traffic could be one of the causes of air pollution. However, the location of industrial installation in the vicinity of these cities has further compounded the problem of increasing pollution levels in some of these cities. The summary results are based on a worst-case scenario, which means that these cities are ranked with very high risk if any one of the monitoring stations falls into this category [7].

6.4.1 Results and Discussion

6.4.1.1 Based on CAQI and Air Quality Using ZD Formalism

Table 6.1 presents the summary of the high-risk pollution levels with numeric DC using ZD formalism and the CAQI approach. The data for the three criteria pollutants have been taken from the Maharashtra Pollution Control Board website for the winter months of November 2014 to January 2015. The ZD formalism describes air quality first linguistically with a numeric DC attached to each description, whereas CAQI gives a numeric value to air quality status and then describes it linguistically.

Table 6.1 Classification of air quality using CAQI and ZD methods.

Station	ZD method with DC	CAQI
Aurangabad	(VP) 0.78	145.87 (P)
Chandrapur	(VP) 0.87	155.06 (VP)
Jalna	(VP) 0.85	228.48 (VP)
Lote	(VP) 0.72	101.00 (P)
Mumbai	(VP) 0.92	167.72 (VP)
Nagpur	(VP) 0.89	100.00 (P)
Nanded	(VP) 0.74	102.00 (P)
Nashik	(VP) 0.86	110.39 (P)
Pune	(VP) 0.76	114.35 (P)
Sangli	(VP) 0.84	152.46 (VP)
Solapur	(VP) 0.65	99.78 (P)
Thane	(VP) 0.87	99.62 (P)

6.4.1.2 Based on Fuzzy C-means Algorithm

The result of the FCM algorithm is computed using PM_{10} concentrations as it is the most important criterion pollutant in the Indian scenario. After running the algorithm through five iterations with varying values of weighting factor ($m = 2.2$), the following clusters were obtained:

Cluster 1 = *Good (G)* = (Lote)
Cluster 2 = *Moderate (M)* = (Aurangabad, Thane)
Cluster 3 = *Unhealthy (UH)* = (Nagpur, Nanded, Solapur, Nashik)

Cluster 4 = *Unhealthy for Sensitive Group (USG)* = (Pune, Sangli, Chandrapur, Mumbai)
Cluster 5 = *Very Unhealthy (VUH)* = (Jalna)

From the clusters obtained using FCM, it is clear that Jalna, Pune, Sangli, Chandrapur, and Mumbai should be given top priority to reduce pollution. In our view, CAQI and FCM methods could be used for reconfirming the results obtained using ZD formalism. We believe that unsupervised learning methods should precede supervised methods.

6.4.1.3 Based on Bellman–Zadeh Formalism

6.4.1.3.1 Assigning Membership Grade

In order to assign a membership grade to various constraints the following criteria have been applied. Though the three criteria pollutants – sulfur dioxide (SO_2), oxides of nitrogen (NO_X), particulate matter (PM_{10}) – were considered in the ZD approach, these three constraints and other constraints, such as population density, wind velocity, wind direction, temperature, humidity, and the capacity/number of TPPs in that particular city, were considered in the BZ method. TPPs rank first in their pollution potential.

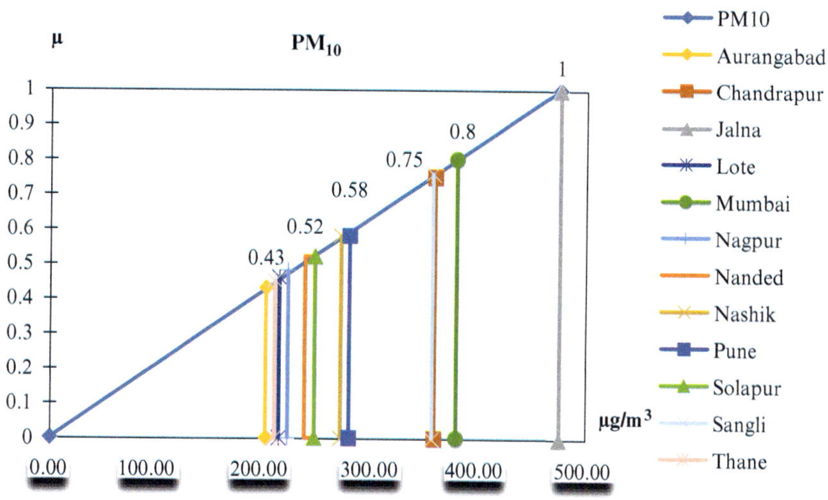

Figure 6.2 Assigning membership grade to PM_{10} pollutant.

The concentration of the criteria pollutants for the winter months has been considered. The maximum concentrations of the pollutants were considered for plotting the membership graph, as shown in Figure 6.2 for PM_{10}. Similar method was applied to SO_2 and NO_X pollutants. Table 6.2 presents the city-wise membership grades for the first three constraints. The DC column obtained by application of ZD formalism now acts as a goal for further computations in BZ formalism.

Table 6.2 Assigning membership grades to SO_2, NO_X, PM_{10} constraints.

Sr. No.	Monitoring station	Goal DC	Constraint 1 SO_2	Constraint 2 NO_X	Constraint 3 PM_{10}
1	Aurangabad	0.78	0.24	0.3	0.43
2	Chandrapur	0.87	0.47	0.48	0.75
3	Jalna	0.85	0.20	0.25	1.00

(Continued)

Table 6.2 (Continued)

Sr. No.	Monitoring station	Goal DC	Constraint 1 SO_2	Constraint 2 NO_X	Constraint 3 PM_{10}
4	Lote	0.72	0.15	0.08	0.46
5	Mumbai	0.92	1.00	1.00	0.80
6	Nagpur	0.89	0.20	0.29	0.48
7	Nanded	0.74	1.00	0.47	0.52
8	Nashik	0.86	0.62	0.23	0.58
9	Pune	0.76	0.6	0.85	0.58
10	Sangli	0.84	0.30	0.49	0.76
11	Solapur	0.65	0.20	0.51	0.52
12	Thane	0.87	0.35	0.18	0.45

For power plants constraint the city with the maximum number of industries/industry with a very high air pollution potential is given highest membership grade (i.e. Chandrapur = 0.8, because of highly polluting super power TPP). For the population constraint wind speed and population along the wind direction are given highest membership grade of 0.9 (i.e. Mumbai). For the temperature constraint, a null hypothesis was used with temp = 32 °C as 0.5 membership grade. For the humidity constraint, humidity ranging from 70 to 79 was assigned $\mu = 0.7$, for 80 to 89 $\mu = 0.8$, and so on. Table 6.3 details the membership grades of these four constraints.

Further the minimum membership grade for the seven constraints, given in Tables 6.2 and 6.3, combined was obtained for all of the cities. Finally, the maximum of the all minimum values obtained for each AQMS was identified as the city with rank = 1, meaning the city with a high risk of pollution levels. Assignment of the membership grade is a subjective measure. High concentration of PM_{10} in air can be attributed to heavy vehicular

Table 6.3 Assigning membership grades to power plants, population, temperature, and humidity constraints.

Sr. no.	Monitoring station	Constraint 4 (power plants)	Constraint 5 (population)	Constraint 6 (temperature)	Constraint 7 (humidity)
1	Aurangabad	0.2	0.5	0.4	0.6
2	Chandrapur	0.8	0.4	0.5	0.8
3	Jalna	0.2	0.3	0.3	0.5
4	Lote	0.5	0.2	0.3	0.7
5	Mumbai	0.4	0.9	0.4	0.7
6	Nagpur	0.4	0.6	0.5	0.7
7	Nanded	0.5	0.5	0.3	0.8
8	Nashik	0.4	0.3	0.4	0.5
9	Pune	0.3	0.7	0.3	0.6
10	Sangli	0.2	0.3	0.4	0.7
11	Solapur	0.3	0.6	0.4	0.6
12	Thane	0.7	0.8	0.5	0.7

traffic, construction activities, cement roads, and untarred roads. In addition, low temperature and less vertical dispersion during the winter months increase the concentration of PM_{10} in the city of Mumbai. The increase in the number of vehicles has eventually led to alarming levels of emissions as they are caught at the intersections and in the usual snarl-ups during peak rush hours. The flyovers constructed to increase traffic speeds only added auto exhaust emissions. Industrial units located near cities add to NOx, SO_2, and PM_{10}. Chandrapur, Nagpur, Thane, and Pune have a large number of vehicles, resulting in high PM_{10} concentrations.

As per the BZ (goal = DC and PM_{10}), the cities of Chandrapur and Mumbai should be given top priority to initiate pollution abatement measures. Constraint 4 (location of TPPs) is important in the case of Chandrapur. Nanded and Pune cities are ranked second due to hazardous industrial installation and excessive concentrations of PM_{10} due to an increase in vehicular traffic.

6.4.1.3.2 Optimal Ranking of AQMS using ZD–BZ and FCM Methods

Table 6.4 finally gives the ranking of AQMS, comparing ZD and BZ (with goal = DC and BZ with goal = PM_{10}) formalism along with FCM results, thus emphasizing that there is slight variation in the ranking of sites using the three methods as ZD formalism is based on concentrations of three pollutants only, whereas ZD–BZ combined formalism takes into consideration a total of seven constraints to arrive at the ranking of cities. The last column of Table 6.4 also shows ranking results using FCM.

As we are interested in optimal ranking on the basis of risk for AQMS, the DC in ZD formalism is considered a goal using the BZ method to finally arrive at a risk-based optimal ranking of AQMS for the identification of pollution mitigation measures in a fuzzy environment. Furthermore, the paper conclusively demonstrates the importance of supervised learning followed by unsupervised learning with a reference group approach in FCM clustering.

Table 6.4 Ranking of air quality monitoring stations.

ZD formalism	BZ formalism with goal = DC	BZ formalism with goal = PM_{10}	Ranking using FCM
Mumbai	Chandrapur/ Mumbai	Chandrapur/ Mumbai	Jalna
Chandrapur	Nanded/Pune	Nanded/Pune	Pune, Sangli, Chandrapur, Mumbai
Nagpur/Thane	Nashik	Nashik	Nagpur, Nanded, Solapur, Nashik
Pune	Aurangabad /Jalna / Nagpur/Sangli/ Solapur	Aurangabad/Jalna/ Nagpur/Sangli/ Solapur	Aurangabad, Thane
Jalna	Thane	Thane	Lote
Sangli	Lote	Lote	—
Nashik/Nanded	—	—	—
Lote	—	—	—
Aurangabad	—	—	—
Solapur	—	—	—

6.5 Case Study II: Optimal Ranking of TPPs

6.5.1 Introduction

In order to meet ever-increasing electric energy demand, developing countries would invariably like to install more power plants, and invariably TPPs. But this comes at the cost of environmental damage subsequently resulting in harm to humankind as well as all living organisms. Twelve out of 40 proposed TPPs in India were considered to demonstrate optimal ranking on the basis of risk for TPPs using the ZD–BZ approach. All the sites whose complete environment impact assessment reports were available were considered in our study [1–9]. Air pollution is of primary concern as air quality is depleted by increases in greenhouse gas (GHS) emissions. Criteria pollutant concentrations were considered in the estimation of the AQI. The ZD approach describes air quality from TPP in linguistic terms as *no risk* (or *very low risk*), *low risk*, *medium risk*, and *high risk* with changing DC. This is considered as the goal and passed on as input to the BZ approach. The decision-maker will naturally like to install *no risk* or *very low risk* and *low risk* TPPs.

6.5.2 Methodology

This section discusses the application of the techniques in order to arrive at the expected outcome summarized in the problem statement.

6.5.2.1 Fuzzy Soft Sets Method

The soft set theory is a generalization of FL that was proposed by Molodtsov [10] in 1999 in order to deal with uncertainty. Mathematically, a soft set is a pair (F, S), where "S" is a set (e.g. a set of parameters) and $F: S \rightarrow P(U)$ is a function such that F(x) is a subset U, where x is an element of the set S, U is the universal set, and P (U) is the power set of set U. Further, each element of F(x) gets mapped to a value in the interval [0, 1], which represents the membership grade of that element of F(x).

A soft set over U is a parameterized family of subsets of the universe U. For $\epsilon \subset A$, F(ϵ) may be considered as the set of ϵ: approximate elements of the soft set (F, A). For two soft sets (F, A) and (G, B) over a common universe U, (F, A) is a soft subset of (G, B) if (i) $A \subset B$ and (ii) if $\epsilon \subset A$, then F(ϵ) and G(ϵ) are identical approximations.

Similarly, (F, A) is said to be a soft super set of (G, B) if (G, B) is a soft subset of (F, A).

If (F, A) and (G, B) are two soft sets then (F, A) AND (G, B) is denoted by

$$(F,A)\wedge(G,B) \tag{6.1}$$

$$(F,A)\wedge(G,B) = (H, A \times B) \tag{6.2}$$

$$H(\alpha, \beta) = F(\alpha) \cap G(\beta) \tag{6.3}$$

where, H is the resultant fuzzy soft set $\alpha \in A$ and $\beta \in B$. The fuzzy soft set method requires formulation of the problem in terms of soft sets.

Figure 6.3 shows the overall flow of application of ZD and BZ techniques. On the basis of the data obtained using bootstrapping method, the probability distribution was transformed into a convex normalized fuzzy number (CNFN). Uncertainty in the expert's observation, termed epistemic uncertainty, is modeled by constructing fuzzy sets and the degree of match (DM) between the fact (parametric data) and the perception of domain experts [11].

Figure 6.4 depicts the plot for possibility distributions of parameter PM_{10} for *no risk* or *very low risk/very good, low risk/good, medium risk/fair*, and *high risk/poor* with expert's knowledge database, and Table 6.5 gives the calculated degree of match for Kolaghat. The Central Pollution Control Board (CPCB) policy is also considered by the experts [12]. The plot of probability distribution is shown as "Kolaghat PM_{10}." This probability distribution is altered to a possibility distribution and shown as a CNFN [13]. The CNFN may overlie with one, two, or three possibility distributions. This coinciding area gives a DM with respective possibility distributions [13]. The above process is repeated for all parameters like SO_2 and NO_2 for all power plants in our study.

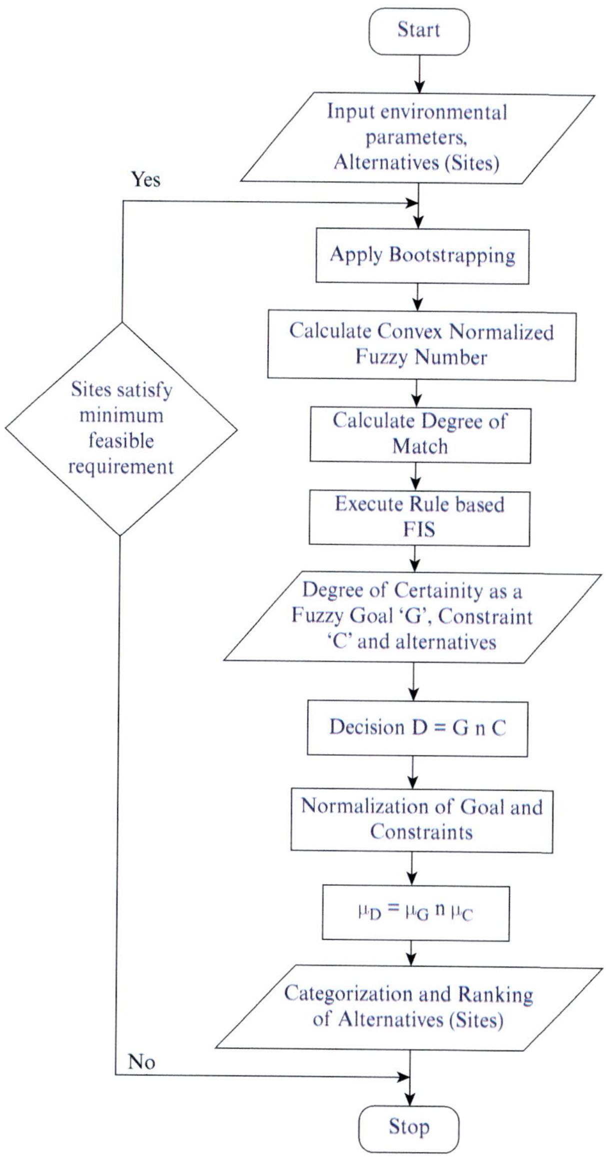

Figure 6.3 Flowchart of ZD–BZ approach.

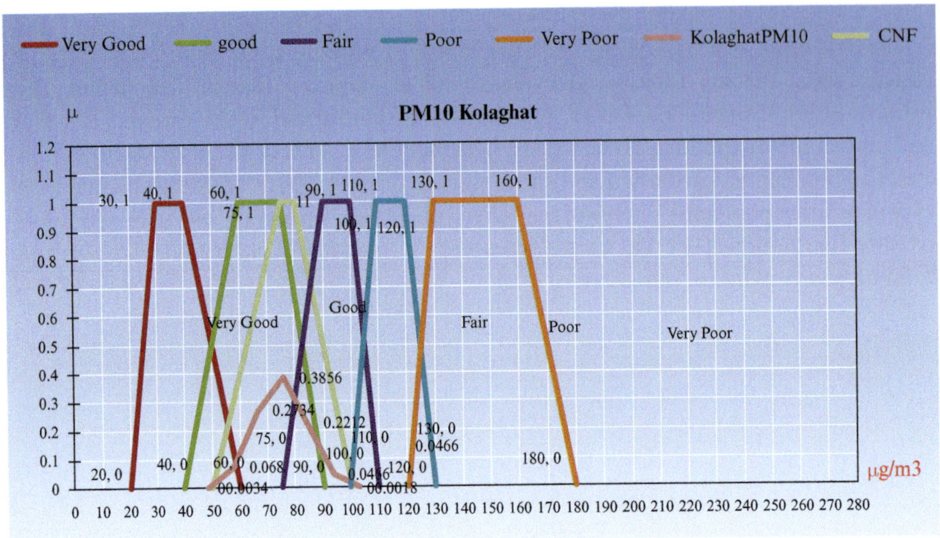

Figure 6.4 Degree of match.

Table 6.5 Degree of match for Kolaghat.

Kolaghat	No risk	Low risk	Medium risk	High risk
SO_2	1	0	0	0
NO_2	0	1	0	0
PM_{10}	0	0.8	0.13	0

The above process was repeated to calculate DM for some other sites.

6.5.2.2 Type-1: Fuzzy Inference System (FIS)

There are 60 fuzzy rules formulated by experts to specify air quality systems. One of the challenges of fuzzy applications is to check the rationality of all these rules. The conjunctive (intersection) operator applies when we want to find the DM of one or more antecedent rules [14]

$$\text{DM(of a rule with steps1, 2, 3, \ldots, } n) = \text{Min}(\text{DM}_1, \text{DM}_2, \text{DM}_3, \ldots, \text{DM}_n)$$

Fuzzy rules which matched for a site at Kolaghat, India

1) If (SO_2 has *no risk* (1) and NO_2 has *low risk* (1) and PM_{10} has *low risk* (0.8) then site has *low risk* with DC(0.8))
2) If (SO_2 has *no risk* (1) and NO_2 has *low risk* (1) and PM_{10} has *medium risk* (0.13) then site has *low risk* with DC (0.13))

Therefore, the site at Kolaghat for a forthcoming TPP is acceptable to have *low risk* with DC (0.8). This process was carried out for all other sites. Similarly, rules were executed for other sites in our study.

A site can be classified based on highest value of DC in a particular class, namely *no risk*, *low risk*, *medium risk*, or *high risk*.

Table 6.6 Degree of certainty for TPPs.

Risk level	Tuticorin	Amreli	Badarpur	Khammam	Kudgi	Kolaghat	Singrauli	Junagadh	Gharghoda
No risk	0.2	—	—	0.3	1	—	—	—	—
Low risk	0.6	0.3	—	0.4	—	0.8	0.4	0.4	0.1
Medium risk	—	0.4	0.3	—	—	0.13	0.6	0.2	0.3
High risk	—	—	0.1	—	—	—	—	—	—
Rank	3	7	9	2	1	4	6	5	8

Source: Adapted from [15].

Plants located at the following have *low risk*: Tuticorin (0.6), Khammam (0.4), Kolaghat (0.8), and Junagadh (0.4).

DC = [0.6, 0.4, 0.8, 0.4]

Normalization: $0.6 + 0.4 + 0.8 + 0.4 = 2.2 \approx 2$

After normalization DC = [(0.6/2), (0.4/2), (0.8/2), (0.4/2)]

Therefore, DC (*low risk*) = [0.3, 0.2, 0.4, 0.2]

This output is passed as a goal to rank sites using the BZ approach, which considers various constraints.

6.5.2.3 Bellman–Zadeh Approach

In the conventional approach to decision-making, the principal components of a decision process are (i) a set of alternatives, (ii) a set of constraints on the choice between alternatives, and (iii) a performance function which associates with each alternative the gain (or loss) resulting from the choice of that alternative [16]. An important aspect of this model is its symmetry with respect to goals and constraints – a symmetry which erases the difference between them and makes it possible to relate in a relatively simple way the concept of a decision to those of goals and constraints of a decision process [16]. Here, the goal considered is to identify power plants that have *low risk* as per the results from the ZD approach. Constraints considered were: SO_2, NO_2, PM_{10}, pH, dissolved oxygen (DO), biochemical oxygen demand (BOD), wind speed, wind direction, and cost of operation and maintenance. Sites with *low risk* were located at Tuticorin, Khammam, Kolaghat, and Junagadh [15]. For the BZ approach the goal was to identify power plant sites with *low risk* constraints: SO_2, NO_2, PM_{10}, pH, DO, BOD, wind direction, wind speed, and cost of operation and maintenance [15].

6.5.2.3.1 Wind Direction Constraint

Table 6.7 provides the results of the evaluation of wind direction and distance from adjoining cities.

For the TPP site located at Junagadh, the common wind direction is north and north-west. Cities/villages such as Jamnagar (113 km), Gandhidham (175 km), Dwarka (174 km), and Upperkot (1.7 km) lie in the direction of the wind. The wind speed is 11 m/s, which is on the slower side and promotes stagnancy and hazardous effects on the population residing near power plants. Most locations are above 100 km, except Upperkot (1.7 km), a membership grade of 0.25 is assigned as the TPP location is not very likely to have harmful effects in consideration of wind speed and wind direction.

For the TPP site located at Tuticorin, the prominent wind direction is north and north-east. The cities/villages which lie in this direction are Ramanathapuram (102 km),

Table 6.7 Details of wind direction and distance from adjoining cities.

TPPsite location	Prominent wind direction	City	Distance (km)	Direction from TPP
Tuticorin, Tamil Nadu	North, north-east	Tirunelveli	46	West
		Ramanathapuram	102	North-east
		Sivaganga	125	North, north-east
		Karaikudi	158	North, north-east
		Pudukkottai	192	North, north-east
		Thanjavur	249	North, north-east
		Karaikudi	158	North, north-east
		Virudhunagar	89	North, north-west
		Madurai	124	North
Khammam, Andhra Pradesh	North, north-east	Bhadrachalam	91	East, north-east
		Malkangiri	223	East, north-east
		Nalgonda	95	West, south-west
		Rajahmundry	176	East
		Warangal	99	North-west
		Guntur	110	South, south-east
		Eluru	123	East, south-east
		Hyderabad	182	West
		Vijayawada	96	South, south-east
		Bhubaneshwar	686	East, north-east
Kolaghat, West Bengal	North-west, west	Bagnan	13	North-east
		Kharagpur	64	South-west
		Amta	25	North-east
		Chaipat	41	North-west
		Trilochanpur	421	North-west
		Daspur	38	North-west
Junagadh, Gujarat	North, north-west	Ahmedabad	—	North-east
		Surat	—	East
		Upperkot	—	North-east
		Jetpur, Navagadh	31	North, north-east
		Gondal	60	North-east
		Jamnagar	113	North, north-west
		Dwarka	174	West, north-west
		Gandhidham	175	North

Sivaganga (125 km), Karaikudi (158 km), Pudukkottai (192 km), Thanjavur (249 km), Karaikudi (158 km), and Pudukkottai (192 km). Distance in kilometers is specified in Table 6.6. As specified in Table 6.8, wind speed is up to 37 m/s. Most of these cities are away from the TPP site at Tuticorin and at a distance of above 100 km in the prominent wind direction to the north-east. Therefore, it is likely to have an adverse effect on the population residing in these cities. So a membership grade of 0 is assigned to the TPP site located at Tuticorin.

For the TPP site located at Khammam, the prominent wind direction is north and north-east and involves cities such as Bhadrachalam (91 km), Malkangiri (223 km), and Bhubaneshwar (686 km). Only one city is located at a distance of 91 km. As specified in Table 6.8, wind speed is not so strong and up to 20 m/s. The remaining cities do not lie in the direction of the wind. Considering these facts, we conclude that this TPP would not have many adverse effects.

For the TPP site located at Kolaghat, the prominent wind direction is north and north-west. Cities or villages such as Chaipat (41 km), Daspur (38 km), and Trilochanpur (421 km) lie in this direction. As specified in Table 6.11, wind speed is up to 4 m/s, which is slow and is harmful as pollutants become stagnant due to this low wind speed. Considering the risk for population staying in the adjoining places, a membership grade of 0.5 is assigned.

Figure 6.5 depicts the fuzzy set for NO_2. Similarly, fuzzy sets were generated for other constraints like SO_2, PM_{10}, pH, DO, and BOD. In Figure 6.5, Tuticorin has a NO_2 value of 24 μ/m^3 then its membership grade is 0.37.

6.5.2.3.2 Cost Constraint

The results of calculations of the operation and maintenance cost of Thermal Power Plants are summarized in Table 6.8 and corresponding cost expressed using the membership grade is given in Table 6.9.

Table 6.8 Operation and maintenance cost of TPPs in India [2, 17].

Rs. Lakh/MW	200/210/250 MW	300/330/350 MW	500 MW	600 MW and above
2009–10	18.20	16.00	13.00	11.70
2010–11	19.24	16.92	13.75	12.37
2011–12	20.34	17.88	14.53	13.08
2012–13	21.51	18.91	15.36	13.82
2013–14	22.74	19.99	16.24	14.62

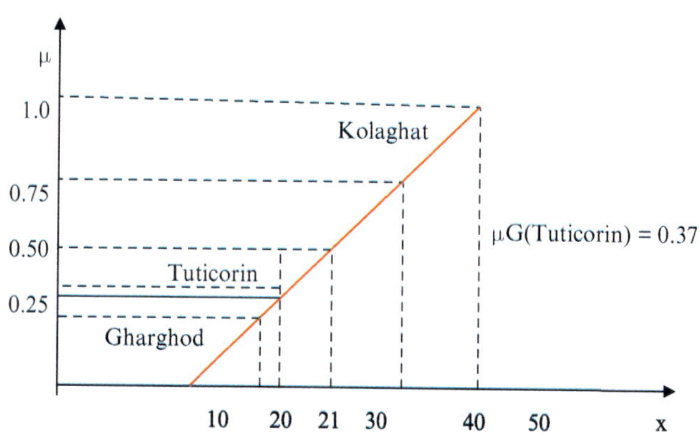

Figure 6.5 Fuzzy set for NO_2 representing the *low risk* goal.

Table 6.9 Calculation of membership grade.

Rs. Lakh/MW	200/210/250 MW	300/330/350 MW	500 MW	>600 MW
Avg	20.40	17.94	14.57	13.11
Membership grade	0.30	0.27	0.22	0.19

The membership grades are calculated by considering the average sum in Table 6.9. So the sum is 66 (i.e. 20.40 + 17.94 + 13.57 + 14.11). Therefore, membership grades are calculated as per power plant capacity as (20.40/66, 17.94/66, 14.57/66, 13.11/66).

The power plants with *low risk* identified in our work with capacity and locations were at Tuticorin (525 MW), Junagadh (660 MW), Khammam (236 MW), and Kolaghat (1260 MW). Therefore, membership grades are assigned depending on the capacity.

Table 6.10 shows the actual data for important constraints considered for ranking, and Table 6.11 includes membership grades from fuzzy sets calculated from data.

Table 6.10 Average values from original data for constraints.

Constraints	Tuticorin	Khammam	Kolaghat	Junagadh
SO_2	16	34	23	20
NO_2	24	36	42	32
PM_{10}	95	66	75	110
pH	7.5	7.0	7.5	8.4
DO	7	8	8	5
BOD	3	2	3	3
Wind speed	37 m/s	20 m/s	4 m/s	11 m/s
Wind direction	North, north-east	North, north-east	North-west, west	North, north-west
Cost (operation and maintenance) with relation to capacity	Rs 14.57 Lakhs/MW	Rs 20.40 Lakhs/MW	Rs 13.11 Lakhs/MW	Rs 13.11 Lakhs/MW

Rs: rupees; Lacks/MW: one hundred thousand per megawatts.

Table 6.11 Membership grades from fuzzy sets (generated from data).

Constraints – C_i	Tuticorin	Khammam	Kolaghat	Junagadh
C_1 – PM_{10}	0.67	0.42	0.5	0.8
C_2 – SO_2	0.1	0.22	0.11	0.1
C_3 – NO_2	0.13	0.21	0.26	0.18
C_4 – pH	0.45	0.35	0.45	0.55
C_5 – DO	0.84	1	1	0.5
C_6 – BOD	0.25	0.25	0.25	0.4
C_7 – Wind speed	0	0.05	0.23	0.13
C_8 – Wind direction	0	0	0.5	0.25
C_9 – Cost	0.22	0.30	0.19	0.19

Table 6.12 Normalized membership grades from fuzzy sets (generated from data).

Constraints – C_i	Tuticorin	Khammam	Kolaghat	Junagadh
C_1 – PM_{10}	0.28	0.17	0.20	0.33
C_2 – SO_2	0.18	0.41	0.20	0.18
C_3 – NO_2	0.16	0.26	0.33	0.23
C_4 – pH	0.25	0.19	0.25	0.30
C_5 – DO	0.25	0.29	0.29	0.14
C_6 – BOD	0.21	0.21	0.21	0.34
C_7 – Wind speed	0	0.12	0.56	0.31
C_8 – Wind direction	0	0	0.66	0.33
C_9 – Cost	0.24	0.33	0.21	0.21

From the normalized values as shown in Table 6.12

$$\text{TPP Tuticorin} = \min\{(C_1,0.28),\ (C_2,0.18),\ (C_3,0.16),\ (C_4,0.25),\ (C_5,0.25),\ (C_6,0.21),\ (C_7,0),\ (C_8,0),\ (C_9,0.24)\}$$

$$\text{TPP Khammam} = \min\{(C_1,0.17),\ (C_2,0.41),\ (C_3,0.26),\ (C_4,0.25),\ (C_5,0.29),\ (C_6,0.21),\ (C_7,0.12),\ (C_8,0.0),\ (C_9,0.33)\}$$

$$\text{TPP Kolaghat} = \min\{(C_1,0.20),\ (C_2,0.20),\ (C_3,0.33),\ (C_4,0.25),\ (C_5,0.29),\ (C_6,0.21),\ (C_7,0.56),\ (C_8,0.66),\ (C_9,0.21)\}$$

$$\text{TPP Junagadh} = \min\{(C_1,0.33),\ (C_2,0.18),\ (C_3,0.23),\ (C_4,0.30),\ (C_5,0.14),\ (C_6,0.34),\ (C_7,0.31),\ (C_8,0.33),\ (C_9,0.21)\}$$

DC for goal: *low risk* after normalization is

$$\text{DC(Low Risk)} = [0.3,\ 0.2,\ 0.4,\ 0.2]$$

$$\text{Decision D} = G_{alt} \cap C_1 \cap C_2 \cap C_3 \cap C_4 \cap C_5 \cap C_6 \cap C_7 \cap C_8 \cap C_9 \tag{6.4}$$

$$\begin{aligned}\text{Decision D} &= \min\{(0.3, 0.0),\ \min(0.2, 0.0),\ \min(0.4, 0.20),\ \min(0.2, 0.14)\}\\ &= \{(\text{Tuticorin, } 0.0),\ (\text{Khammam, } 0.0),\ (\text{Kolaghat, } 0.20),\ (\text{Junagadh, } 0.14)\}\end{aligned}$$

The constraints are obtained as (0, 0, 0.20, 0.14). Therefore, the site at Kolaghat is ranked 1. Removing the details of the first ranked site's constraints and goals we acquire the goals for Tuticorin, Khammam, and Junagadh as {(0.3, 0.2, 0.2)} and the corresponding constraints as {(0.0, 0.0, 0.14)}. This process is called re-normalization. Decision D is given below. Therefore, Junagadh is ranked next (i.e. second) and Tuticorin and Khammam are ranked third

$$\begin{aligned}D &= \min\{(0.3, 0.0), (0.2, 0.0), (0.2, 0.14)\}\\ &= \{(\text{Tuticorin}, 0.0), (\text{Khammam}, 0.0), (\text{Junagadh}, 0.14)\}\end{aligned}$$

As shown in Table 6.9, the ranks acquired by the ZD and the combination of the ZD with the BZ approach has considered many important constraints other than the air and water qualities, like wind speed and wind direction, that play an important role in identifying pollution risk. The cost is also an important constraint when setting up a power plant. Both these approaches are found to be the best for evaluation. The ZD approach has been further extended by combining it with the BZ approach for performing risk-based optimal ranking. It was found to be essential to first apply the ZD approach to acquire the membership grade of the goal set to identify sites with *low risk* as everyone would prefer to set up a power plant in a *low risk* area.

Table 6.13 Ranks with ZD and ZD–BZ.

Approach	Tuticorin	Khammam	Kolaghat	Junagadh
ZD	2	1	3	4
ZD–BZ	3	3	1	2

Source: Adapted from [15].

As shown in Table 6.13, the ranks acquired by ZD and ZD–BZ could consider important constraints other than wind speed and wind direction that play an important role in identifying pollution risk. Cost is also an important constraint when setting up a power plant.

Figure 6.6 shows the TPP location (sites) in India. The symbol "V" indicates sites which have *very low risk* (actually meaning *no risk*). The sites identified for TPP with very low risk

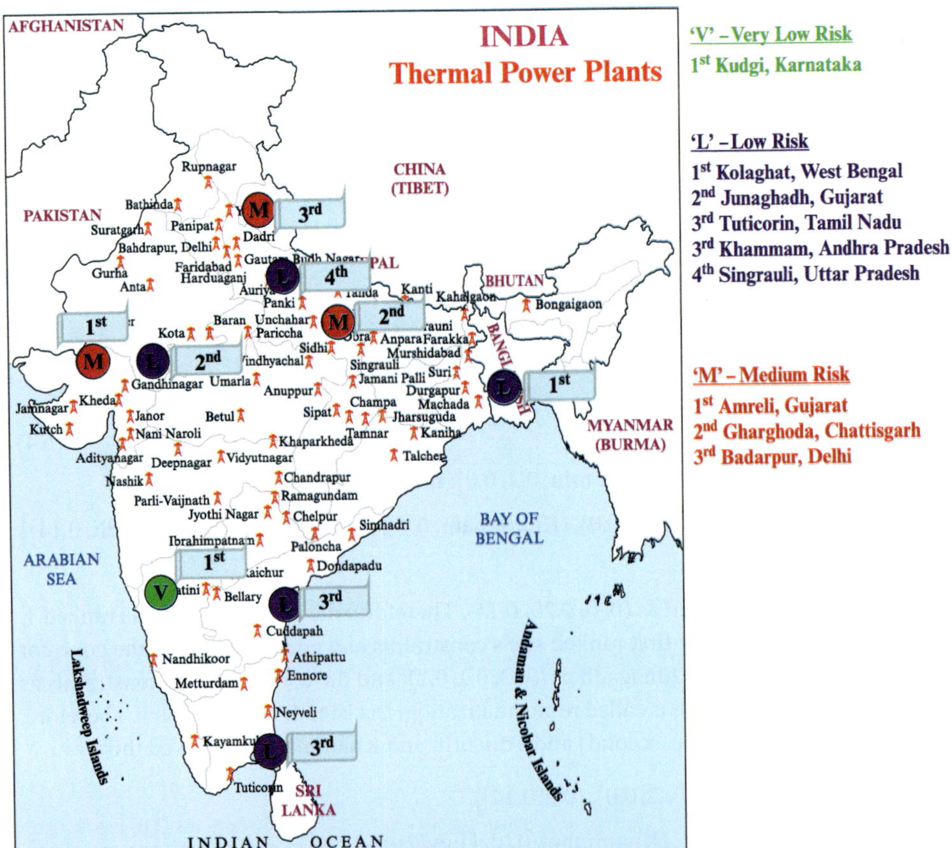

Figure 6.6 Power plant locations in India. *Source:* Adapted from [15].

are located at Kudgi, in the southern state of Karnataka. The symbol "L" indicates sites with *low risk* out of some of the sites considered in the ZD and BZ approaches. They are: Kolaghat, West Bengal (first rank); Junagadh, Gujarat (second rank); Tuticorin, Tamil Nadu (third rank); and Khammam, Andhra Pradesh (third rank). Another TPP site considered in our study was located at Singrauli was ranked fourth. The symbol "M" indicates sites with *medium risk*, which are ranked as: Amreli, Gujarat (first); Gharghoda, Chhattisgarh (second rank); and Badarpur, Delhi (third rank).

The rank of Tuticorin remains more or less the same with both the approaches. For the Junagadh site, where the wind direction is north-west, only two cities lie in this direction. These two cities, namely Jamnagar (113 km) and Dwarka (174 km), lie more than 100 km away. The cost of operation and maintenance is the lowest compared to the other cities. These two constraints (wind direction and cost) are dominant in lifting the rank of Junagadh from fourth by ZD to second by combining both approaches. A site at Singrauli, Uttar Pradesh, was also identified as having *low risk*. The symbol "M" indicates sites with *medium risk*. The sites identified to have *medium risk* are located at: Amreli, Gujarat; Gharghoda, Chhattisgarh; and Badarpur, Delhi.

For the site location Khammam, which was ranked first by the ZD approach, has come down to rank three by a combined ZD and BZ approach, as it has been observed to have the highest cost for operation and maintenance amongst the four sites. Kolaghat was ranked as third by the ZD approach and first by the combined ZD and BZ approach. The cost of operation and maintenance was a dominant constraint. The Kolaghat site has fewer cities/villages around it. Also, it does not have very good air quality. Therefore, it was given a lower rank by the ZD approach because it considers air quality constraints important.

6.6 Concluding Remarks

Decision-making is a complex issue and needs careful study as it involves the participation of all stakeholders [18]. Site selection would become an almost irreversible decision after its installation of a TPP due to heavy financial investment and labor. Therefore, there is a need to devise a method to avoid harmful effects to the environment and subsequently to humankind [19]. Soft computing techniques with focus on fuzzy set theory in deciding risk-based optimal ranking of TPPs is demonstrated in this chapter.

The current scenario and possible risks are considered in this study during the ranking process. The ZD formalism helps describe the site suitability linguistically with a DC attached to each term. This approach also helps to define certain goals, such as the identification of sites with *no risk* (or *very low risk*), *low risk*, *medium risk*, or *high risk*. A combination of the ZD approach with the BZ approach has proved a useful tool in risk-based optimal ranking of power plant sites. In summary, the visible conclusion of the research is to assist policymakers to take decisions based on the risk-based optimal ranking of TPPs. A list of TPPs considered in our study is given in Table 6.14.

It is our belief that the ZD–BZ combination and reference level approach in fuzzy clustering may have wider practical applications in some of the areas of science and technology wherein the goal and constraints are imprecise or fuzzy and where policymakers are confronted with estimating a risk-based optimal strategy for a defined problem.

Table 6.14 List of thermal power plants considered in our study.

Sr. no.	Power plant	Location
1.	Badarpur TPP	Mayapuri Industrial Area, Delhi
2.	Amreli TPP	Lunsapur, Gujarat
3.	Bharuch TPP	Bharuch District, Gujarat
4.	Tuticorin TPP	Fisheries College, Tamil Nadu
5.	Khammam TPP	Khammam, AP
6.	Nagapattinam TPP	Nagapattinam District, Tamil Nadu
7.	Thiruvallur TPP	Gummidipoondi, Tamil Nadu
8.	Junagadh TPP	Gujarat
9.	Kudgi TPP	Karnataka
10.	Gharghoda TPP	Mohali, Chhattisgarh
11.	Nalgonda TPP	Peddaraveedu, Andhra Pradesh
12.	Ajmer TPP	Andheri Deori, Rajasthan
13.	Sipat TPP	Sipat, Chhattisgarh
14.	Khancheepuram TPP	Chitharkadu, Tamil Nadu
15.	Kamrup TPP	Assam
16.	Sagardighi TPP	Murshidabad, West Bengal
17.	Raipur TPP	Raigarh, Chhattisgarh
18.	Janjgir-Champa TPP	Bhilai, Chhattisgarh
19.	Bargawan	Sidhi District, Madhya Pradesh
20.	Sri Damodaran TPP	Nellore, Andhra Pradesh
21.	Kolaghat TPP	Purba, West Bengal
22.	Captive TPP	Ariyalur, Tamil Nadu
23.	Chandrapur TPP	Wardha, Maharashtra
24.	Kutch TPP	Kutch, Gujarat
25.	Sundargarh TPP	Angul, Odisha
26.	Singrauli TPP	Sonebhadra, Uttar Pradesh

Acknowledgment

We are immensely grateful to our mentor and guide Dr. Ashok Deshpande for motivating us to write this chapter. His contribution in writing this chapter is commendable and worthy.

References Part I (Case study I)

1 Bellman, R.E. and Zadeh, L.A. (1970). Decision making in a fuzzy environment. *Management Science* 17 (4): B141–B164.

2 Ekel, P.Y., Kokshenev, I.V., Parreiras, R.O., Alves, G.B., and Souza, P.M.N. (2013). Fuzzy set based models and methods of decision making and power engineering problems. *Engineering* 5: 41–51.

3 Popov, V.A. and Ekel, Y.P. (1987). Fuzzy set theory and problems of controlling the design and operation of electric power systems. *Soviet Journal of Computer and System Sciences* 25 (4): 92–99.

4 Fang, S. and Hu, C.F. (1999). Linear programming with fuzzy coefficients in constraints. *Computers and Mathematics with Applications* 37: 63–76.

5 Fang, S.C. and Li, G. (1999). Solving fuzzy relation equations with a linear objective function. *Fuzzy Sets and Systems* 103: 107–113.

6 Sakawa, M. and Yano, H. (1989). An interactive fuzzy satisficing method for multi-objective nonlinear programming problems with fuzzy parameters. *Fuzzy Sets and Systems* 30 (10): 221–238.

7 Chanas, S. and Kuchta, D. (1998). Discrete fuzzy optimization. In: *Fuzzy Sets in Decision Analysis Operations Research and Statistics: The Handbook of Fuzzy Sets Series* (ed. E. Slowinski), 249–276. Dordrecht, Netherlands: Kluwer Academic.

8 Kacprzyk, J. and Esogbue, A.O. (1996). Fuzzy dynamic programming: main developments and applications. *Fuzzy Sets and Systems* 81 (1): 31–46.

9 Liu, J. and Fang, S.C. (2001). Solving nonlinear optimization problems with fuzzy relation equation constraints. *Fuzzy Sets and Systems* 119: 1–20.

10 Molodtsov, D.A. (1999). Soft set theory-first results. *Computers and Mathematics with Applications* 37: 19–31.

11 Deshpande, A.W. and Raje, D.V. (2003). *Fuzzy Logic Applications to Environmental Management Systems: Case Studies. IEEE.*

12 Mckone, T. and Deshpande, A. (2005). Can fuzzy logic bring complex environmental problems into focus? *Environmental Science and Technology* 39 (2): 42A–47A.

13 Jyoti, Y., Kharat, V., and Deshpande, A. (2011). Fuzzy description of air quality: a case study. *6th International Conference on Rough Sets and Knowledge Technology (RSKT)* (9–12 October). Banff, Canada, 420–427.

14 Jyoti, Y., Kharat, V., and Deshpande, A. (2013). Evidence theory and fuzzy relational calculus in estimation of health effects due to air pollution. *International Journal of Intelligent Systems* 22 (1): 9–22.

15 Goenka, D. and Guttikunda, S. (2011). Coal kills: an assessment of death and disease caused by India's dirtiest energy sources. *Urban Emissions in partnership with the Conservation Action Trust and Greenpeace India.*

16 Matrix Thermal Power Pvt. (2010). *Environment impact assessment report on coal based washery based thermal power plant, Khammam, Andhra Pradesh, India.*

17 Narmada Thermal Power Limited. (2012). Draft environmental impact assessment report for thermal power plant. *National Environmental Engineering Research, Jaitapur.*

18 NTPC Limited. (2010). *Badarpur combined cycle power project, Badarpur, Delhi.*

19 Patel Energy Limited. (2011). *Draft environmental impact assessment for coal based thermal power plant, Amreli, Gujarat.*

20 R.S. Envirolink Technologies. (2010). *GMR Bajoli Holi hydro power plant, environmental impact assessment report.*

21 R.S. Envirolink Technologies. (2010). *TT energy, environmental impact assessment of H.E. Project.*

22 Smec India. (2009). *SEW Nafra Corporation, environmental impact assessment and environmental management plan for NAFRA hydro electric power project.*

23 SPIC Electric Power corporation Pvt., Ltd. (2010). *HW Tuticorin thermal power project.* Tuticorin, Tamilnadu.

24 Ross Timothy, J. (1995). *Fuzzy Logic with Engineering Application.* New Mexico: McGraw Hill.

25 Ministry of Environment and Forest (MoEF). (2011). *Environmental guidelines for industries and impact assessment*. https://moef.gov.in/wp-content/uploads/2017/06/moef_gov_in_citizen_specinfo_enguin_html.pdf (accessed 4 October 2022).

26 Deshpande, A.W., Yadav, J.Y., and Kharat, V. (2014). Zadeh Deshpande approach for fuzzy description of air and water quality. *International Journal of Information Technology* 6: 677–682.

27 McKone, T.E. and Deshpande, A.W. (2011). Can fuzzy logic bring complex environmental problems into focus? *Environmental Science & Technology*. 39 (2): 42A–47A.

28 Deshpande, A.W. and Salla, K.R. (2017). Can fuzzy logic bring complex issues in risk based optimal ranking of hazardous plant sites into focus? a case study. *International Journal of System Assurance Engineering and Management* 8 (1): 1–11.

29 Soong, T. (1987). Stochastic processes and stochastic differential equations. In: *The Bellman Continuum* (ed. N. Bellman and R.S. Roth), 551–666. World Scientific Publishing.

30 Motghare, P. and Vilas, C.R. (2014). Generation cost calculation for 660 MW thermal power plants. *International Journal of Innovative Science Engineering and Technology* 1 (10): 660–664.

31 Sambhoo, K., Kadam, S., and Deshpande, A. (2013). Rule base for grading of sites. *International Journal of Computer and Technology* 10 (7): 1836–1847.

32 Zadeh, L.A., Abbasov, A.M., Yager, R.R., Shahbazova, S.N., and Reformat, M.Z (2014). Efficacy of selected soft computing techniques in ranking of sites for hazardous industrial installation. In: *Recent Developments and New Directions in Soft Computing, Studies in Fuzziness and Soft Computing* (ed. K. Salla, S. Kadam, and A. Deshpande), 317: 345–359. Switzerland: Springer International Publishing.

Part III

Fuzzy Logic Application in Healthcare Decision-making

7

Evaluation of Health Effects Due to Environmental Pollution Based on Belief and Possibility

Jyoti Yadav[1] *and Kedar Rijal*[2]

[1] *Department of Computer Science, Savitribai Phule Pune University, Pune, India*
[2] *Tribhuvan University, Department of Environmental Science, Kathmandu, Nepal*

7.1 Case Study I: Effects of Air Pollution on Human Health Using Fuzzy Relational Calculus and Evidence Theory

7.1.1 Introduction

It is a proven fact that air pollution affects human health. Human beings living in a pollution-free environment lead a better quality of life, but are we really aware of how bad air affects us and which organs of the human body are damaged by air pollutants? Different kinds of pollution have different effects on various organs of the human body. Emissions from different transport modes, industrial pollution, forest fires, the burning of fossil fuels, and huge construction projects are some of the major causes of air pollution. These sources of emissions liberate toxic gases and substances that are dangerous to human health, resulting in a number of respiratory diseases depending upon pollutant type; its magnitude, length, and frequency of exposure; along with its toxicity. Miniscule air pollutants easily penetrate into the respiratory system, damaging lungs, heart, and brain. These pollutants include some common criteria pollutants, namely carbon monoxide, oxides of sulfur, particulate matter, ground-level ozone, oxides of nitrogen, lead, and some harmful air pollutants, namely hydrochloric acid, formaldehyde, asbestos, benzene, carbon tetrachloride, chlordane, chloroform, heptachlor, mercury, methanol, phenol, and toluene. Many researchers are involved in recognizing various diseases and accompanying risk factors on human health due to polluted air. Air pollution is aggravated by increasing levels of traffic, urbanization, and industrialization.

The primary source of air pollution is vehicular transportation. Trucks and cars with diesel engines contribute to vehicular pollution in metropolitan cities. The major toxic components emitted by vehicular traffic are carbon monoxide, ozone, particulate matter, and other smog-forming emissions. Bad air causes a serious health risk. Research studies have shown that the increase in vehicular traffic is one of the causes of respiratory diseases. There are two types of pollution discharged by petrol vehicles, like evaporative emissions, that occur when vapors of fuel are released into the atmosphere, without being burnt, and exhaust emissions, including dangerous gases such as carbon monoxide, oxides of nitrogen, hydrocarbons, and particulates. Respiratory diseases due to air pollution are growing day by day. People are suffering from various respiratory diseases including asthma, chronic obstructive pulmonary disease (COPD), cardiovascular disease, and lung cancer. Small children and the elderly are especially at risk.

Fuzzy Systems Modeling in Environmental and Health Risk Assessment, First Edition. Edited by Boris Faybishenko, Rehan Sadiq, and Ashok Deshpande.

Pope et al. [1] have tried to associate day-to-day exposure to particulate air pollution and adverse health effects, using statistical analysis for around 16 years' worth of data using the Bartlett test and the Cox proportional hazards model. In their study, 500 000 adults were served with a questionnaire about individual risk factor data such as age, sex, race, weight, height, smoking history, education, marital status, diet, alcohol consumption, and occupational exposures. The risk factor data were linked with air pollution data throughout the United States and combined with vital status and cause of death data. The authors concluded that elevated fine particulate air pollution exposures were associated with significant increases in lung cancer and cardiopulmonary mortality even after controlling for cigarette smoking, body mass index (BMI), diet, occupational exposure, other individual risk factors, and also after controlling for regional and spatial differences. Some of the prominent researchers who have studied the facets of air pollution and its health effects, especially from the outdoor air pollution perspective, include [2–4]. To estimate the health risks of chronic exposure a cohort study is typically used. For example, Dockery et al. [5] examined the output of a cohort study in which over 8000 adults in six US cities were followed for a period of 14–16 years. The association between total daily mortality and air pollution was investigated for one year from September 1985 through August 1986 to evaluate the relative importance of various measures of particulate and gaseous air pollution as predictors of daily mortality [6].

Air pollution studies often analyze data from a number of cities [7, 8], using a variety of statistical approaches. This variation in statistical methodology is partly responsible for the heterogeneity observed in the pollution–health associations which have been estimated. This heterogeneity could be reduced by implementing large multicity studies, including air pollution and health.

Epidemiological data for over a decade were collected by the researchers for establishing the association between respiratory diseases and polluted air. In the majority of cases, such data may not be reliable and also not available. The hospital staff might not have collected the information for drawing inference using statistical methods. This is known as happenstance data (collected for something and used for something else). Also, the collation of data is labor intensive and time consuming. Realizing the limitations of data collection used by the researchers, we have approached the problem of the association between respiratory diseases and inhaling polluted air using a different kind of formalism, one based on the knowledgebase of pulmonologists/chest physicians.

The proposed formalism applies fuzzy relational calculus and evidence theory. The Dempster–Shafer theory (DST) can be viewed as a method for reasoning under epistemic uncertainty. Reasoning under epistemic uncertainty refers to logically arriving at decisions based on available knowledge. The most important part of this theory is Dempster's rule of combination, which combines evidence from two or more sources to form inferences. According to Shafer, DST is based on two ideas: the idea of obtaining degrees of belief for one question from subjective probabilities for a related question and Dempster's rule for combining such degrees of belief when they are based on independent items of evidence. In traditional probability theory, evidence is associated with only one possible event. In DST, evidence can be associated with multiple possible events (e.g. sets of events). The significant innovation of this framework is that it allows for the allocation of a probability mass to sets or intervals. This is a potentially valuable tool for evaluation when knowledge is obtained from expert elicitation. An important aspect of this theory is the combination of evidence obtained from multiple sources and the modeling of conflict between them.

7.1.2 Study Methodology

As the number of vehicles on the road is growing enormously, vehicular pollution has been the main cause of respiratory diseases. Figure 7.1 shows the growth in the number of vehicles registered in Pune and Mumbai cities. Karve Road in Pune city is one of the busiest locations (Figure 7.2), where diesel-driven three-wheelers plying their trade on narrow roads are around 60% of the total number of vehicles registered, emitting around a total pollution load of 36 tonnes/day in a congested area with high rise buildings occupying very high population density. The approximate number of vehicles passing along Karve Road, consisting mainly of two- and three-wheelers, were 64 965 and 13 523 vehicles/day, respectively, during the study carried out by the authors in January 2012.

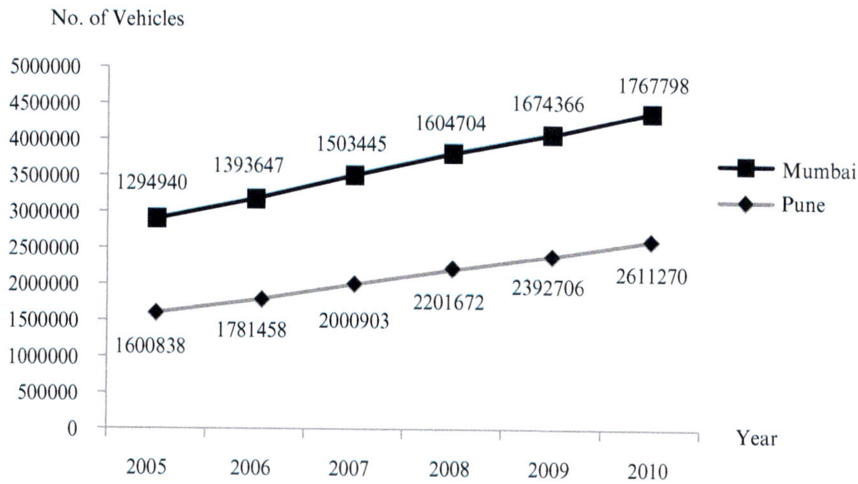

Figure 7.1 Growth in number of registered vehicles.

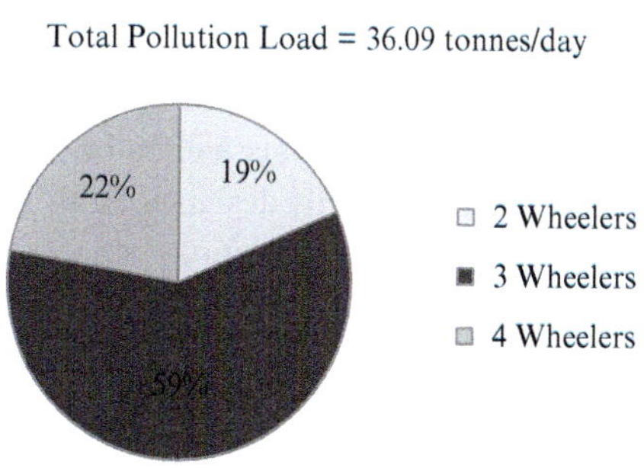

Figure 7.2 Total pollution load caused due to vehicles at the Karve Road hotspot.

Figure 7.3 depicts the number of vehicles/hours that pass through this intersection. With reference to the data collected from a renowned hospital in Pune, there is an increase in the number of patients suffering from respiratory illness (Table 7.1).

In order to establish the correlation between different respiratory diseases and air pollution, the perceptions of chest physicians were collected. Based on their expertise and years of experience, the respiratory diseases were classified into four major groups:

A) Simple bronchitis and COPD with lung cancer as the subset of COPD.
B) Asthma.

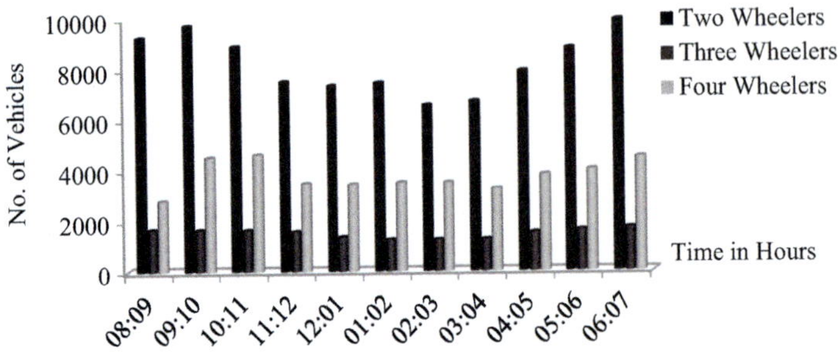

Figure 7.3 Total number of vehicles running per hour at the Karve Road hotspot.

Table 7.1 Fuzzy tolerance relation.

X1	X2	X3	X4	X5	X6	X7
1.00	0.99	0.97	0.89	0.96	0.96	0.92
0.99	1.00	0.97	0.90	0.96	0.97	0.94
0.97	0.97	1.00	0.94	0.96	0.96	0.95
0.89	0.90	0.94	1.00	0.90	0.93	0.95
0.96	0.96	0.96	0.90	1.00	0.96	0.95
0.96	0.97	0.96	0.93	0.96	1.00	0.97
0.92	0.94	0.95	0.95	0.95	0.97	1.00

C) Allergic rhinitis and conjunctivitis.
D) Heart attack.

The domain experts (chest physicians/pulmonologists) related to Case Study I on air pollution were interviewed for their views on air-borne/respiratory diseases caused by an increase in concentrations of air pollutants like PM_{10}, NO_X, and SO_X as these pollutants were monitored in India till 2011. Each domain expert wrote down their evidence/belief for the diseases (termed basic evidence assignment) in the questionnaire specially designed for this purpose. The mathematical framework for building the possible association between inhaling polluted air and respiratory diseases is shown in Figure 7.4.

The three cornerstones of determining the cause–effect relationship between polluted air and respiratory diseases are the parametric data of pollutants, epidemiological study, and the perceptions of chest physicians that play a central role. Conservationists and doctors have associated vehicular pollution with respiratory diseases like asthma, conjunctivitis, heart attack, allergic rhinitis, simple bronchitis, and COPD with lung cancer as a subset of COPD. Many researchers have worked on establishing this relationship. The current chapter addresses this research problem using the perception of pulmonologists who are experts in diagnosing respiratory diseases caused by polluted air. Figure 7.4 presents the overall diagnostic model for air pollution control and its associated health effects, while Figure 7.5 estimates the combined belief in connection with air pollution and respiratory disease.

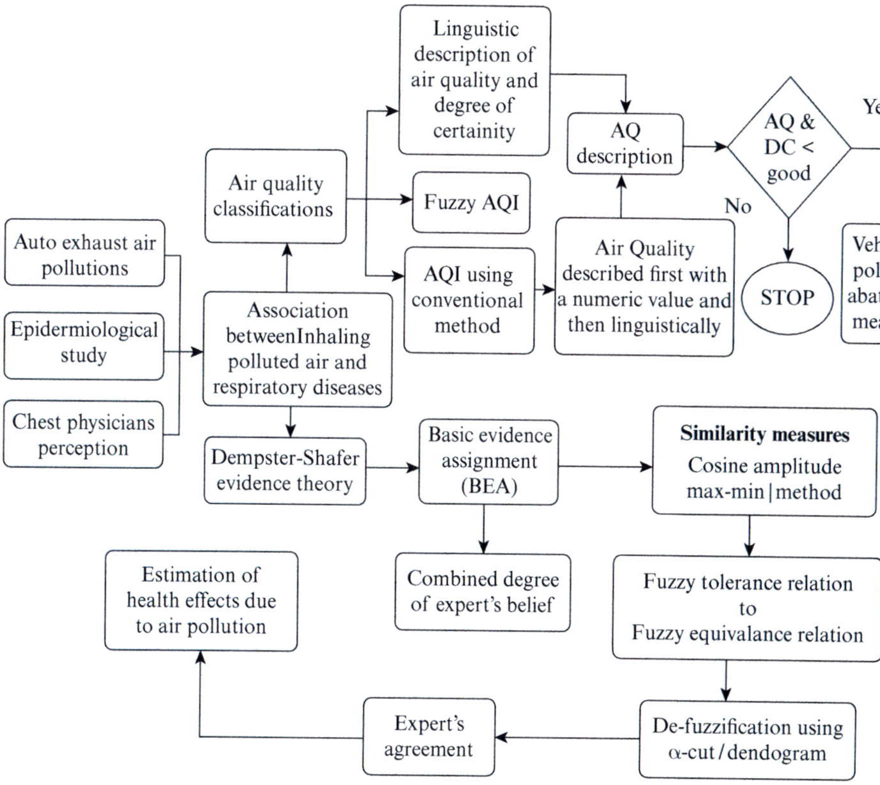

Figure 7.4 Diagnostic model for air pollution control and associated health effects.

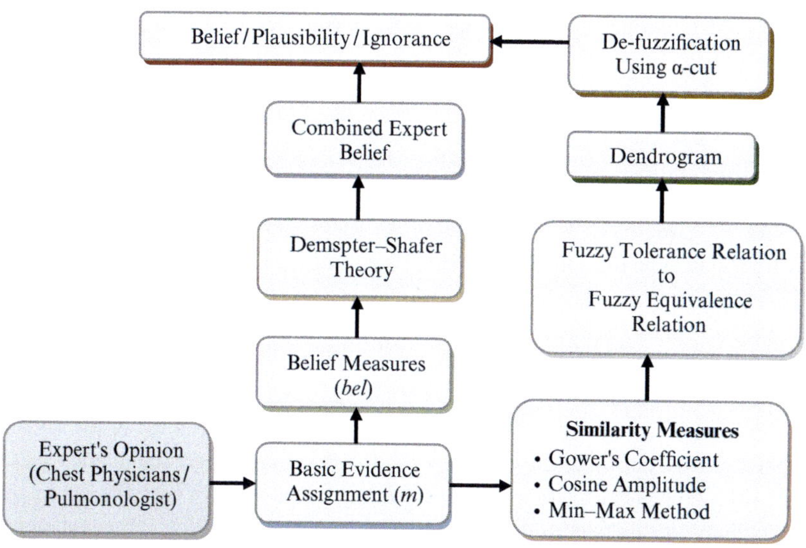

Figure 7.5 Estimating the combined belief of chest physicians for associating air pollution with respiratory diseases.

7.1.3 Establishing Association Between Inhaling Polluted Air and Respiratory Diseases

The dendrograms (Figures 7.6a and 7.6b) present similarity between the domain experts, in this case the chest physician's opinion on the hypothesis concerning the association between inhaling polluted air and respiratory diseases using the concept of α-cut as a measure of possibility of accepting the hypothesis. In Figure 7.6a the possibility of belief (α-cut) that experts 1 and 2 agree with the hypothesis is 0.99 and the possibility of belief for all the experts except expert 4 (X_4) is 0.96 where $X_1, X_2, \ldots, X_7$ are chest physicians/pulmonologists. In Figure 7.6b, the possibility of belief that experts 1 and 2 agree with the hypothesis is 0.92 and the degree of agreement of the experts in both the methods varies from 0.80 to 0.99. To summarize, the seven identified proficient pulmonologists believe that vehicular pollution could be the cause of respiratory diseases or could be aggravated or triggered in the already infected individuals.

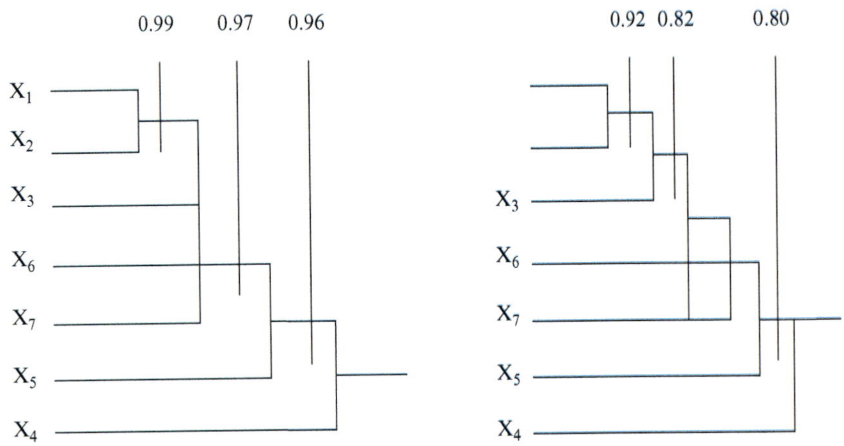

Figure 7.6 (a) Cosine amplitude method. (b) Max–min method (dendrogram).

7.1.4 Results and Discussion

The vagueness in experts' perception when stating their belief or assigning basic evidence (*bea*) is modeled using DST. The combined belief (evidence) of two experts has been computed for the power sets (and not fuzzy sets) of the respiratory diseases, namely (A) simple bronchitis and COPD with lung cancer as the subset of COPD, (B) asthma, (C) allergic rhinitis and conjunctivitis, and (D) heart attack and its variants (e.g. $A \cup B$, $A \cup B \cup C$, and so on). The experts' opinion was ranked on a scale from 0 to 100 which was further normalized to arrive at *bea* denoted as *m*. The belief measure is denoted by *bel*. In order to find out the similarity between the domain experts' knowledge base, the *bea* of all the experts was considered for the computation of similarity measures using the cosine amplitude method. A fuzzy tolerance relation was applied while examining the similarity matrix. The fuzzy tolerance relation was defuzzified to obtain the classical relation by transforming it to the fuzzy equivalence relation using the transitivity closure operation. Tables 7.1 and 7.2 give the fuzzy tolerance relation obtained by using the cosine amplitude method and transformed fuzzy equivalence relation after two iterations, respectively. The dendrogram presents the similarity between the experts in their opinion. It can be inferred for α-cut = 0.99 that the opinion of experts 1 and 2 regarding their *bea* is almost the same, with α-cut = 0.96. Except expert 4, all the other six experts have a similar basic evidence/belief assignment (Figure 7.1).

Table 7.2 Fuzzy equivalence relation.

X1	X2	X3	X4	X5	X6	X7
1.00	0.99	0.97	0.95	0.96	0.97	0.97
0.99	1.00	0.97	0.95	0.96	0.97	0.97
0.97	0.97	1.00	0.95	0.96	0.97	0.97
0.96	0.95	0.95	1.00	0.95	0.95	0.95
1.00	0.96	0.96	0.95	1.00	0.96	0.96
1.00	0.97	0.97	0.95	0.96	1.00	0.97
1.00	0.97	0.97	0.95	0.96	0.97	1.00

The combined expert's belief for disease A is 0.12, for B it is 0.17, for C it is 0.18, and for D it is 0.11. A typical computational procedure for belief (*bel*) and plausibility (*Pl*) is

$$\begin{aligned} Bel(A \cup B \cup C) &= m_1(A) + m_1(B) + m_1(C) + m_1(A \cup B) + m_1(A \cup C) + m_1(B \cup C) \\ &\quad + m_1(A \cup B \cup C) \\ &= 0.04 + 0.07 + 0.08 + 0.06 + 0.07 + 0.08 + 0.08 = 0.48 \end{aligned}$$

$$\begin{aligned} Pl(A) &= m_{1,2}(A) + m_{1,2}(A \cup B) + m_{1,2}(A \cup C) + m_{1,2}(A \cup D) + m_{1,2}(A \cup B \cup C) \\ &\quad + m_{1,2}(A \cup B \cup D) + m_{1,2}(A \cup C \cup D) + m_{1,2}(A \cup B \cup C \cup D) \\ &= 0.12 + 0.05 + 0.06 + 0.05 + 0.03 + 0.02 + 0.02 + 0.01 = 0.36 \end{aligned}$$

$$\begin{aligned} Pl(A \cup B) &= m_{1,2}(A) + m_{1,2}(B) + m_{1,2}(A \cup B) + m_{1,2}(A \cup C) + m_{1,2}(A \cup D) \\ &\quad + m_{1,2}(B \cup C) + m_{1,2}(B \cup D) + m_{1,2}(A \cup B \cup C) + m_{1,2}(A \cup B \cup D) \\ &\quad + m_{1,2}(A \cup C \cup D +) + m_{1,2}(B \cup C \cup D) + m_{1,2}(A \cup B \cup C \cup D) \\ &= 0.12 + 0.17 + 0.05 + 0.06 + 0.05 + 0.07 + 0.06 + 0.03 + 0.02 + 0.02 \\ &\quad + 0.02 + 0.01 \\ &= 0.68 \end{aligned}$$

$$\begin{aligned} Pl(A \cup B \cup C) &= m_{1,2}(A) + m_{1,2}(B) + m_{1,2}(C) + m_{1,2}(A \cup B) + m_{1,2}(A \cup C) \\ &\quad + m_{1,2}(A \cup D) + m_{1,2}(B \cup C) + m_{1,2}(B \cup D) + m_{1,2}(C \cup D) \\ &\quad + m_{1,2}(A \cup B \cup C) + m_{1,2}(A \cup B \cup D) + m_{1,2}(A \cup C \cup D +) \\ &\quad + m_{1,2}(B \cup C \cup D) + m_{1,2}(A \cup B \cup C \cup D) \\ &= 0.12 + 0.17 + 0.18 + 0.05 + 0.06 + 0.05 + 0.07 + 0.06 + 0.05 \\ &\quad + 0.03 + 0.02 + 0.02 + 0.02 + 0.01 \\ &= 0.91 \end{aligned}$$

We can see that the combined belief of two experts is 0.18 for diseases, namely allergic rhinitis and conjunctivitis, which is the maximum for the four diseases groups (Table 7.3). The second largest combined evidence/belief measure for asthma is 0.17. The computed combined belief/evidence of two experts for A or B or C is 0.68, while the computed plausibility is 0.91. However, there remains ignorance in the domain experts' perception for the diseases A or B or C, which is computed using evidence theory as 0.23.

Table 7.3 Combined evidence measures of two experts about respiratory diseases.

	Expert 1		Expert 2		Combined evidence			
Disease	m_1	bel_1	m_2	bel_2	$m_{1,2}$	$bel_{1,2}$	**Plausibility**	**Ignorance**
A	0.04	0.04	0.05	0.05	0.12	0.12	0.36	0.24
B	0.07	0.07	0.08	0.08	0.17	0.17	0.43	0.26
C	0.08	0.08	0.08	0.08	0.18	0.18	0.44	0.26
D	0.03	0.03	0.03	0.03	0.11	0.11	0.33	0.22
$A \cup B$	0.06	0.17	0.06	0.19	0.05	0.34	0.68	0.34
$A \cup C$	0.07	0.19	0.06	0.19	0.06	0.36	0.67	0.31
$A \cup D$	0.06	0.13	0.06	0.14	0.05	0.27	0.59	0.32
$B \cup C$	0.08	0.23	0.08	0.24	0.07	0.43	0.74	0.31
$B \cup D$	0.08	0.18	0.07	0.18	0.06	0.34	0.65	0.31
$C \cup D$	0.06	0.17	0.07	0.18	0.05	0.34	0.67	0.33
$A \cup B \cup C$	0.08	0.48	0.09	0.50	0.03	0.68	0.91	0.23
$A \cup B \cup D$	0.07	0.41	0.06	0.41	0.02	0.57	0.83	0.26
$A \cup C \cup D$	0.06	0.40	0.06	0.41	0.02	0.58	0.84	0.26
$B \cup C \cup D$	0.08	0.48	0.07	0.48	0.01	0.66	0.90	0.24
$A \cup B \cup C \cup D$	0.08	1.00	0.08	1.00	0.01	1.00	1.00	0.00
Total	1.00		1.00		1.00			

The dendrogram obtained in Figure 7.7 represents the similarity between the experts in their opinion using the cosine amplitude similarity measure. The possibility of belief (α-cut) that experts 1 and 2 agree with the hypothesis is 0.99 and the possibility of belief for all the experts except expert 4 (X_4) is 0.96 where $X_1, X_2, \ldots, X_7$ are chest physicians/pulmonologists.

The other significant observations include:

- The expert chest physicians believe that diseases A or B or C can be a cause of vehicular pollution and could be aggravated or triggered.
- Particularly, in the case of asthma, the chest physicians believe that the disease gets aggravated in patients who are already suffering from that disease, and healthy people are likely to become asthmatic because of long-term exposure to vehicular pollution.
- Furthermore, they justify the disease as pathophysiological and relate asthma to the functional changes that can be caused in the respiratory tract.
- The chest physicians are of the view that PM_{10} triggers asthma. Untreated asthma gets converted to COPD and might lead to lung cancer. This is in line with the results obtained by [9] in their exhaustive study.

The toxic pollutants emitted from diesel- or petrol-driven two/three/four-wheel vehicles mainly act as allergens, which may lead to irritation of eyes and resulting in conjunctivitis. Figures 7.1 and 7.2 infer the spectacular vehicular growth – primarily diesel-driven two/three-wheelers.

7.1.5 Concluding Remarks

The case study presents a new formalism on perception-based modeling in order to estimate the association between toxic air pollutants and human health.

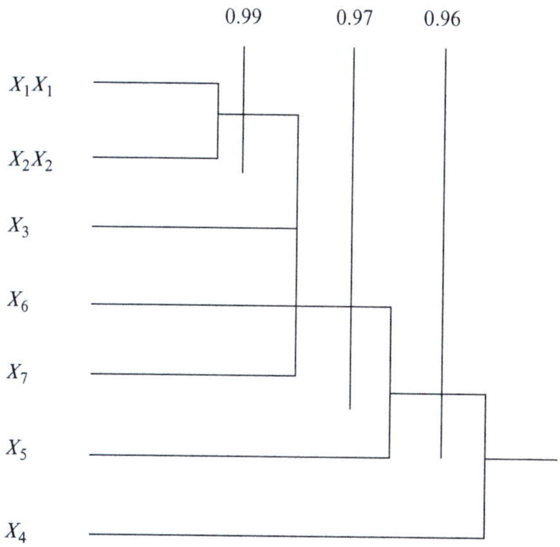

Hypothesis: Association between polluted air and respiratory diseases
α cut is the possibility of accepting the hypothesis
Possibility of belief (α cut) that experts 1 and 2 agree to the hypothesis: 0.99
Possibility of belief for all experts except 4 (X_4) is 0.96
X_1, X_2, ...,X_6 are domain expert Chest Physician / Pulmonologists

Figure 7.7 Dendrogram with different α-cuts for the agreement.

Epidemiological data for over a decade were collected by the researchers for establishing the association between respiratory diseases and polluted air. In the majority of cases, such data may not be reliable and also not available. The hospital staff might not have collected the information for drawing inference using statistical methods. This is known as happenstance data (collected for something and used for something else). Also, the collation of data is labor intensive and time consuming. Realizing the limitations of data collection used by the researchers, we have approached the problem of the association between respiratory diseases and inhaling polluted air using a different kind of formalism which is based on the knowledge base of pulmonologists/chest physicians.

7.2 Case Study II: Association Between Polluted Water and Human Health

7.2.1 Introduction

Hindus and Buddhists in India and Nepal regard bathing in holy rivers as sacred bathing that washes away their sins. But bathing in such places that contain contaminated water may lead to water-borne diseases (WBDs). In spite of much effort, there has been no solution available to prove/disprove the cause–effect relations between bathing in holy river waters and WBDs. This case study aims to establish the association between bathing in polluted river water and WBDs, and also models domain experts' epistemic uncertainty in supporting their evidence for various WBDs.

7.2.2 Literature Review

No significant research has been reported on bathing in polluted river water and WBDs. In the past, efforts were made to establish a linkage between health risk and bathing in polluted waters in Lake Michigan, the Ohio River, and Long Island Sound [10]. A bacteriological study of the water of River Ganga in the stretch of Bhagalpur (a town in the state of Bihar) clearly indicated that water was fit for neither drinking nor for bathing purposes either for human beings or cattle due to a high level of contamination [11]. Several studies on beaches have failed to establish a relationship between bathing and WBDs [12–18].

7.2.3 Study Methodology

Understanding the importance and complexity in associating polluted water with human health, a formalized method has been developed. The multifaceted approach includes: interviewing student community ranging between ages 8–31 years with a structured questionnaire and collecting belief measure of the resident medical practitioners about bathing in polluted river and WBDs. Furthermore, we also model epistemic uncertainty resident in the perception of the domain experts in supporting their evidence for WBDs. The study investigates the health status of the bathers during festive (*mela*) periods.

7.2.3.1 Cause–Effect Relationship Between Bathing in Polluted Water and WBDs

The case study is related to water quality classification and religious bathing in the highly polluted sacred Indrayani River at Alandi, a town around 30 km from Pune, Maharashtra state, India. In spite of significant development in science and technology, there has been no scientific justification for religious bathing. Religious belief is the only reason that can be cited for bathing in holy rivers!

Though there are no industrial (treated/untreated) wastewater discharges into the water course, the river still remains polluted due to the washing of clothes, the throwing of flowers, defecation, the dumping of animal waste, etc. The Maharashtra Pollution Control Board has been regularly monitoring the water quality parameters and keeping the concerned authorities informed. It should be mentioned here that the present location for abstracting water for the town of Alandi is located just half a kilometer away from the bathing ghats and the water quality at this location is far from satisfactory. The biochemical oxygen demand (BOD) level is as high as 12 mg/L, and also the coliform count is over 100 000 MPN/100 ml of water sample. During festive periods, devotees from the various parts of Maharashtra state come to Alandi in large numbers to take a holy dip at these bathing places located along the river, resulting in a further deterioration of the river's water quality. In the absence of the strict implementation of pollution control strategies, the authorities adopt the principle of dilution in order to reduce bacterial load in the river, especially in these festive periods (June/July and November/December). The State Irrigation Department has a policy to discharge a sizeable quantity of fresh water in the river, in these months, from the barrage about 6 km from Alandi, so as to bring down the coliform count.

Figure 7.8 portrays the most important activities of the multifaceted formalism for finally describing river water quality for bathing in the Indian context. The salient features of the structured approach are:

- Health status study to find out the relation between WBDs and bathing activity in polluted river water, and the use of DST for combining belief measures for modeling domain experts' perceptions.
- Random sampling from the identified stretch of river and laboratory analysis for the selected parameters. To increase precision in water quality data, estimation of bootstrap mean values for the defined parameters.

- Estimation of the selected water quality indices and its corresponding linguistic description along with a commentary on the results obtained.
- Application of fuzzy logic to describe water quality linguistically with some degree of certainty attached to each linguistic term for the purpose of bathing.
- The perception of the end-user about the river water quality for bathing in linguistic terms.
- Group discussion: formal and informal meetings with different stakeholders, public health experts, water quality experts, bathers, and the public at large.
- Development and use of software for water quality index and combined belief measures.

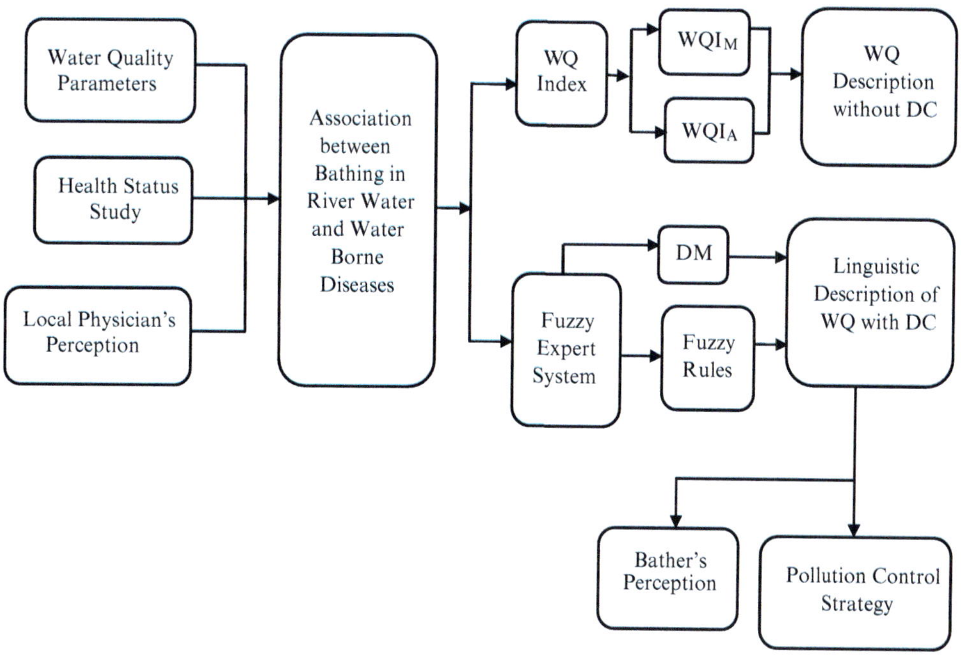

Figure 7.8 Multifaceted approach to river water classification.

The procedure formulated for establishing the relation between bathing in river water and the incidence of WBDs is detailed here. Water samples at the defined locations along the river were collected and analyzed, which showed a high fecal count (the known indicative organism for defining water pollution). The local physicians, based on several years' experience, are of the opinion that they receive complaints of WBDs from the residents almost daily, and could possibly be due to bathing in the Indrayani River. The activities include: a series of discussions with the domain experts including the resident physicians, data collation for epidemiological study, and water quality analysis (left-hand part of Figure 7.8).

7.2.3.2 Data Sampling, Health Status Study, and Resident Physicians' Perception

Students from various parts of the state, apart from their regular schooling, undergo training in religious literature and traditional instruments. The students are generally from economically poorer sections of society and stay in the hostels built by charitable organizations and individuals. The student population from three different hostels was selected for the study and the groups were formed on the basis of bathing activity. The first group consisted of those students who took a bath in the river almost regularly (bathers), while the second group was designated as nonbathers (do not use river water, meaning they were using treated municipal water supply for bathing). Out of the 220 students selected as samples, 83 were regular bathers and 137 were nonbathers. It is important to mention that all other activities, such as daily food intake, water consumption, and daily movement except bathing, were almost identical amongst all the hostel students. Similarly, the students were from the same age group. The only effective mechanism considered applicable in order to draw meaningful conclusions in such investigations is through personal dialogue. The students were interviewed at different times in a period of six months. The inputs from the social scientists in this regard were extremely helpful wherein uncertainty level could be very high. The selected students from the three hostels were asked relevant questions by respective hostel in charge, relating to the symptoms, if any, after taking a bath in the river. The resident physicians, on preliminary examination, opined that some of the students showed symptoms of WBDs. This was the first but very important step in the study (Table 7.4). Those students who did not show any symptoms of WBD are also portrayed in Table 7.4.

Table 7.4 Number of students with and without WBDs.

Hostel		Water-borne Diseases		
		Yes	No	Total
A	S_wWBD	5	18	23
	$S_{wo}WBD$	1	32	33
	Subtotal	6	50	56
B	S_wWBD	5	24	29
	$S_{wo}WBD$	1	41	42
	Subtotal	6	65	71
C	S_wWBD	4	27	31
	$S_{wo}WBD$	1	61	62
	Subtotal	5	88	93
Grand total		17	203	220

S_wWBD and $S_{wo}WBD$ stand for students with and without water-borne diseases, respectively.

7.2.3.3 Application of Dempster–Shafer Theory and Combined Belief Measures

The study was aimed at investigating health status of the bathers primarily during festive period. The first task in this approach was to gather information on the belief of the resident medical practitioners about bathing in polluted river and WBDs, based on their expert knowledge. DST models epistemic uncertainty of the domain experts in supporting their perceptions about various WBDs. Initially, only one expert was consulted for assigning this belief about bathing in the river and the occurrence of WBDs. The belief measure (m) of the expert about the occurrence of diarrheal diseases after bathing in the river was worked out as 0.41, and that for cholera as 0.04, and so on (Table 7.5). The expert also indicated this belief for diarrheal diseases and cholera together as 0.03; he also opined that cholera and giardiasis cannot occur together, which means that the bather cannot have these two diseases at the same time. A domain expert's knowledge is the key to success in subjective information.

Table 7.5 Belief measure of expert one for different WBDs.

Focal element (A_i)	Power set	Exp. $m(A_i)$
Diarrhea	a	0.41
Cholera	b	0.04
Giardiasis	c	0.15
Diarrhea and cholera	$a \cup b$	0.03
Diarrhea and giardiasis	$a \cup c$	0.02
Diarrhea, cholera, and giardiasis	$a \cup b \cup c$	0.01
Others	d	0.34

Exp. $m(A_i)$: expert measure of the element A_i.

7.2.4 Results and Discussion

The computed values of degree of belief and plausibility (*bel* and *Pl*, respectively) for each focal element are presented in Table 7.6. The typical computational procedure for *bel* and *Pl* is given as

$$bel(a) = m(a) = 0.41, bel(a \cup b) = m(a) + m(b) + m(a \cup b) = \{0.41 + 0.04 + 0.03\} = 0.48$$

$$pl(a \cup b) = m\{(a \cup b) \cup (a)\} + m\{(a \cup b) \cup (b)\} + m\{(a \cup b) \cup (a \cup b)\}$$
$$= \{0.41 + 0.04 + 0.03\} = 0.48$$

The expert believes that evidence supporting diarrheal diseases is at least 0.41 and 0.04, 0.15, 0.34 are cholera, giardiasis, and others, respectively. It is also necessary to have a second opinion, and therefore one more expert (second expert) was consulted for assigning his belief about bathing in the river and the occurrence of WBDs. The details are shown in Table 7.6.

The second expert believed that the evidence supporting diarrheal diseases is at least 0.30 and plausibly as high as 0.78. The computed plausibility value signifies that occurrences of diarrheal disease is very high if people bathe in polluted water. It is also inferred that there is a strong association between bathing in a polluted river and the incidence of diarrheal diseases.

Table 7.6 Measures of evidence given by second expert on WBDs.

Focal element (diseases)	Power set	Expert		
		Belief measure $m_1(A_i)$	Degree of belief $bel_1(A_i)$	Plausibility(A_i)
Diarrhea	a	0.39	0.39	0.78
Cholera	b	0.8	0.08	0.16
Giardiasis	c	0.16	0.16	0.32
Diarrhea and cholera	$a \cup b$	0.01	0.46	0.48
Diarrhea and giardiasis	$a \cup c$	0.02	0.54	0.57
Diarrhea, cholera, and giardiasis	$a \cup b \cup c$	0.04	0.59	0.67
Others	d	0.30	0.30	0.30

7.2.5 Concluding Remarks

It is practical to calculate the combined belief of the two experts, using DST formalism. The variability in basic belief assignment (BBA) values of two domain experts (m_1 [A_i] and m_2 [A_i]) for each focal element was worked out. Table 7.7 presents the details of the computed values of the combined evidence for different WBDs. Finally, the combined belief works out to $m_{12} = 0.58$ for the focal element A (diarrheal diseases).

Table 7.7 Focal elements and the combined belief measures for WBDs.

Focal elements (A_i)	Expert 1		Expert 2		Combined evidence	
	$m_1(A_i)$	$bel_1(A_i)$	$m_2(A_i)$	$bel_1(A_i)$	$m_{12}(A_i)$	$bel_{12}(A_i)$
a	0.41	0.41	0.39	0.39	0.58	0.58
b	0.04	0.04	0.08	0.08	0.02	0.02
c	0.15	0.15	0.16	0.16	0.10	0.10
$a \cup b$	0.03	0.43	0.01	0.36	0.00	0.60
$a \cup c$	0.02	0.54	0.02	0.54	0.00	0.68
$a \cup b \cup c$	0.01	0.59	0.04	0.59	0.00	0.71
d	0.w34	0.34	0.30	0.30	0.28	0.28

a: diarrhea; b: cholera; c: giardiasis; $a \cup b$: diarrheal or cholera; $a \cup c$: diarrheal or giardiasis; $a \cup b \cup c$: diarrheal or cholera or giardiasis; and d: others.

The combined degree of belief (bel_{12}) for the focal element such as diarrheal diseases is equal to 0.58 as against 0.41, based on a single expert's belief measure. In the case of diarrhea and giardiasis, the degree of belief, based on a single expert's perception, was 0.53, while the combined degree of belief of two experts was 0.68. Based on the combined evidence of the two domain experts, it can be inferred that the evidence supporting diarrheal diseases is 0.58.

Although there is a need for guidance on issues involving risk due to bathing in polluted river water, there is no well-defined approach that focuses on this complex health problem. The measurements and predication of these risks are technically difficult and the evaluation of the detriment is controversial. The quantification of epistemic uncertainty of the experts in their degree of belief for specific WBDs, such as diarrheal diseases (which include dysentery), cholera, etc., is the uniqueness of the suggested formalism. It could be stated that the computation of degree of belief of at least two experts, based on their combined evidence, could be an appropriate computational procedure, as it helps to model ambiguity in the basic probability assignment of the domain experts. In sum, the application of fuzzy measures and DST, exhaustive field investigations with the support of domain experts, and water quality data could go a long way to resolve, at least to some extent, the most difficult issue of bathing risk in polluted river water.

The case study presents an approach for establishing the relationship between bathing in polluted river and WBDs, supported by quantifying epistemic uncertainty of the experts in their degree of belief for specific WBDs, such as diarrhea and giardiasis, etc. Computing degree of belief of at least two experts, based on their combined evidence, could be an appropriate computational procedure, as it helps to model ambiguity in the basic probability assignment of the domain experts.

The two case studies stress the importance of fuzzy-logic-based formalism on environmental issues such as water quality, air quality, environmental impact assessment, etc. Since fuzzy-logic-based methods rely on the perceptions of domain experts, it is necessary to have experts who have experience and expertise in the relevant domain limiting the performance of the proposed studies.

Acknowledgment

Jyoti and Kedar would like to sincerely acknowledge Dr. Ashok Deshpande for the efforts and guidance provided for framing this chapter into an intelligent product.

References

1 Pope, A.C. et al. (2002). Lung cancer, cardiopulmonary mortality, and long-term exposure to fine particulate air pollution. *Journal of the American Medical Association* 287: 1132–1142.
2 Brauer, M., Hoek, G., Van Vliet, P., Meliefste, K., Fischer, P.H. et al. (2002). Air pollution from traffic and the development of respiratory infections and asthmatic and allergic symptoms in children. *American Journal of Respiratory and Critical Care Medicine* 166 (8): 1092–1098.
3 Goldberg, A.S. et al. (2003). Association between ambient air pollution and daily mortality among persons with congestive heart failure. *Environmental Research* 91: 8–20.
4 Kwon, H.J., Cho, S.H., Nyberg, F., and Pershagen, G. (2001). Effects of ambient air pollution on daily mortality in a cohort of patients with congestive heart failure. *Epidemiology* 12: 413–419.
5 Dockery, D., Pope, C., Xu, X., Spengler, J., Ware, J., Fay, M., Ferris, B., and Speizer, F. (1993). An association between air pollution and mortality in six US cities. *New England Journal of Medicine* 329: 1753–1759.
6 Dockery, D., Schwartz, J., and Spengler, J.D. (1992). Air pollution and daily mortality: associations with particulates and acid aerosols. *Environmental Research* 89: 362–373.
7 Spix, C., Heinrich, J., Dockery, D., Schwartz, J., Volksch, G., Schwinkowski, K., Collen, C., and Wichmann, H. (1993). Air pollution and daily mortality in Erfurt, East Germany, 1980–1989. *Environmental Health Perspectives* 101: 518–526.
8 Braback, L. and Forsberg, B. (2009). Does traffic exhaust contribute to the development of asthma and allergic sensitization in children: findings from recent cohort studies. *Environmental Health* 8: 17–28.
9 Brauer, M. et al. (2002). Air pollution from traffic and the development of respiratory infections and asthmatic and allergic symptoms in children. *American Journal of Respiratory and Critical Care Medicine* 166: 1092–1098.
10 Sinton, L.W., Donnison, A.M., and Hastie, C.M. (1993). Faecal streptococci as faecal pollution indicators: a review: part II: sanitary significance, survival, and use. *New Zealand Journal of Marine and Freshwater Research. Royal Society of New Zealand* 27: 17–137.
11 Bilgrami, K.S. and Kumar, S. (1998). Bacterial contamination in water of the River Ganga and its risk to human health. *International Journal of Environmental Health Research* 8: 5–13.
12 Brown, J.M., Campbell, E.A., Rickards, A.D., and Wheeler, D. (1987). Sewage pollution of bathing water. *Lancet* 2: 1208–1209.
13 Cabelli, V.J. (1989). Swimming associated illness and recreational water quality criteria. *Water Science and Technology* 21: 13–21.
14 Cabelli, V.J., Dufour, A.P., McCabe, L.J., and Levin, M.A. (1983). A marine recreational water quality criterion consistent with indicator concepts and risk analysis. *Journal of the Water Pollution Control Federation* 55: 1306–1324.

15 Pond, K. (2005). *Water Recreation and Disease, Plausibility of Associated Infections: Acute Effects, Squeal and Mortality*. Padstow: TJ International.

16 Public Health Laboratory Service (1959). Sewage contamination of costal bathing waters in England and Wales: a bacteriological and epidemiological study. *Journal of Hygiene* 57: 435–472.

17 Stevenson, A.H. (1953). Studies of bathing water quality and health. *American Journal of Public Health* 53: 529–534.

18 World Health Organization (2000). *The World Health Report 2000*. https://apps.who.int/iris/bitstream/handle/10665/42281/WHR_2000-eng.pdf (accessed 4 October 2022).

8

Respiratory Disease Risk Assessment Among Solid Waste Workers Using a Fuzzy Rule Based System Approach

Namrata Jariwala and R.A. Christian

SV National Institute of Technology, Surat, India

8.1 Introduction

The prediction of any disease is a complex process affected by several factors. Medical science uses the symptoms, laboratory tests, and medical history of a person as diagnostic tools for detecting diseases. In addition, studies of many biochemical parameters of healthy people in various age groups, body mass index (BMI), and habit data are required to predict diseases. Theoretically, several years of delayed onset might be observed for the possible development of diseases among the study cohort. Medical science considers the criteria of odds ratio and relative risk for the study of disease occurrence pattern, in which the exposed and nonexposed groups are compared.

Respiratory disease is the main cause of mortality in developed countries as well as India. The risk assessment of respiratory disease is a sophisticated process as it is influenced by several factors, such as age, BMI, diet, genetic predispositions, personal habits (smoking and alcohol drinking), and working environment condition. The development of respiratory disease takes a long time as the first symptoms appear after a prolonged period; occasionally, the patient misses the chance of being able to take any preventive measures. Thus, it is important to identify the factors responsible for the main cause of respiratory disease and also determine their individual risk value.

Reportedly, solid waste workers are at high risk of developing respiratory disease. This study's main objective was to determine the individual solid waste worker's risk of respiratory disease under selected parameters. The factors responsible for the development of respiratory disease were identified and measured among solid waste workers. Hence, the study population consisted of door-to-door waste-collecting workers of the city of Surat in India. In order to determine the individual worker's risk without any test or diagnosis in scalar value 0 to 1, fuzzy logic (FL) and the fuzzy rule based system (FRBS) were applied. In the FRBS approach, available information from the medical experts was considered, and the real patient's data were used to predict the individual's risk on a 0 to 1 scale without any test. The determined risk value was used to modify the factors responsible for the occurrence of diseases in an individual. Preventive steps can prevent the disease to some extent but definitely cannot cure it.

A number of studies have used FL and FRBS with various soft computing techniques in medical science for the prediction of health risks related to different diseases.

Fuzzy Systems Modeling in Environmental and Health Risk Assessment, First Edition. Edited by Boris Faybishenko, Rehan Sadiq, and Ashok Deshpande.

Carreno and Jani [1] developed a rule-based system approach for insurance risk assessment using fuzzy CLIPS (C-Language Integrated Production System, version 5.1 developed by NASA). To assess the degree of health risk associated with each person's physical characteristics (such as height, weight, age), additional information (such as exercise, smoking, drinking, and eating habits) was considered in the model. The FRBS numerical value represents the putative risk of a person due to their physical characteristics and eating habits.

McCauley-Bell and Badiru [2] used the FRBS approach and analytical hierarchy process to predict occupational injury, such as cumulative trauma disorders (CTDs). The present study suggested that the development of a preliminary linguistic rating scale for qualitative assessment of risk condition mainly involves task-, personal-, and occupational-related risk factors.

Mure et al. [3] developed a risk assessment of occupational accidents using a fuzzy approach. Frequency, severity, contact factor, and degree of protection have been selected as input parameters for the fuzzy inference system, and the output was termed as risk. Industrial work environmental data of an Italian tire company were used, and the risk value in the manufacturing section was obtained on a 0 to 1 scale.

Keshwani et al. [4] developed two Mamdani-type (three inputs–one output and two inputs–one output) fuzzy models to predict the permeability of compounds through human skin. The models were derived from multiple data sources, including laboratory data, published databases, published statistical models, and expert opinion. The input to the model includes information about the compound (molecular weight and octanol/water partition coefficient) and the application temperature. One model included all three parameters as inputs, and the other model only included information about the compound. Each model was a collection of rules that expressed the correlation between each input and the permeability of the compound through human skin. The quality of the model was determined by comparing the predicted and actual fuzzy classification and defuzzification of the predicted outputs to correlating the estimates with the published values.

Simić and Simić [5] considered FRBS as a successful technique for knowledge-based decisions in many domains, including medicine. It is also used as a tool to diagnose migraine diseases. The diagnostic criteria developed have been used in epidemiology research. A total of 80 individuals' responses were assessed based on the FRBS model. The system rules were implemented by the application of International Classification of Headache Disorders (ICHD) criteria for headache type, and hence the system could be considered to provide optimal output.

Khatibi et al. [6] developed a fuzzy-based hybrid inference framework using the Dempster–Shafer theory (DST) of evidence and fuzzy sets theory for the prediction of coronary heart disease. The output revealed 91.58% accuracy of the model.

Adeli and Neshat [7] designed a fuzzy expert system using Mamdani inference with 13 inputs and one output to determine coronary heart disease risk. Anooj [8] developed clinical decision support systems (DSSs) to determine the risk level of heart disease using weighted FRBS. In this model, weighted fuzzy rules were developed using the data mining technique.

8.2 Research Framework

Figure 8.1 is a flowchart of the sequence of stages followed in the present study to predict the risk of respiratory disease among solid waste workers using the FRBS approach.

8.2.1 Experts' Perception and Opinions

The health data and subjective judgment of experts were used to develop the FL system. Healthy discussions were made with medical practitioners and dietitians. The solid waste management system and the socioeconomic condition of solid waste workers were also studied in depth.

8.2.2 Identify the Factors for the Development of Respiratory Disease

Based on the experts' opinions and perceptions, the factors responsible for respiratory disease were considered. The factors responsible for computing respiratory disease were considered as an input variable, and the output of the system was termed as risk. The rate of change in these factors was neither linear nor incremental but complex. However, the system had two types of inputs: base and incremental. The base input variables for the system were age and BMI, and the data were variable in individuals. BMI is a measure that indicates if a person is overweight. It is calculated by dividing the weight in kilograms by the square of height in meters (BMI = weight/height^2). Incremental input variables addressed the specific habits and typical characteristics of solid waste workers, such as tobacco chewing, alcohol drinking, smoking, diet, personal hygiene, and working environment.

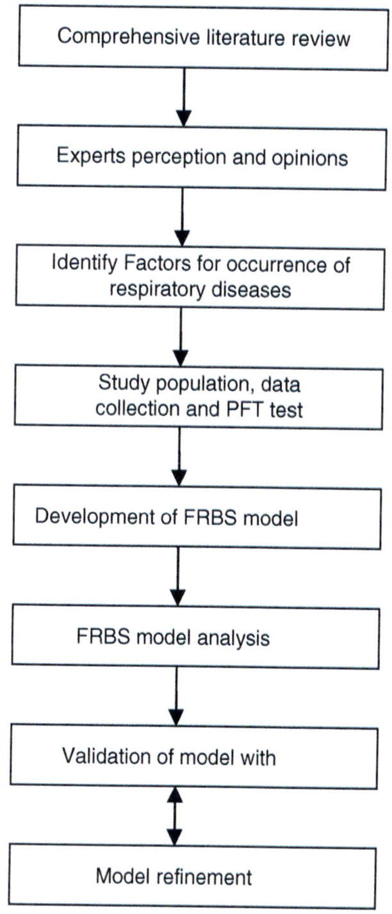

Figure 8.1 Research framework for the study.

8.2.3 Study Population and Data Collection

A cross-sectional study was conducted on a group of 292 males working at seven transfer stations in Surat. Male employees of door-to-door waste collection facilities were selected. Surat is divided into seven zones for administration. Solid waste workers, including door-to-door waste collection workers at six transfer stations like Pal, Bhatar, Umarpada, Anjana, Pandesara, and Vastadevdi transfer stations, were considered in this study. The workers were selected randomly during the morning shift from 8 a.m. to 2 p.m. Data were collected using questionnaires that were prepared referring to the ATS-DLD (American Therapeutic Society – Diffuse Lung Disease) questionnaire. Occupation, family, and personal medical history were recorded, as well as smoking habits, drinking habits, and tobacco chewing were included, in the questionnaire in addition to diet and personal hygiene details.

The factors selected to assess the occurrence of respiratory disease were measured among the study population of solid waste workers. These included age and BMI data. For the quantification of personal habits, a normal approach of collecting data from experts and medical practitioners was considered. However, it was difficult to quantify the personal hygiene, working environment, and diet of solid waste workers. Thus, the relative weight/age ratio of individual solid waste workers was calculated based on certain critical criteria. For personal hygiene, diet, and working environment, a value for the relative weight/age ratio of 0–9 was assigned to the workers. Their personal hygiene was judged based on criteria such as using protective material: mask, earplugs, gloves, and gumshoes at the site, washing hands after duty, changing clothes after duty, and nail cutting and hair cutting regularly. The criteria selected for environmental conditions included the type of transfer station at which the workers worked, daily working hours, type of work, shift patterns, and involvement in other physical activities. Good personal hygiene and working environment lowered the risk of occurrence of respiratory disease. For diet, weightage was related to poor, moderate, and good nutrition. If a worker was categorized as having a good nutritious diet, the relative risk for the occurrence of respiratory disease was low.

8.2.4 Pulmonary Function Test of Solid Waste Workers

The PFT was performed for each solid waste worker, and the data were recorded using a spirometer by Maestro midline version 2.1. The measurements were taken during morning hours. The forced vital capacity (FVC), FEV_1 (forced expiratory volume in 1 second), the maximum flow rate at 50%, and the last 25% of the vital capacity were measured. The lung function was evaluated according to the recommendation of the American Thoracic Society. Based on the diagnosis of the PFT test, the workers were classified into four categories: normal abnormalities, obstructive type abnormalities (O type), restrictive type abnormalities (R type), and combined type abnormalities (O + R type). The diagnosis of O and R type abnormalities is essential as they are both curable if proper care taken. However, if neglected, and no preventive measures are taken, the risk of respiratory disease is elevated. Both O and R type abnormalities are considered as high risk for the occurrence of respiratory disease. Table 8.1 represents the PFT abnormalities observed among solid waste workers of Surat.

Table 8.1 PFT data of solid waste workers of Surat.

	Normal type	O type abnormalities	R type abnormalities	O + R type abnormalities	Total
No. of workers	96 (32.8%)	47 (16.10%)	73 (25%)	76 (26.10%)	292

8.3 Development of the FRBS Model to Determine Respiratory Risk

The present study deals with different kinds of knowledge acquisition to gain information of the body that could be beneficial in developing fuzzy linguistic parameters and their associated members' function to quantify the risk of occurrence of respiratory disease. A schematic fuzzy decision system to identify the risk of respiratory disease among solid waste workers is illustrated in Figure 8.2.

8.3.1 FRBS Approach

The rule-based system utilized a model that represents human knowledge in the form of *if–then* rules. This conventional approach had been adopted to build FRBS for this study. Relevant and available input variables were selected, and their domain was partitioned into several fuzzy sets or linguistic variables. In the FL system, an expert defines the rule to describe the characteristics

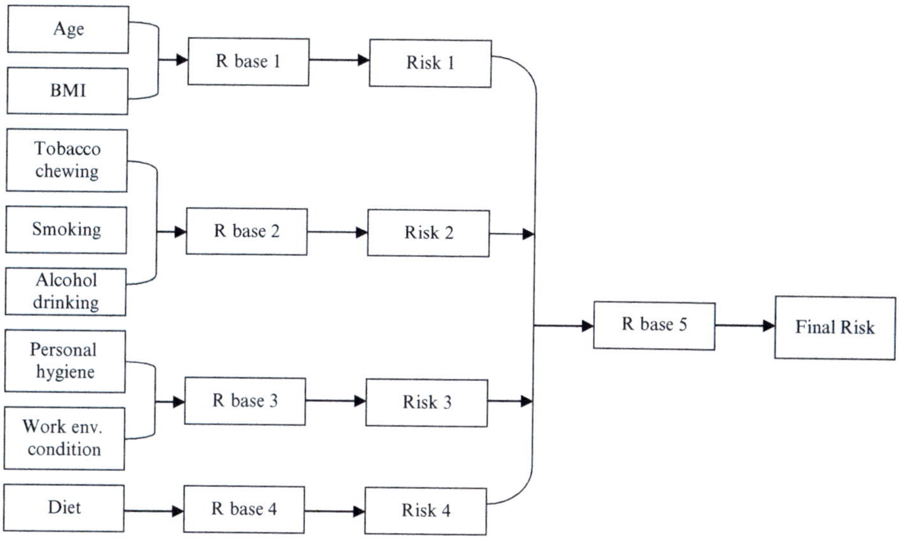

Figure 8.2 Fuzzy logic risk assessment.

of risk assessment for each factor. The input variables were processed by rules to generate an appropriate output. The fuzzy rules represented the logical correlation between the input and output variables that were derived from previous data or the experience of the analyst or the experts' judgment or perception.

After obtaining the fuzzy numbers with their corresponding degree of match, the rules of the system were defined. A specific set of rules was constructed for the risk prediction model. Each rule had an antecedent proposition connected together using an *AND* operator, resulting in some consequence. The assertions related to its antecedent part obtained from the experts were imprecise or fuzzy. Thus, an FRBS could be developed for the knowledge representation or reasoning process. Herein, partial matching was allowed, and the analyst could estimate the extent to which the assertion satisfied the antecedent part of the rule contrary to the rule-based system, which examined whether the antecedent part is satisfied.

8.3.1.1 Development of Fuzzy Sets

Subjective judgment, gathered data, and linguistic variables were employed to develop fuzzy membership functions of all input and output parameters. When human experts are asked to evaluate a variable, they describe it in words. Furthermore, fuzzy linguistic variables are extensions of numerical variables such that they represent the condition of an attribute at a given interval by considering the fuzzy sets as their values. The input and output parameters of the system are discussed in Figure 8.2, and R base 1–5 is defined using the fuzzy toolbox of MATLAB. Also, the implication, aggregation, and defuzzification are calculated using the same program.

8.3.1.2 Fuzzy Rules

Figure 8.2 is the representation of the FRBS model to evaluate the risk of respiratory disease in solid waste workers. Herein, a simple FRBS model consisting of five steps of R base 1 to R base 5 was utilized. The first step had two inputs and one output with 16 inference rules. The second step had three inputs and one output with 27 inference rules. The third step had two inputs and one output with nine inference rules. The fourth step had one input and one output with four inference rules. Finally, the fifth step had four inputs and one output with 81 inference rules.

8.3.1.3 Defuzzification

The final step in the fuzzy controller is the defuzzification process that converts all fuzzy outputs into an equivalent crisp value. The fuzzy input values entering the inference engine are converted to fuzzy output values after the evaluation of rules in the rule base and whilst considering the membership function values. The fuzzy output values are defuzzified using the widely accepted and effective centroid method.

8.3.2 Determination of Risk 1

The quantitative solution of the fuzzy system is made convenient by the following steps: defining input and output parameters of the system using MATLAB's Fuzzy Logic Toolbox™ formulation of the logical rule, and defuzzification operation performed using MATLAB 6.5 (R2008). Here, the FL system was used to evaluate Risk 1 for solid waste workers based on two input parameters: age and BMI. The output of the system Risk 1 was evaluated by the Mamdani FRBS using MATLAB 6.5 (released in 2008).

Figure 8.3 shows the FL system used to evaluate Risk 1 for solid waste workers with age and BMI as two input parameters. The output of the system Risk 1 is evaluated by the Mamdani FRBS.

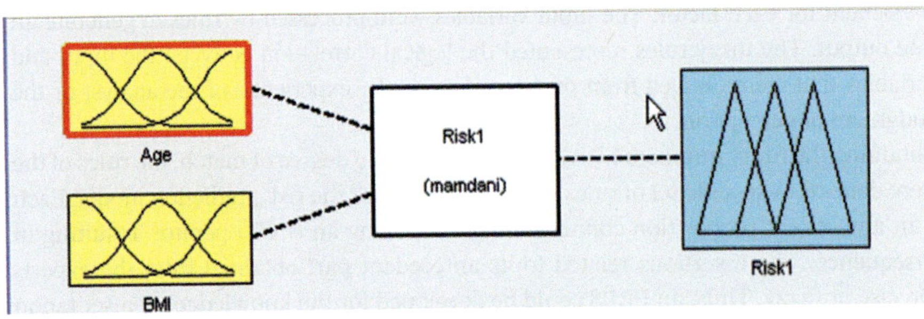

Figure 8.3 Fuzzy inference system editor for the determination of Risk 1.

8.3.2.1 Defining Input and Output Parameters for Risk 1

The input variable age (years) has a range of 0–100 and has been divided into four fuzzy sets as follows: very young (0, 0, 15, 35), young (15, 35, 55), middle age (35, 55, 75), and old age (55, 80, 100, 100). Figure 8.4 represents the membership function selected for the input variable age.

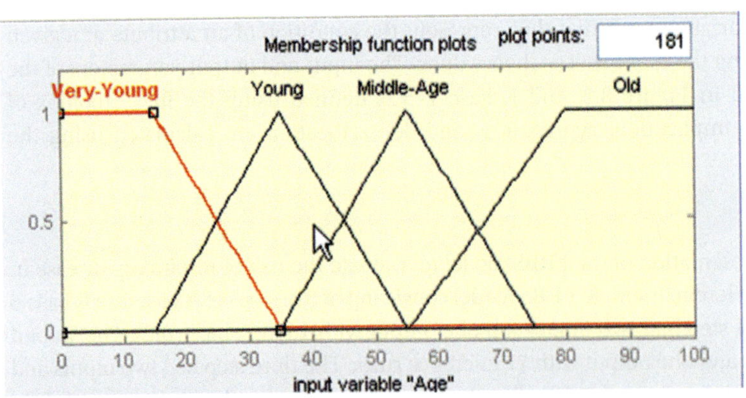

Figure 8.4 Membership function for age (Sivanandam et al. [9]).

The input variable BMI (kg/m^2) has a range of 0–40 and has been divided into four fuzzy sets: underweight (0, 0, 18.5, 19.5), ideal (18.5, 21.5, 24.9), overweight (25, 27.5, 29.9), and obese (28.5, 29.9, 40, 40). Figure 8.5 shows the membership function for BMI.

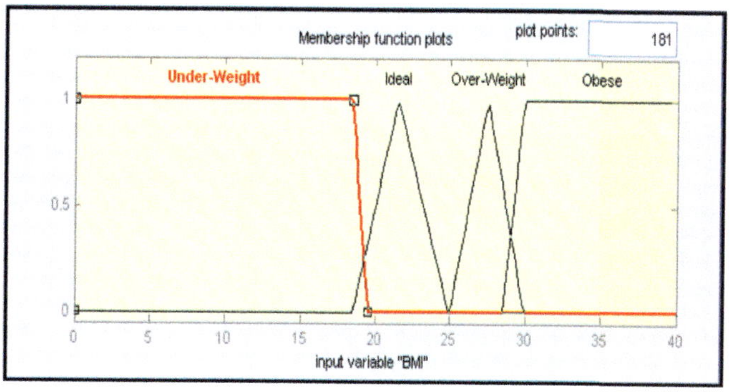

Figure 8.5 Membership function for BMI. *Source:* [1].

The output variable Risk 1 is a numerical value with a range of 0–1 and has been divided into three fuzzy sets: low (0, 0, 0.15, 0.33), medium (0.34, 0.44, 0.54, 0.67), and high (0.68, 0.8, 1,1). Figure 8.6 represents the membership function for output variable Risk 1.

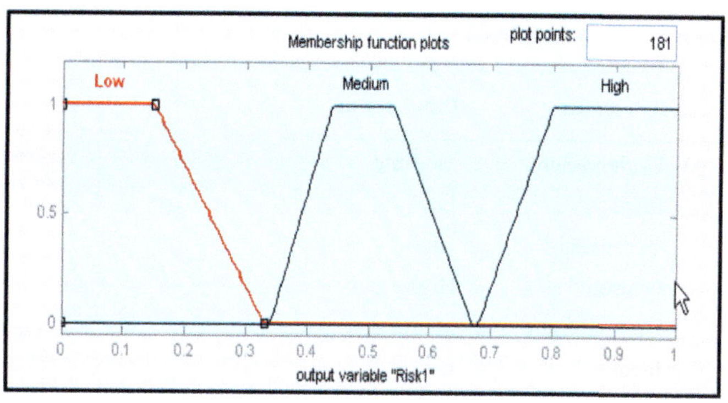

Figure 8.6 Membership function for output Risk 1.

8.3.2.2 Defining Rules for Risk 1

The rule editor of Risk 1 in the MATLAB fuzzy toolbox is shown in Figure 8.7. The rule editor is a platform on which the correlation between input and output has been established. The R base 1 is the system to determine Risk 1. The Risk 1 system consists of two input variables, age and BMI, and each has four linguist variables; $4 \times 4 = 16$ inference rules to evaluate the system.

The evaluated rules for Risk 1 are based on if–then statements and established using knowledge, experience, and guidance given by medical experts. This is represented in Table 8.2.

8.3.2.3 Defuzzification Process for Risk 1

Figure 8.8 shows the defuzzification steps to determine Risk 1. The input data of age 50 years and BMI 20 has been considered, and the output Risk 1 value was 0.141 after the defuzzification process, according to the fuzzy rule processed in MATLAB.

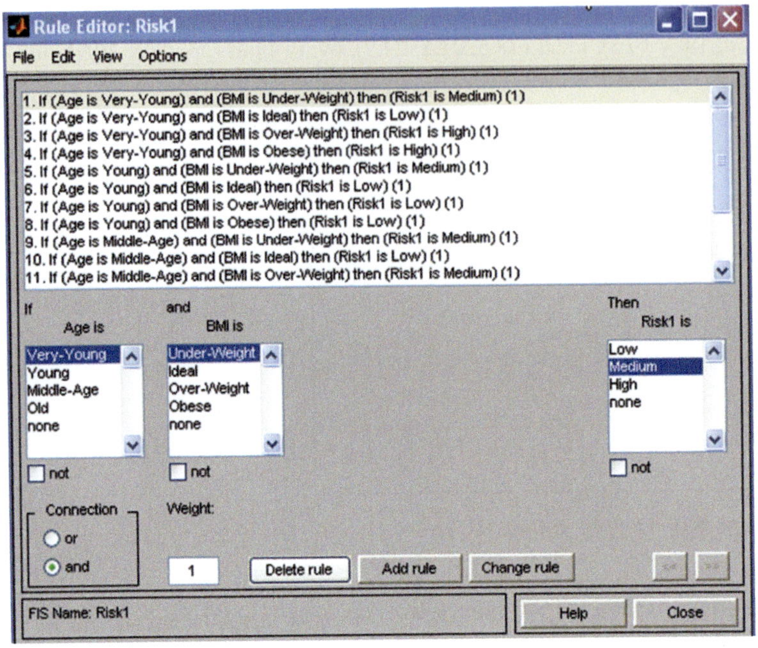

Figure 8.7 Rule editor for Risk 1.

Table 8.2 Fuzzy rules considered for evaluation of Risk 1.

Station number	Age status	BMI status	Risk 1 status
1	Very young	Underweight	Medium
2	Very young	Ideal	Low
3	Very young	Overweight	High
4	Very young	Obese	High
5	Young	Underweight	Medium
6	Young	Ideal	Low
7	Young	Overweight	Low
8	Young	Obese	Low
9	Middle age	Underweight	Medium
10	Middle age	Ideal	Low
11	Middle age	Overweight	Medium
12	Middle age	Obese	High
13	Old	Underweight	Medium
14	Old	Ideal	Medium
15	Old	Overweight	High
16	Old	Obese	High

Risk 2 mainly occurs due to personal habits, such as tobacco chewing, smoking, and alcohol drinking. Risk 3 can be attributed to personal hygiene and working environment, and Risk 4 may be based on the diet pattern.

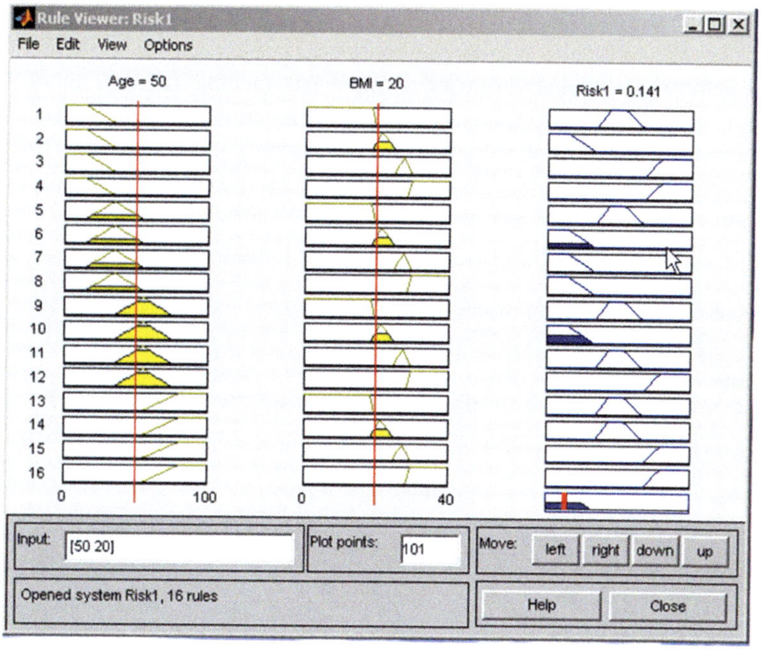

Figure 8.8 Defuzzification process for Risk 1.

8.3.3 Linguistic Outputs

After the inference is completed, the obtained output is the linguistic value that represents the risk of respiratory disease associated with solid waste workers. The output value is considered on a scale of 0–1 scale very low risk (0–0.2), low risk (0.21–0.4), medium risk (0.41–0.6), high risk (0.61–0.8), and very high risk (0.81–1). The obtained result risk and PFT result observation have been compared based on an expert's perception. Table 8.3 represents the PFT test comments and the corresponding linguistic output value for the study.

Table 8.3 Linguistic output and PFT result consideration.

Crisp output	Linguistic output	PFT comments considered
0–0.2	Very low risk	Individual worker may not have any symptoms that can be considered respiratory abnormalities
0.21–0.40	Low risk	Individual worker may have irregular symptoms but these cannot be considered respiratory disease
0.41–0.60	Medium risk	Individual may experience minor respiratory abnormalities on irregular basis, like O type and R type, but these are not considered respiratory disease
0.61–0.80	High risk	Individual is expected to experience regular minor or major O type, R type, or a combination of these
0.81–1.0	Very high risk	Individual is expected to experience regular O type, R type, and O + R type, which is noted as a severe respiratory disease

8.4 Model Analysis

The collected data of solid waste workers are used as input parameters. After the model application, the output value of final respiratory disease risks may be obtained on a scalar value on a range of 0–1. The obtained risk value for the present study varied based on the data of the study population in the range of 0.08–0.92. Furthermore, the obtained risk value was classified as a linguistic output group and compared to the PFT abnormalities. Table 8.4 represents the obtained output classification according to the PFT results.

Table 8.4 Final output classification with PFT results.

Crisp output	Linguistic output	N type (96)	O type (47)	R type (73)	O + R type (76)
0–0.2	Very low risk	37 (12.7%)	1 (0.3%)	—	—
0.21–0.40	Low risk	45 (15.4%)	3 (1.0%)	2 (0.7%)	—
0.41–0.60	Medium risk	10 (3.4%)	37 (12.7%)	63 (21.6%)	7 (2.4%)
0.61–0.80	High risk	1 (0.3%)	6 (2.1%)	4 (1.4%)	47 (16.1%)
0.81–1.0	Very high risk	3 (1%)	—	4 (1.4%)	22 (7.5%)

From the developed method in the present study, the risk was estimated as a scalar value, which varied from 0.08 to 0.92 in the present study. Moreover, it was categorized as very low risk (0–0.2), low risk (0.21–0.4), medium risk (0.41–0.6), high risk (0.61–0.8), and very high risk (0.81–1).

A total of 120/292 (41.1%) workers exhibited O or R type PFT abnormalities. Among these, 76 (26.1%) had severe abnormalities, recognized as possible obstruction with concurrent restriction, as observed in the PFT result. The remaining 96 (32.8%) workers showed normal lung function.

According to the observed value of the final risk out of 292 workers 82 (28%) workers were classified into the low risk category, where, as per the PFT results, 96 (32.8%) workers reported with regular abnormalities, showing only 4.8% deviations. The model predicted that 110 (37.6%) workers were classified into the medium risk category and, according to the observed PFT value, 120 (41%) workers were identified with Pulmonary O or R type diseases with a 3.4% deviation. Subsequently, the observed value for high risk was 69 (23.6%) obtained from the model, and according to the test results 76 (26.1%) workers were recognized to have O + R type abnormalities with only a 2.5% deviation.

8.5 Validation of the Model

The system was validated by comparing the results of FRBS modeling and the PFT. To check the adequacy of FRBS modeling, the potential outputs were categorized with PFT abnormalities. For example, if the crisp numerical output of the system was 0.37 for individual workers, the overall risk associated with the condition was low. Interestingly, in the PFT, the same person should be recognized as having no PFT abnormalities or with a normal condition. The results of Tables 8.1 and 8.3 can be compared to check the adequacy of the FRBS model.

Cases were evaluated by a 2 × 2 contingency table to check whether the results of the FRBS model were similar to the PFT results. Table 8.5 summarizes the cases in various cells based on FRBS modeling and PFT analysis in a 2 × 2 cell.

Table 8.5 Results in (2 × 2) contingency table with FRBS model prediction.

Respiratory abnormalities (PFT analysis)				
FRBS model prediction		**Yes (76)**	**No (216)**	**Total (292)**
	Yes	69 (90%)	18 (8%)	87
	No	7 (10%)	198 (92%)	205
	Total	76	216	292

The FRBS model predicted that 69 solid waste workers were possessed the risk of developing respiratory abnormalities. This prediction was confirmed by a sensitivity of 0.90. The model also predicted that 18 individuals were at high risk for developing respiratory abnormalities, and none of these individuals was confirmed by the PFT results. This was considered as a probability of a type-1 error of 0.08.

Seven solid waste workers were already experiencing high respiratory abnormalities, but in the FRBS model prediction they were categorized as medium risk with a probability of a type-2 error of 0.10. A total of 198 individuals were predicted to have a minimum number of risks for the development of respiratory disease.

Cases were also evaluated by a 3×3 contingency table to check whether the results of the FRBS model were similar to the PFT results. Table 8.6 summarizes the cases in various cells based on FRBS modeling and PFT analysis in a 3×3 cell.

Table 8.6 Results in (3 × 3) contingency table with FRBS model prediction.

Respiratory abnormalities (PFT analysis)				
FRBS model prediction	**Low (76)**	**Medium (120)**	**High (76)**	**Total (292)**
Low	82 (28%)	6 (2.1%)	0	88
Medium	10 (3.4%)	100 (34.2%)	7 (2.4%)	117
High	4 (1.4%)	14 (4.8%)	69 (23.7%)	87
Total	96	120	76	292

FRBS model prediction and PFT result with 2×2 and 3×3 contingency tables suggested that the predictions are representative of the condition of respiratory disease.

8.6 Sensitivity Analysis

Sensitivity analysis evaluated the most influencing factor for the risk of respiratory disease among the study population. Consequently, we found that Risk 2 was dominating, followed by Risk 3, Risk 1, and Risk 4 among the study population of solid waste workers. Personal habits, such as tobacco chewing, smoking, and alcohol drinking, were the major influencing factors for the risk prediction model up to 47.8%, followed by personnel hygiene at 23.5%. Further analysis also revealed that workers in a high risk category were middle age and had a high BMI.

Finally, sensitivity analysis is essential to inform the workers about the change in the most influencing parameter to drive toward respiratory disease. Thus, if the worker modified the influencing parameter in daily life, the chance to get the disease risk may vary accordingly.

8.7 Conclusion

The present study was designed to determine the risk for the occurrence of respiratory disease among solid waste workers. A cross-sectional study was carried out with a pulmonary lung function test of 292 solid waste workers of the city of Surat.

In this study, a new risk assessment approach was proposed and utilized. The present method evaluated the risk in scalar value based on an analyst's judgment, expert human knowledge, experience, and available literature. However, the quantification of risk in scalar value is subject to uncertainties for many reasons, including difficulties in defining input and output parameters, consequences of severity, and the mathematics of combining them.

The risk assessment for solid waste workers' working environment conditions ensured that uncertainties are inherent in every condition. FRBS provides linguistic expressions rather than metrics to assess the risks. Moreover, precise data are not available, and nothing is documented by private or government agencies for solid waste management workers. The workers are unable to maintain health data, although their exposure level and frequency of coming in contact with waste are remarkably high. Since data compilation is insufficient for the risk determination, the present study proposed the FRBS method to determine the risk for the occurrence of respiratory disease in solid waste workers.

The proposed methodology has new outcomes with respect to risk assessment. The model prediction results with a probability of a type-1 error as 0.08 and a probability of a type-2 error as 0.10 show that FRBS modeling is an effective approach for predicting respiratory risk levels with PFT comparison.

About 90 and 92% overall prediction of the system from 2×2 and 3×3 contingency tables, respectively, indicates that this system can successfully predict respiratory abnormalities, and that the PFT is an effective tool for the validation of the system. The model can be redefined for different types of the study population, and a slight modification in the factor could be considered depending on the study population.

Based on the above observations, it is concluded that the developed model with FRBS can determine the respiratory risk if the influencing parameters could be defined correctly. For any health risk assessment, the medical data and ethics of the data are essential. Typically, it is rather challenging to obtain the above data; then, the proposed methodology serves as the basic guideline for risk determination. The proposed approach may be implemented in different fields for risk assessment. However, in this system genetic disorders, diabetes measurements, and some difficulty in the quantification of certain factors were not considered in the study.

References

1 Carreno, L.A. and Jani, Y. (1993). A fuzzy expert system approach to insurance risk assessment using FuzzyCLIP. In: *WESCON/'93 Conference Record*, pp.IEEE, 28–30.

2. McCauley-Bell, P. and Badiru, A.B. (1996). Fuzzy modeling and analytic hierarchy processing: means to quantify risk levels associated with occupational injuries: part 2: the development of a fuzzy rule-based model for the prediction of injury. *Scopus Export 1990s 2656*. https://stars.library.ucf.edu/scopus1990/2656 (accessed 4 October 2022).

3 Mure, S., Deminchela, M., et al. (2006). Assessment of the risk of occupational accidents using a fuzzy approach. *Cognition, Technology & Work* 8: 103–112.

4 Keshwani, D.R., Jones, D.D., Meyer, G.E., and Brand, R.M. (2008). Rule-based Mamdani-type fuzzy modeling of skin permeability. *Applied Soft Computing* 8: 285–294.

5 Simić, S., Simić, D., Slankamenac, P., and Simić-Ivkov, M. (2008). Rule-based fuzzy logic system for diagnosing migraine. In: *Artificial Intelligence: Theories, Models and Applications* (ed. J. Darzentas, G.A. Vouros, S. Vosinakis, and A. Arnellos). Berlin: Springer. doi: 10.1007/978-3-540-87881-0_37.

6 Khatibi, V. and Montazer, G.A. (2010). A fuzzy-evidential hybrid inference engine for coronary heart disease risk assessment. *Expert Systems with Applications* 37: 8536–8542.

7 Adeli, A. and Neshat, M. (2010). A fuzzy expert system for heart disease diagnosis. *Proceedings of the International Multi-Conference of Engineers and Computer Scientists*, 1–6.

8 Anooj, P. (2011). Clinical decision support system: risk level prediction of heart disease using weighted fuzzy rules. *Journal of King Saud University: Computer and Information Sciences* 24 (1): 27–40.

9 Sivanandam, S.N., Sumathi, S., and Deepa, S.N. (2007). *Introduction to Fuzzy Logic Using MATLAB*. Springer.

9

Risk Analysis for Indoor Swimming Pools

A Fuzzy-based Approach

Sana Saleem, Haroon R. Mian, Manjot Kaur, Roberta Dyck, Guangji Hu, Kasun Hewage, and Rehan Sadiq

School of Engineering, University of British Columbia (Okanagan), Kelowna, BC, Canada

9.1 Introduction

Active leisure helps people stay fit and healthy, both physically and mentally. It can reduce health risk factors specifically linked to obesity. An active lifestyle has long been associated with health benefits and healthcare cost savings [1]. A significant portion of the Canadian population has a preference for indoor swimming over other physical activities [2, 3]. According to Statistics Canada, 25% of the population had gone swimming at least once in the last three months with an average of five visits per month [4]. Due to this increasing trend, swimming pool water can be a host to various pathogens that cause water-borne diseases. In British Columbia, municipalities are required to adopt disinfection techniques (i.e. chlorination) to make pool water safe for swimming [5]. While it neutralizes the pathogens, chlorine also reacts with organic content (i.e. urine, skin squama, cosmetics, sweat, hair, sunblock, and body lotion) introduced by people into the pool [6]. This results in the formation of disinfection by-products (DBPs) [7–12] that can impact human health.

Exposure to these DBPs impacts not only swimmers but also the surrounding nonswimmers [13]. Exposure to DBPs has been associated with several health problems, including cancer [2], upper airway respiratory symptoms, nose and eye irritation, voice extinction, and asthma [13–19] among swimmers, lifeguards, pregnant women, children, and infants [13]. In swimming pools, DBPs were first reported in the 1980s [20]. Since then, more than 100 DBPs have been identified in the existing literature [21]. Among all identified DBPs, chloroform is one of the most frequently occurring DBP species in swimming pools [22]. Hence, it is necessary to control the concentrations of DBP precursors in pool water as well as the risks associated with the formation of chloroform in water. It is difficult to quantify the risk due to the complexity of pool systems (e.g. different pool types and related various operations) and imprecise inputs (i.e. failure to account for subjectivity of small samples in variable selections). The inputs and outputs of environmental systems can be highly complex, variable, and difficult to measure accurately [23]. Indeed, for any environment system uncertainty is an unavoidable element as it represents partial information about a particular subject [24].

Fuzzy Systems Modeling in Environmental and Health Risk Assessment, First Edition. Edited by Boris Faybishenko, Rehan Sadiq, and Ashok Deshpande.

In environment modeling and analysis, when calculating the risk, there may exist both aleatory and epistemic uncertainty [25]. Aleatory uncertainty arises during the natural sampled system, whereas epistemic uncertainty arises due to imperfect knowledge of the system. More precisely, in swimming pools aleatory uncertainty arises due to swimming frequency, swimming duration, swimmer weight, and surface area; epistemic uncertainty arises due to incomplete or conflicting information, uncertainty in sampling, and limited data (i.e. data are not disclosed or available for analysis) [22]. These uncertainties must be included and characterized properly to facilitate the most accurate risk interpretation.

Several techniques and methods have been used to estimate the formation risk of DBPs. The literature mentions that, to describe aleatory uncertainty, probabilistic or stochastic methods are more effective because the major source of uncertainty is randomness or natural variability [26]. Moreover, possibilities rather than probabilities describe epistemic uncertainty arising from imperfect human understanding, due to the measurement of data and expert opinion [27]. However, data can be constrained in a complex variable environmental system, and in such systems fuzzy logic (FL) is used to deal with uncertainty more often than traditional probabilistic representations of uncertainties [28]. This chapter focuses on addressing the uncertainty involved in the estimation of formation risk of DBPs considering the different types of pools in a swimming facility.

9.2 Methodology

This section explains the analysis of DBP formation risk in four different types of pools in an indoor swimming pool facility, using a fuzzy-based approach.

9.2.1 Fuzzy-based Risk Analysis

In complex environmental problems, uncertainties can be encountered during the assessment that lead to an imprecise understanding of risk [29]. These uncertainties are due to incomplete data, sampling, and frequency, which can be addressed by various techniques. However, the application of soft computing techniques such as FL has been used often in risk analysis. Fuzzy is a language with syntax and semantics used to convert qualitative knowledge into numerical reasoning [30]. However, for the risk analysis of complex systems, such as indoor swimming pools, it is difficult to communicate the potency of risk on the basis of numerical values. Hence, engineers, environmentalists, decision-makers, regulators, managers, and stakeholders represent the risk linguistically (i.e. *very low*, *low*, *medium*, *high*, and *very high*). This helps to improve the decision-making process by identifying those areas that are at high risk and require mitigation measures. FL uses different types of fuzzy numbers to represent linguistic variables. However, trapezoidal fuzzy numbers and triangular fuzzy numbers (TFNs) are most commonly used. In this chapter, TFNs have been used. The formation risk in an indoor swimming pool facility can be defined as

$$\text{Risk (R)} = \text{Likelihood (L) Consequences (C)} \tag{9.1}$$

The likelihood and consequence of the formation risk are represented by TFNs (i.e. triangular fuzzy likelihood (TFL) (a_L,b_L,c_L) and triangular fuzzy consequence (TFC) (a_C,b_C,c_C), respectively) to incorporate the uncertainty, as illustrated in Table 9.1.

Table 9.1 Linguistic definitions for five-grade scale for likelihood, consequence, and risk using triangular fuzzy numbers.

Granular (G)	Likelihood (L)/ consequence (C)/risk (R)	Triangular fuzzy number (TFL, TFC, or TFR)	Centroid(Gr)*
1	Very low	[0, 0, 0.25]	0.08
2	Low	[0, 0.25, 0.5]	0.25
3	Medium	[0.25, 0.5, 0.75]	0.5
4	High	[0.5, 0.75, 1]	0.75
5	Very high	[0.75, 1, 1]	0.92

*Gr is the centroid of TFR defined as $\dfrac{\int x.\mu_A^x dx}{\int \mu_A^x dx}$ [31].

Using Eq. (9.1), the triangular fuzzy risk (TFR) can be calculated as

$$\text{TFR} = \text{TFLÄTFC} = (a_L, b_L, c_L) \otimes (a_C, b_C, c_C)(a_L * a_C, b_L * b_C, c_L * c_C) \tag{9.2}$$

Table 9.1 describes the five-grade (or granular, G_1 to G_5) qualitative scaling: very low (VL), low (L), medium (M), high (H), and very high (VH), for both likelihood and consequence. This five-grade scaling provides more flexibility in expressing the linguistic variables for likelihood and consequence comprehensively, to the decision-maker. Additionally, the fuzzy risk will be described with the help of a five-scale system denoted by G_1 to G_5 for VL to VH, respectively, as shown in Table 9.1. The corresponding membership function (μ) is defined as

$$\mu_1^L(x_L) \; or \; \mu_1^C(x_C) = \begin{Bmatrix} 1-4x & 0 \le x \le 0.25 \\ 0 & 0.25 \le x \le 1 \end{Bmatrix}$$

$$\mu_{P,}^L(x_L) \; or \; \mu_p^C(x_C) = \begin{Bmatrix} 0 & 0 \le x \le \dfrac{p-2}{4} \\ 4x-(p-2) & \dfrac{p-2}{4} \le x \le \dfrac{p-1}{4} \\ p-4x & \dfrac{p-1}{4} \le x \le \dfrac{p}{4} \\ 0 & \dfrac{p}{4} \le x \le 1 \end{Bmatrix} (p = 2,3,4)$$

$$\mu_5^L(x_L) \; or \; \mu_5^C(x_C) = \begin{Bmatrix} 0 & 0 \le x \le 0.75 \\ 4x-3 & 0.75 \le x \le 1 \end{Bmatrix}$$

First, the TFR is mapped on the G_1 to G_5 scale, as illustrated in Figure 9.1, to obtain the fuzzy risk. The points where TFR intersects with the defined scale VL to VH will be identified. Next, a non-normalized five-tuple set of all the intersection points (intersection at more than one point uses a maximum operator) will be created, which shows the membership of non-normalized fuzzy risk (NFR) to the grades from VL to VH. Further, the NFR will be normalized to obtain fuzzy risk, dividing the membership value of each by the cardinality C (sum of all the membership values). The risk will be obtained by defuzzification of the normalized fuzzy risk (FR) using centroid method (i.e. $G_r \times FR$).

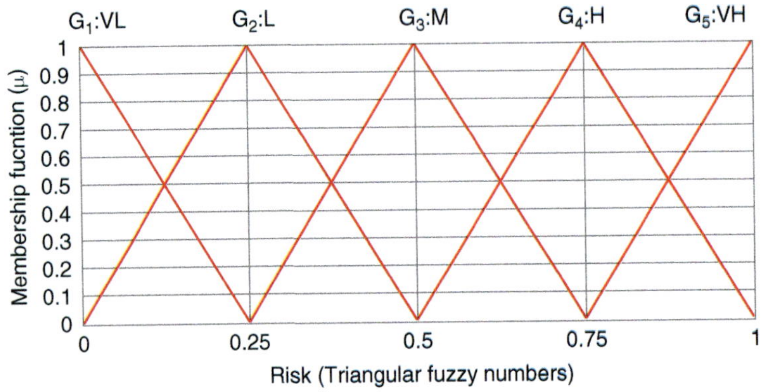

Figure 9.1 Risk evaluation scale.

9.2.2 Application: An Indoor Swimming Pool Facility Case Study

In general, indoor pool facilities can be broadly classified based on (i) use (e.g. sports pools, therapy and thermal baths, and competition pools) [32, 33]; (ii) types of user (e.g. public pools, private pools, and semipublic pools) [33]; (iii) temperature (e.g. hot water pools and cold water pools) [33]; (iv) size (50 m lap pool and 25 m lap pool); and (v) buildings (e.g. residential pools, hotel pools, gymnasium pools, school pools, and university pools) [33, 34].

For this case study, the indoor swimming pool facility used consists of four types of pools: a 50 m lap pool (PT-1), a 25 m lap pool (PT-2), a leisure pool (PT-3), and a hot tub (PT-4). All these pool types face potential risk in terms of DBP formation. The formation of DBPs is highly influenced by the organic content that has been introduced in the pool water by anthropogenic sources such as humans [35]. The presence of these DBPs varies by pool type due to variation in number of users, operational factors, and organic content input. These DBPs are believed to cause serious health problems, as mentioned in Section 9.1. In addition to organic content, previous studies demonstrate many operational and social factors that contribute to the deterioration of pool water.

The amount of organic content is associated with bathing load. For example, a high amount of organic content represents a high bathing load [10, 33, 36]. The bathing load is also related to the number of swimmers visiting the pool facility on a daily basis and their unhygienic behaviors (i.e. not taking showers, urinating in the pool causing deterioration of the pool environment, and having no knowledge about hygienic rules of the facility) [36, 37]. According to British Columbia guidelines for pool operation,[1] the formation risk can be avoided if the amount of free chlorine is maintained between 0.5 ppm (parts per million) and 5 ppm for water temperature less than 30°C [38]. If the water temperature is higher than 30°C then the amount of free chlorine should be maintained at 1.5 ppm [38]. In addition, the pool water temperature should not exceed 37°C for all types of pool except hot tubs; the water temperature in a hot tub should not exceed 30°C [38]. To protect the health of swimmers it is essential to estimate the risk for each pool type. This will help pool managers and operators to take the necessary actions for risk mitigation and to identify the pool type with the highest risk.

1 B.C. Guidelines for Pool Operations is designed to help operators and regulators interpret the pool regulation with respect to the operation of pools. These guidelines represent generally accepted standards of safe practices in British Columbia.

To assess the formation risks associated with these pools, data were abstracted from Saleem et al. [39] for four types of pools. The data related to number of swimmers, water temperature, free residual chlorine chloroform, and total organic carbon (TOC) were selected, as listed in Table 9.2 [39].

Table 9.2 Factors impacting indoor swimming pools and chloroform.

Pool type	Number of swimmers	Temperature °C	Free chlorine (ppm)	TOC (ppm)	Chloroform (ppm)
50 m* (PT-1)	12	25.6	1.43	1.68	35.7
	16	25.9	1.23	1.48	35.7
	29	25.5	1.14	1.5	30.9
	6	25.9	0.83	1.55	38.3
	9	25.7	1.14	1.42	30.4
	27	25.6	1.55	1.39	32.8
	19	24.7	1.35	1.42	34.3
25 m* (PT-2)	2	27.4	1.12	1.51	41.5
	0	25.8	0.95	1.39	46.4
	10	26.4	0.91	1.38	46.4
	10	25.6	0.88	1.42	48.1
	5	27.6	0.87	1.53	42.1
	10	26.9	1.22	1.34	26.3
	9	26.4	1.02	1.38	41.1
Leisure (PT-3)	5	29.5	1.56	3.95	63.6
	17	30.7	1.53	3.76	69.5
	16	29.5	0.94	3.6	63.6
	0	28.4	0.78	3.4	86.4
	4	31.0	1.35	3.32	64.7
	9	30.0	1.09	3.3	80.1
	0	31.6	1.2	3.14	91.8
Hot tub (PT-4)	0	30.2	9.0	2.0	87.1
	3	32.1	1.24	5.66	82.4
	7	31.6	1.46	6.0	72.6
	0	32.9	2.18	5.75	51.2
	4	36.0	3.0	5.9	57.0
	9	32.7	2.2	6.19	95.5
	2	34.6	1.02	6.24	65.9

*Lap pool.

9.3 Results and Discussion

For indoor swimming pools, the likelihood of chloroform concentration has been calculated considering the factors mentioned in Table 9.2. Due to uncertainty in these factors, the likelihood is converted into TFNs for each pool type. TFL is calculated by categorizing the data for all pool types separately, and individually taking the minimum, average, and maximum value of each pool type data. All values are normalized on the scale of 0–1, and the results of TFL are shown in Table 9.3.

Table 9.3 Triangular fuzzy likelihood (TFL).

Pool type	TFL		
	Minimum	Average	Maximum
PT-1	0.050	0.204	0.372
PT-2	0.001	0.061	0.110
PT-3	0.002	0.251	0.749
PT-4	0.010	0.346	1.000

The consequence of formation in each type of indoor swimming pool is represented through the concentration of chloroform that has been generated due to likelihood (factors). The approach used for the calculation of TFL is used for the calculation of TFC to incorporate the uncertainty in collection and analysis of samples, as described in Table 9.4.

Table 9.4 Triangular fuzzy consequence (TFC).

Pool type	TFC		
	Minimum	Average	Maximum
PT-1	0.30	0.34	0.38
PT-2	0.26	0.42	0.48
PT-3	0.64	0.74	0.92
PT-4	0.51	0.73	0.96

TFR of each pool was calculated with parameters given in Table 9.5. For example, TFR for PT-1 is the fuzzy multiplication of (0.05, 0.204, 0.372⊗) (0.30, 0.34, 0.38) obtained as (0.0152, 0.0694, 0.1423). The obtained TFR is mapped on the scale G_1 to G_5 to attain a five-tuple set of NFR, as shown in Figure 9.2. The five-tuple set of NFR of PT-1 is $\left\{\frac{0.78}{VL}, \frac{0.44}{L}, \frac{0}{M}, \frac{0}{H}, \frac{0}{VH}\right\}$ and the NFR for the remaining pools, PT-2, PT-3, and PT-4 is illustrated in Table 9.6.

Table 9.5 Triangular fuzzy risk (TFR).

Pool type	TFR		
	Minimum	Average	Maximum
PT-1	0.0152	0.0694	0.1423
PT-2	0.0002	0.0253	0.0528
PT-3	0.0012	0.1867	0.6877
PT-4	0.0053	0.2529	0.9550

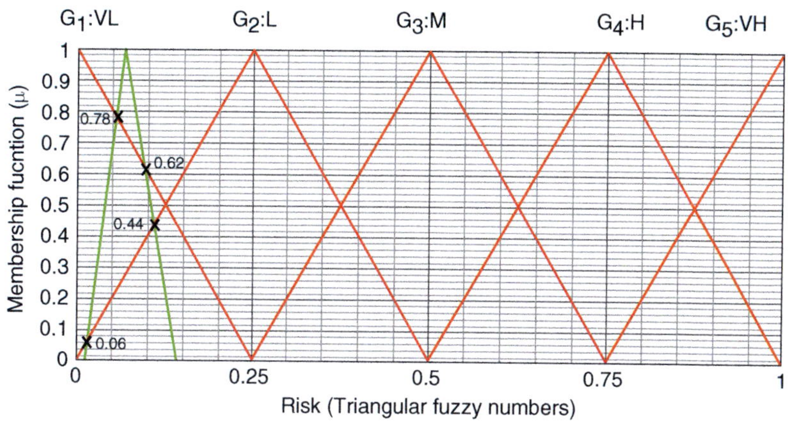

Figure 9.2 Estimating five-tuple set non-normalized fuzzy risk (NFR).

The NFR is then normalized using $\frac{\mu}{C}$ (i.e. membership divided by cardinality) to obtain FR, as listed in Table 9.7. For example, $C = 1.22$ for PT-1, FR will be $\left\{\frac{0.64}{\text{VL}}, \frac{0.36}{\text{L}}, \frac{0}{\text{M}}, \frac{0}{\text{H}}, \frac{0}{\text{VH}}\right\}$.

Table 9.6 Five-tuple set non-normalized fuzzy risk (NFR).

Pool type	NFR
PT-1	$\left\{\frac{0.78}{VL}, \frac{0.44}{L}, \frac{0}{M}, \frac{0}{H}, \frac{0}{VH}\right\}$
PT-2	$\left\{\frac{0.92}{VL}, \frac{0.2}{L}, \frac{0}{M}, \frac{0}{H}, \frac{0}{VH}\right\}$
PT-3	$\left\{\frac{0.55}{VL}, \frac{0.92}{L}, \frac{0.59}{M}, \frac{0.33}{H}, \frac{0}{VH}\right\}$
PT-4	$\left\{\frac{0.48}{VL}, \frac{0.98}{L}, \frac{0.74}{M}, \frac{0.48}{H}, \frac{0.22}{VH}\right\}$

Table 9.7 Normalized fuzzy risk (FR).

Pool type	FR
PT-1	$\left\{\frac{0.64}{VL}, \frac{0.36}{L}, \frac{0}{M}, \frac{0}{H}, \frac{0}{VH}\right\}$
PT-2	$\left\{\frac{0.82}{VL}, \frac{0.18}{L}, \frac{0}{M}, \frac{0}{H}, \frac{0}{VH}\right\}$
PT-3	$\left\{\frac{0.23}{VL}, \frac{0.38}{L}, \frac{0.25}{M}, \frac{0.14}{H}, \frac{0}{VH}\right\}$
PT-4	$\left\{\frac{0.17}{VL}, \frac{0.34}{L}, \frac{0.26}{M}, \frac{0.17}{H}, \frac{0.08}{VH}\right\}$

Finally, the centroid method is used to obtain the final risk by defuzzifying the FR (i.e. multiplying the FR with the centroid (*Gr*), as described in Table 9.8). For instance, the risk of PT-1 is 0.14 (i.e. low risk of formation). The PT-4 represents the hot tub as having the highest value of formation risk, whereas the PT-2 has the lowest formation risk value. Although the PT-4 has a relative higher risk compared to the other pools, on the developed fuzzy risk scale it comes under the medium risk category, which is 0.25, 0.5, and 0.75 (Table 9.1). The risk value estimated for PT-4 is close to PT-3 as both come under the medium risk category. There are several factors that can be associated with relatively higher formation risk values for PT-4 and PT-3, including high temperature, increased number of swimmers, and increased duration of swimmers in the pool water, which is indirectly linked with the increase in the organic content in pool water.

Table 9.8 Final risk of each pool type.

Pool type	Final risk
PT-1	0.14
PT-2	0.11
PT-3	0.34
PT-4	0.43

Recent studies have demonstrated various pool design and management factors that have an influence on pool facilities' overall performance with respect to both water and air quality. Moreover, the formation of DBPs has always been a prime source of concern for pool managers as these DBPs are linked with various human health effects, as mentioned in Section 9.1. The formation of these DBPs is also linked to factors such as source water quality, temperature, chlorine concentration, water renewal frequency, and proximity of bathers [40]. The temperature and number of swimmers are generally high in leisure (PT-3) and hot tub (PT-4) pools according to some studies [39]. This might be a reason for the high formation risk estimated for PT-4 and PT-3. Compared to 25 m and 50 m lap pools (i.e. PT-1 and PT-2), the estimated formation risk in PT-4 is around three to four times higher, whereas the risk estimated in PT-3 is two to three times higher. Some previous studies have also shown that, after some time, water in the hot tub is two to four times more mutagenic compared to both the original water used to fill them [41] and other lap pools. This is due to the higher temperature of hot tub water providing conditions for bacteria to grow. These bacteria can be the source of various infections or disease. Bacteria such as *Legionella* and *Pseudomonas* have been reported to cause numerous infections including severe skin rashes, eye and ear infections, and pneumonia. These bacteria can be controlled and minimized by maintaining the residual disinfectant level of the water [42]. Moreover, the PT-3 (leisure pool) has relatively high formation risk compared to the lap pool. This is possibly due to bathers being at much closer proximity and staying for a longer period of time in the leisure pool. This increases the chances of reaction between organic matters from humans and pool disinfectant, indirectly causing an increase in the possibility of risks [40].

There are several ways to minimize the associated formation risk in pools. Some studies have shown a relationship between user behavior and human health risks. According to the World Health Organization (WHO), all stakeholders, including pool facility workers, local authorities, pool managers, and families, could play an important role in minimizing unhealthy behaviors and ensuring pool safety through public education and appropriate and targeted information. An increase in knowledge/awareness of health risks associated with swimming pools is required. Pool facility management and lifeguards could play a key role if, in addition to technical skills training, they received training in educational and preventive aspects of human health effects. Moreover, the managers of pool facilities should aim to promote the benefits of healthy behaviors through proper training of staff as well as educational projects that target different groups of swimming pool users according to age, frequency of exposure, and other factors that can help minimize risks [43].

9.4 Conclusion

The model created in this chapter is based on FL, which gives a risk assessor the ability to solve complex problems plagued with uncertainty and vagueness. There are several risks associated with indoor swimming pools, which are relatively intangible in nature since exact values cannot be given. However, in this chapter formation risk has been studied through main variables, including likelihood and consequences. For both likelihood and consequences, membership functions have been established. The method adopted allows for a realistic preliminary assessment of the risk of formation in an indoor swimming pool. This method can be used by swimming pool facility managers to evaluate the risk in all pool types and ensure that safety measurements are satisfactorily based on the given data. This method can be used as a preliminary risk assessment tool, able to highlight critical situations and the need for more in-depth and complete analysis. Further, it can also be used to aid in decision-making for risk reduction.

References

1 Chen, J. and Millar, W.J. (1999). Health effects of physical activity. *Health Reports: Statistics Canada* 11 (1): 21–30.

2 Hang, C., Zhang, B., Gong, T., and Xian, Q. (2016). Occurrence and health risk assessment of halogenated disinfection byproducts in indoor swimming pool water. *Science of the Total Environment* 543: 425–431.

3 Bougault, V., Turmel, J., Levesque, B., and Boulet, L.P. (2009). The respiratory health of swimmers. *Sports Medicine* 39 (4): 295–312.

4 Gilmour, H. (2007). Physically active Canadians. *Health Reports / Statistics Canada, Canadian Centre for Health Information = Rapports Sur la Sante / Statistique Canada, Centre Canadien D'information Sur la Sante* 18 (3): 45–65.

5 Olsen, K. (2007). Clear waters and a green gas: a history of chlorine as a swimming pool sanitizer in the United States. *Bulletin for the History of Chemistry* 32 (2): 129–140.

6 Florentin, A., Hautemanière, A., and Hartemann, P. (2011). Health effects of disinfection by-products in chlorinated swimming pools. *International Journal of Hygiene and Environmental Health* 214 (6): 461–469.

7 Bureau, G., Lévesque, B., Dubé, M., Gauvin, D., Lépine, F., and Laliberté, D. (2017). Indoor swimming pool environments and self-reported irritative and respiratory symptoms among lifeguards. *International Journal of Environmental Health Research* 27 (4): 306–322.

8 Aggazzotti, G., Fantuzzi, G., Righi, E., and Predieri, G. (1998). Blood and breath analyses as biological indicators of exposure to trihalomethanes in indoor swimming pools. *Science of the Total Environment* 217 (1–2): 155–163.

9 Aggazzotti, G. and Predieri, G. (1986). Survey of volatile halogenated organics (VHO) in Italy: levels of VHO in drinking waters, surface waters and swimming pools. *Water Research* 20 (8): 959–963.

10 Chu, H. and Nieuwenhuijsen, M.J. (2002). Distribution and determinants of trihalomethane concentrations in indoor swimming pools. *Occupational and Environmental Medicine* 59 (4): 243–247.

11 Judd, S.J. and Jeffrey, J.A. (1995). Trihalomethane formation during swimming pool water disinfection using hypobromous and hypochlorous acids. *Water Research* 29 (4): 1203–1206.

12 Zwiener, C., Richardson, S.D., De Marini, D.M., Grummt, T., Glauner, T., and Frimmel, F.H. (2007). Drowning in disinfection byproducts? Assessing swimming pool water. *Environmental Science and Technology* 41 (2): 363–372.

13 Parrat, J., Donzé, G., Iseli, C., Perret, D., Tomicic, C., and Schenk, O. (2012). Assessment of occupational and public exposure to trichloramine in Swiss indoor swimming pools: a proposal for an occupational exposure limit. *Annals of Occupational Hygiene* 56 (3): 264–277.

14 Massin, N., Bohadana, A.B., Wild, P., Hery, M., Toamain, J.P., and Hubert, G. (1998). Respiratory symptoms and bronchial responsiveness in lifeguards exposed to nitrogen trichloride in indoor swimming pools. *Occupational and Environmental Medicine* 55 (4): 258–263.

15 Hery, M., Hecht, G., Gerber, J.M., Gender, J.C., Hubert, G., and Rebuffaud, J. (1995). Exposure to chloramines in the atmosphere of indoor swimming pools. *Annals of Occupational Hygiene* 39 (4): 427–439.

16 Thickett, K.M., McCoach, J.S., Gerber, J.M., Sadhra, S., and Burge, P.S. (2002). Occupational asthma caused by chloramines in indoor swimming-pool air. *European Respiratory Journal* 19 (5): 827–832.

17 Jacobs, J.H., Spaan, S., van Rooy, G.B.G.J., Meliefste, C., Zaat, V.A.C., Rooyackers, J.M., and Heederik, D. (2007). Exposure to trichloramine and respiratory symptoms in indoor swimming pool workers. *European Respiratory Journal* 29 (4): 690–698.

18 Chu, T.S., Cheng, S.F., Wang, G.S., and Tsai, S.W. (2013). Occupational exposures of airborne trichloramine at indoor swimming pools in Taipei. *Science of the Total Environment* 461–462: 317–322.

19 Lévesque, B., Vézina, L., Gauvin, D., and Leroux, P. (2015). Investigation of air quality problems in an indoor swimming pool: a case study. *Annals of Occupational Hygiene* 59 (8): 1085–1089.

20 Beech, J.A. Estimated worst case trihalomethane body burden of a child using a swimming pool. *Medical Hypotheses* 6 303–307.

21 Richardson, S.D., DeMarini, D.M., Kogevinas, M., Fernandez, P., Marco, E., Lourencetti, C., Ballesté, C., Heederik, D., Meliefste, K., McKague, A.B., Marcos, R., Font-Ribera, L., Grimalt, J.O., and Villanueva, C.M. (2010). What's in the pool? A comprehensive identification of disinfection by-products and assessment of mutagenicity of chlorinated and brominated swimming pool water. *Environmental Health Perspectives* 118 (11): 1523–1530.

22 Dyck, R.A. and Sadiq, R. Fuzzy physiologically based pharmacokinetic (PBPK) model of chloroform in swimming pools.

23 Wen, M. (2015). *Uncertain Data Envelopment Analysis*. Berlin: Springer.

24 Sadiq, R. and Tesfamariam, S. (2009). Environmental decision-making under uncertainty using intuitionistic fuzzy analytic hierarchy process (IF-AHP). *Stochastic Environmental Research and Risk Assessment* 23 (1): 75–91.

25 Chhipi-Shrestha, G., Hewage, K., and Sadiq, R. (2017). Microbial quality of reclaimed water for urban reuses: probabilistic risk-based investigation and recommendations. *Science of the Total Environment* 576: 738–751.

26 Dyck, R., Sadiq, R., Rodriguez, M., Simard, S., and Tardif, R. (2017). A comparison of membership function shapes in a fuzzy-based fugacity model for disinfection byproducts in

indoor swimming pools. *International Journal of Systems Assurance Engineering and Management* 8 (S4): 2051–2063.

27 Reza, B., Sadiq, R., and Hewage, K. (2013). A fuzzy-based approach for characterization of uncertainties in energy synthesis: an example of paved road system. *Journal of Cleaner Production* 59: 99–110.

28 Mays, M.D., Bogardi, I., and Bardossy, A. (1997). Fuzzy logic and risk-based soil interpretations. *Geoderma* 77 (2–4): 299–315.

29 Phillis, Y.A., Kouikoglou, V.S., Andriantiatsaholiniaina, L.A., and Zhu, X. Environmental applications of fuzzy logic. In: *2004 IEEE International Conference on Fuzzy Systems (IEEE Cat. No.04CH37542)*, 1, 341–346. IEEE.

30 Sadiq, R., Kleiner, Y., and Rajani, B. (2007). Water quality failures in distribution networks: risk analysis using fuzzy logic and evidential reasoning. *Risk Analysis* 27 (5): 1381–1394.

31 Yager, R.R. (1980). On a general class of fuzzy connectives. *Fuzzy Sets and Systems* 4 (3): 235–242.

32 ISO. (2012). Risk management information public swimming pools. https://amtrustfinancial.com/getmedia/43dfcde1-0b40-4b25-9b4e-c788dea2e828/Public-Swimming-Pools_1.pdf (accessed 4 October 2022).

33 WHO. (2006). Guidelines for safe recreational water environments: volume 2: swimming pools and similar environments.

34 Buchtal, A. (2010). The great book of swimming pools.

35 Lakind, J.S., Richardson, S.D., and Blount, B.C. (2010). The good, the bad, and the volatile: can we have both healthy pools and healthy people? *Environmental Science and Technology* 44 (9): 3205–3210.

36 Pasquarella, C., Veronesi, L., Napoli, C., Castaldi, S., Pasquarella, M.L. et al. (2014). What about behaviours in swimming pools? Results of an Italian multicentre study. *Microchemical Journal* 112: 190–195.

37 Nemery, B., Hoet, P.H.M., and Nowak, D. (2002). Indoor swimming pools, water chlorination and respiratory health. *European Respiratory Journal* 19 (5): 790–793.

38 Health Protection Branch MOH. (2014). BC guidelines for pool operations.

39 Saleem, S., Dyck, R., Hu, G., Hewage, K., Rodriguez, M., and Sadiq, R. (2019). Investigating the effects of design and management factors on DBPs levels in indoor aquatic centres. *Science of the Total Environment* 651 (Pt 1): 775–786.

40 Anses. (2016). Health risks associated with swimming pools for collective use. https://www.anses.fr/en/content/health-risks-associated-swimming-pools-collective-use (accessed 21 September 2022.

41 Daiber, E.J., DeMarini, D.M., Ravuri, S.A., Liberatore, H.K., Cuthbertson, A.A. et al. (2016). Progressive increase in disinfection byproducts and mutagenicity from source to tap to swimming pool and spa water: impact of human inputs. *Environmental Science and Technology* 50 (13): 6652–6662.

42 British Columbia HealthLinkBC. (2017). Residential hot tubs and pools: health and safety tips. https://www.healthlinkbc.ca/healthlinkbc-files/residential-hot-tubs-and-pools-health-and-safety-tips (accessed 21 September 2022).

43 Gallè, F., Dallolio, L., Marotta, M., Raggi, A., Di Onofrio, V., Liguori, G., Toni, F., and Leoni, E. (2016). Health-related behaviors in swimming pool users: influence of knowledge of regulations and awareness of health risks. *International Journal of Environmental Research and Public Health* 13 (5).

Part IV

Fuzzy Logic Applied to the Management of Water Distribution Networks

10

Fuzzy Parameters in the Analysis of Water Distribution Networks

Rajesh Gupta[1] *and Lindell Ormsbee*[2]

[1] *Visvesvaraya National Institute of Technology, Nagpur, India*
[2] *Kentucky Water Resources Research Institute, University of Kentucky, Lexington, KY USA*

10.1 Introduction

The analysis and design of water distribution networks (WDNs) involves several parameters, like length, diameter, pipe roughness, pipe discharge and head loss, nodal demands, reservoir water levels, etc. Some of the parameters are known and used to determine others in the analysis and design of WDNs. The values of some of the parameters, like pipe lengths, are precisely known and remain constant over time. The precise value of other parameters, like pipe diameter and pipe roughness coefficients, can be obtained when new; however, due to the aging process and deposition on the inside surface of pipes, the diameters and pipe roughness coefficient change with time and their future value cannot be obtained accurately. Uncertainty creeps in due to the unknown magnitude of the change. Further, there are some parameters, like nodal demands, that are dependent on various factors, such as population served by node, per capita consumption, peak factor, etc., which themselves have inherent uncertainty. Therefore nodal demands cannot be obtained accurately. Even though the diameter of a pipe varies with time, it is common practice to assume that the diameter remains constant over time, but its pipe roughness coefficient value does change. Thus, pipe roughness coefficients, reservoir water levels, and nodal demands are taken as imprecisely known. Besides, imprecise knowledge of the values of parameters, there are other sources of uncertainties, such as from the head loss formula and the coefficients taken in that formula. The Darcy–Weisbach formula, the Hazen–Williams (HW) formula, the modified HW formula, the Manning or Strickler formula are all used in different countries. Different formulas give different values of head loss. Moreover, there are uncertainties in the coefficients in these formulas. For example, in the HW formula the constant 10.68 is taken to be as low as 10.4516 to as high as 10.9031 in the literature [1, 2], and the value of the exponent of discharge is 1.85 or 1.852, giving different values of head loss in pipes. However, once the formula is chosen and the coefficients fixed, the answers we get are acceptable. However, inherent uncertainty is present. In order to explore how the uncertainties in the independent or basic parameters are propagated to dependent or derived parameters we will employ fuzzy set theory.

Fuzzy Systems Modeling in Environmental and Health Risk Assessment, First Edition. Edited by Boris Faybishenko, Rehan Sadiq, and Ashok Deshpande.

Several shapes of membership functions, such as triangular, trapezoidal, Gaussian, etc., are available. Fuzzy parameters of WDNs are represented through membership functions, such as triangular and trapezoidal. A triangular membership function is most common for any parameter as it shows likely variation on either side of most likely (normal) value of the parameter, and can be used for both nodal demands and pipe roughness coefficient values. However, when the pressure requirement is to be checked for both minimum and maximum daily demands with associated uncertainties, a trapezoidal membership function is used. A triangular membership function for the HW coefficient, C, and trapezoidal function for nodal demand, q, are shown in Figure 10.1 (a) and (b), respectively.

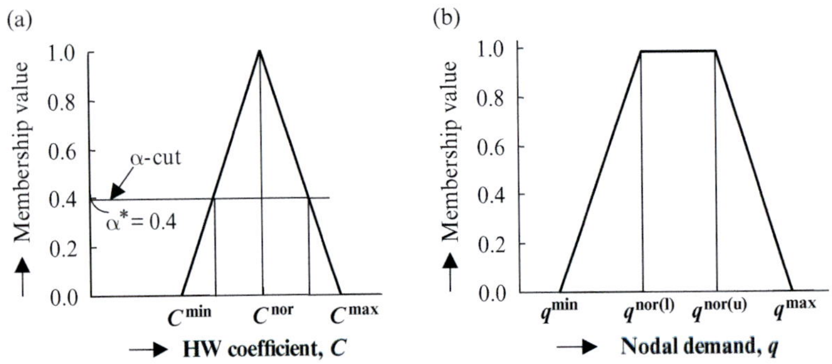

Figure 10.1 Membership functions: (a) triangular for the HW coefficient, C, and (b) trapezoidal for nodal demand, q.

In Figure 10.1, C^{min} = minimum value of C; C^{nor} = most-likely (i.e. normal value of C); and C^{max} = maximum value of C; q^{min} = minimum value of q; $q^{nor(l)}$ = lower limit of the range of most-likely values (i.e. lower limit of the range of normal value of q); $q^{nor(u)}$ = upper limit of the range of the normal value of q; and q^{max} = maximum value of q. It is worthwhile to remember that the membership function shows possibility distribution. The possibility of a value increases as one move from zero to one. A cut parallel to the x-axis is known as an alpha-cut (α-cut), or lambda-cut (λ-cut). Thus, for an α-cut with $\alpha = \alpha^*$ $(0 < \alpha^* < 1)$, we get two values of the parameter, both with a membership value of α^*. Considering the precise and fuzzy known parameters as independent parameters, we will obtain membership functions for the unknown parameters as dependent parameters. The base of fuzzy independent parameter (i.e. the maximum and minimum values) are fixed based on the past experience.

10.2 Pipe Network Hydraulic Principles and Relationships

Considering the fuzziness of the independent parameters, for an α-cut, $\alpha = \alpha^*$, the head loss h_x in pipe x is given by

$$(h_x)_{\alpha=\alpha^*} = \frac{KL_xQ_x^n}{D_x^r} = (R_xQ_x^n)_{\alpha=\alpha^*} \tag{10.1}$$

in which L_x, D_x, and Q_x are the length, diameter, and discharge in pipe x, respectively. K, n, and r have values depending on the formula used and the units of different parameters. For the Darcy–Weisbach formula, $K = f/12.1$, $n = 2$, and $r = 5$ (f is friction factor); for the HW formula, $K = 10.68 / C_{HW}^{n}$, $n = 1.85$, and $r = 4.87$ (C_{HW} is the HW coefficient); for the Strickler formula, $K = 10.3/S^2$, $n = 2$, and $r = 5.33$ (S is the Strickler coefficient); and for the Manning formula $K = 10.3\ N^2$, $n = 2$, and $r = 5.33$ (N is the Manning coefficient). The independent parameter pipe resistance constant, R_x, is fuzzy due to the fuzziness of pipe roughness characteristic such as coefficient of friction f in the Darcy–Weisbach formula, HW coefficient C_{HW}, expressed here as C, Manning coefficient N or Strickler coefficient S.

For α-cut, $\alpha = \alpha^*$, the node flow continuity relationship is given by

$$\sum_{x \text{ cnnected to } j} (Q_x)_{\alpha=\alpha^*} + (q_j)_{\alpha=\alpha^*} = 0 \tag{10.2}$$

in which nodal flow q_j is also fuzzy.

For α-cut, $\alpha = \alpha^*$, the loop head loss relationship is given by

$$\sum_{x \in c} (h_x)_{\alpha=\alpha^*} = \sum_{x \in c} (R_x Q_x^n)_{\alpha=\alpha^*} = 0 \tag{10.3}$$

Fuzzy analysis using α-cut methodology involves determining the extreme values of dependent parameters at various α-cut levels by selecting appropriate values of independent fuzzy parameters and solving Eqs. (10.2) and (10.3). The extreme values of any dependent parameters at various α-cut are then used to develop its membership functions.

10.3 Monotonic and Non-monotonic Change in Dependent Parameters

The nodal heads and the pipe discharges (or pipe velocities) are the usual dependent parameters of concern in the network analysis. The nodal heads and pipe discharges in the network may increase or decrease depending on the uncertainty in nodal demands and pipe roughness coefficients. The following can be interpreted from network hydraulic principles described by Eqs. (10.1) to (10.3).

1) With the increase in demand at any node in the network having a constant supply head, the head loss in the network increases. Therefore, nodal head decreases at all the demand nodes.
2) With the increase in resistance of any pipe (due to change in the pipe roughness coefficient), the head loss in that pipe increases, thereby decreasing the discharge in that pipe. Therefore, discharge will increase in other connected pipes to this node to balance the continuity equations. The effect will continue to other nodes also. Thus, pressure will increase at some nodes and decrease at others.

Monotonic change implies continuous increase or decrease of dependent parameters with the change in independent parameters in the space of α-cut interval, as shown in Figure 10.2(a) [3]. In the non-monotonic change there may exist global or local extreme values in the space of α-cut intervals of independent parameters, as shown in Figure 10.2(b). If the pressure head decreases due to an increase in demand or an increase in pipe resistance, this will continue to do so. The pressure head may be observed less than the minimum required pressure at some of the nodes. Similarly, if the discharge (or velocity) in any pipe reduces due to the change in nodal demands and pipe resistances, it may happen that flow direction in that reverses. Figure 10.3(a) shows a graph of HW coefficient for the case of a monotonic pipe discharge, when negative sign of velocity is considered [4, 5]. In reality, the negative sign of pipe discharges (or velocities) is actually reflecting a change in the direction of flow. Therefore, the minimum value pipe discharges (or velocities) will be zero, as shown in Figure 10.3(b) [3, 6–8]. As we will see, monotony in dependent parameters helps to simplify the process of fuzzy analysis.

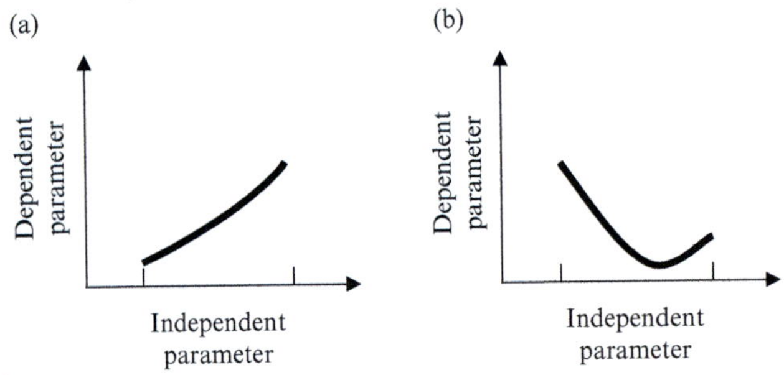

Figure 10.2 Change in dependent parameter with independent parameter: (a) monotonic and (b) non-monotonic.

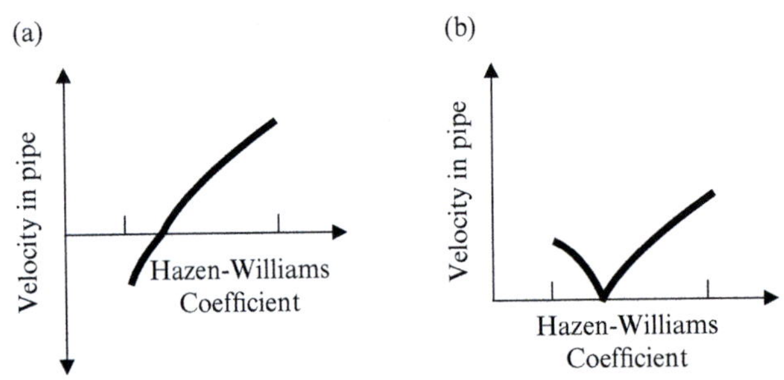

Figure 10.3 Change in velocity with HW coefficient: (a) monotonic, when negative sign of velocity is considered and (b) non-monotonic, when only magnitude of velocity is considered.

10.4 Methods of Fuzzy Analysis

Fuzzy analysis methods can be classified broadly into two categories as: (i) methods based on optimization and (ii) methods based on analysis. While methods based on optimization solve several optimization problems to obtain the extreme values of different dependent parameters at various α-cut levels, methods based on analysis obtain such values based on hydraulic analysis only.

10.4.1 Optimization-based Methods

Consider a looped network having M source nodes, labelled 1, ..., M, and N demand nodes, labelled $M + 1, ..., M + N (= J)$, so that the node labels are $j = 1, ..., J$. Let the pipe labels be x, $x = 1, ..., X$; and the loop labels, including pseudo loops be c, $c = 1, ..., C + M - 1$. The membership function of a dependent parameter, say Q_x, for an α-cut at $\alpha = \alpha^*$, requires the maximum and minimum values of Q_x, i.e. Max $(Q_x)_{\alpha=\alpha^*}$ and Min $(Q_x)_{\alpha=\alpha^*}$. This optimization problem must satisfy the node flow continuity equality constraints given by Eq. (10.2) and loop head loss equality constraints of Eq. (10.3). Further, the selected value of the independent parameter must lie within the boundaries of the minimum and maximum values at $\alpha = \alpha^*$. The optimization problem can be formulated for obtaining the membership functions for pipe discharges as follows by considering the resistance constant of each pipe as imprecise.

$$Max\ or\ Min(Q_x)_{\alpha=\alpha^*}, x = 1,\ldots,X \tag{10.4}$$

Subject to

$$\sum_{x\ \text{cnnected to}\ j} (Q_x)_{\alpha=\alpha^*} + q_j = 0, j = 1,\ldots,J \tag{10.5}$$

$$\sum_{x\in c}(h_x)_{\alpha=\alpha^*} = \sum_{x\in c}(R_x Q_x^n)_{\alpha=\alpha^*} = 0, c = 1,\ldots,C+M-1 \tag{10.6}$$

$$(R_x^{\min})_{\alpha=\alpha^*} \le R_x \le (R_x^{\max})_{\alpha=\alpha^*}, x = 1,\ldots,X \tag{10.7}$$

This is a constrained nonlinear optimization problem. Similarly, problem can be written for obtaining membership functions of nodal heads. Also, the constraints can be added for other uncertain parameters, such as water levels in the reservoir or nodal demands.

10.4.1.1 Solution Using Sequential Quadratic Programming

Revelli and Ridolfi [5] were the first to address the problem of fuzzy pipe network analysis. They used the sequential quadratic programming (SQP) method to obtain a solution for the problem defined by Eqs. (10.4) to (10.7). They found the method to be both efficient and robust. The method uses a Newton-type iterative procedure to search for the maximum and minimum values of the objective function. The method consists of three steps:

Step 1: Combine the objective function of Eq. (10.4) and the constraints of Eqs. (10.5) and (10.6) to obtain a Lagrange function, thus converting the problem into an unconstrained one. (The boundary-value constraint (10.7) is treated separately.) It is then approximated by means of its Hessian at each iteration.
Step 2: Solve the problem by quadratic programming.
Step 3: Use the solution to identify the directions along which the independent fuzzy parameters should move to reach the extreme values of the objective function.

To improve the convergence characteristics, they suggested to start with $\alpha^* = 1$ (the normal value), and then decrease α^* from time to time and use the extreme values of the previous α^*-cut level as the tentative values in the next step corresponding to the new α^*-cut level. This makes the algorithm to locate the extreme values of the new α^*-cut level more efficiently and with more certainty. Thus, in this method, to obtain $Q_x^{\max}$ and $Q_x^{\min}$ values for an α^*-cut for all X pipes of the network, $2X$ fuzzy optimization problems, one for the maximum and the other for the minimum for each of the X pipes are solved. The exact shape of the membership function of a dependent parameter depends on the number of α-cut levels, which can be chosen depending upon the required accuracy. For nf number of α-cuts, the number of fuzzy optimization problems will be $(nf - 1) \times 2X + 1$.

Once the intervals $(Q_x^{\min}, Q_x^{\max})_{\alpha=\alpha^*}$ for the pipe discharges are known, using these values it is possible to obtain the intervals for pipe head loses. Thus

$$(h_x^{\min}, h_x^{\max})_{\alpha=\alpha^*} = (R_x^{\min}, R_x^{\max})_{\alpha=\alpha^*} \times (Q_x^{\min}, Q_x^{\max})^n_{\alpha=\alpha^*} \tag{10.8}$$

in which the extreme values of h_x are obtained by combining the extreme combinations $(R_x^{\min}, Q_x^{\min})$ and $(R_x^{\max}, Q_x^{\max})$. Thus, starting from the node with a known hydraulic gradient level (HGL) value, it is possible to obtain the membership functions of the HGL values at all nodes of a network.

The method is observed to work well for small networks. However, for the medium-sized network having one source node, 179 pipes, 141 demand nodes, and 38 loops of Gittikhadan Zone of Nagpur, India, introduced by Gupta and Bhave [4], Moosavian and Lence [9] (also 10) observed that the SQP method did not converge to the global optimum, and a suboptimal, narrower range of nodal pressures was obtained for all nodes.

10.4.1.2 Solution Using Genetic Algorithm

Branisavljevic and Ivetic [3] considered uncertainty in pipe roughness coefficients for fuzzy analysis. The suggested genetic algorithm (GA) was used to obtain the membership functions of nodal heads and pipe discharges. Thus, this method requires solving $2X$ fuzzy optimization problems to obtain $Q_x^{\max}$ and $Q_x^{\min}$ values, and further $2N$ fuzzy optimization problems to obtain $H_j^{\max}$ and $H_j^{\min}$ for any α*-cut. They used GA components code written in C++ from the library [11] and linked EPANET [12] as a hydraulic solver in the code written in C++. GA is relatively simple to use when dependent parameters change monotonically with independent parameters as one of two boundary values are to be selected for each uncertain parameter. Thus, decision variables will be discrete and limited to $2X$ in this case. However, in non-monotonic change, the independent variable can take any value in the space between α*-cut interval. Thus, in this case the decision variable will be continuous, and obtaining its exact value will be relatively difficult. Sivakumar et al. [13], who also linked GA and EPANET, admitted this fact by stating that "variables are generated as continuous function. This causes slight variation in the result particularly in the lower range of roughness value."

Haghighi and Asl [6] discussed that the optimization problem that arises in the fuzzy analysis of pipe networks is neither a single objective problem nor a common multi-objective problem. It belongs to a special class of optimization in which there are many independent objective functions to be simultaneously optimized with the same set of constraints. A modified nondominated sorting genetic algorithm (NSGA-II) was developed to find all extreme values of the dependent parameters in a single optimization run. A new closeness/distance operator was introduced to the NSGA-II instead of its standard crowding distance operator since no tradeoff between the problem objective functions is required in the fuzzy analysis. The new operator helped in quickly obtaining the extreme values for the several independent objective functions. Although multi-objective optimization provided the entire fuzzy analysis in a single run, the number of evaluations and time required were significant as discussed in Section 10.4.1.3.

10.4.1.3 Solution Using Other Evolutionary Methods

Shibu and Reddy [14] considered the monotonic behavior of dependent parameters with independent parameters. In such cases, extreme values of independent parameters provide extreme values of dependent parameters. Thus, the constraint equation, Eq. (10.7), is replaced by boundary value constraint as

$$(C_x^{\min})_{\alpha=\alpha^*} \leq C_x \leq (C_x^{\max})_{\alpha=\alpha^*}, x = 1, \ldots, X \tag{10.9}$$

Shibu and Reddy [14] used the cross-entropy method to select the proper combination of pipe resistance constants to obtain the membership functions of nodal heads and pipe discharges.

Sabzkouhi and Haghighi [7, 8] developed many-objective particle swarm optimization (MO-PSO) to solve fuzzy analysis problems involving several independent objective functions to be solved with the same set of constraints. The MO-PSO starts in a similar way to standard PSO, by flying particles with randomly assigned positions and velocities in the search space. Each particle is analyzed using EPANET and evaluated for different objective functions. Particles are ranked based on the different objectives and grouped as per the objectives. New positions of the particles are generated considering global best and historical best position of particles in the individual group. The optimization process continues till no further improvement in the solution is observed. However, the number of EPANET runs required in MO-PSO as compared to that in NSGA-II is relatively less. The network introduced by Haghighi and Asl [6] of the city of Khorramshahr in southwest Iran had 45 demand nodes, 65 pipes, and 20 loops, required 590 040 EPANET runs against 2 620 000 by Haghighi and Asl [6]. However, this approach is also computationally burdensome for the fuzzy analysis of large WDNs.

Gupta and Sreetu [15] suggested the Jaya algorithm [16] to obtain membership function of pipe velocities considering non-monotonic behavior. By application on the Khorramshahr city network, they concluded that the Jaya algorithm was capable of providing minimum velocity as zero in pipes with a reversal of flow but was also computationally burdensome.

10.4.2 Analysis-based Methods

These methods assumed monotonic change of dependent parameters with independent parameters.

10.4.2.1 Solution Using Impact Table

Gupta and Bhave [4] suggested a fuzzy analysis method that avoided using any optimization model. Their methodology first determined the impact of change of all uncertain independent parameters on all dependent parameters numerically by increasing/decreasing all of the parameters, one by one, by a small amount through hydraulic analysis. Considering the monotonic change of the dependent parameter with the independent parameter, proper extreme values of independent parameters were selected to determine the extreme values of the dependent parameters. Thus, if the impact of increase in the independent parameter was observed to be positive (i.e. there was an increase in the value of the dependent parameter) then the maximum value of this independent parameter was selected during the maximization of the dependent parameter. If the impact of increase in the independent parameter was observed to be negative (i.e. there was a decrease in the value of the dependent parameter) then the minimum value of this independent parameter was selected during the maximization of the dependent parameter.

This method requires (i) a normal analysis using the most likely values of uncertain parameters, one analysis for each of the uncertain parameters to obtain the impact of change, and (ii) an analysis at each selected α^*-cut, except for $\alpha^* = 1$ for obtaining minimum and maximum values of dependent parameters. Thus, considering X pipe resistance constants (or pipe roughness coefficients) and N nodal demands as uncertain, there will be $X + N + 1$ hydraulic analyses for constructing an impact table. Further, considering the requirement of the membership function for nodal heads and pipe discharges (or velocities), $2 \times (X + N)$ hydraulic analysis will be required at each α^*-cut except for $\alpha^* = 1$ for minimum and maximum values of dependent parameters. Considering nf number of α^*-cuts, the number of hydraulic analyses for minimum and maximum values of dependent parameters for all α^*-cut except $\alpha^* = 1$ will be $2 \times (X + N) \times (nf - 1)$. Thus, the total hydraulic analysis for an entire fuzzy analysis will be $(2 \times nf - 1) \times (X + N) + 1)$. It is to be noted that hydraulic analysis can be reduced substantially if membership functions are required for only nodal heads and not required for pipe discharges (or velocities). In such cases, impact is to be determined only for the change in pipe roughness on nodal heads, as the impact of a change in nodal demands on nodal heads can be simply predicted from hydraulic principles as discussed above. In this case, the number of hydraulic analysis for the nodal heads will be $2 \times (nf - 1) \times N + X + 1$. Further, if nodal demands are uncertain and membership functions for nodal heads are required, no analysis is necessary for obtaining an impact table. The number of hydraulic analysis for the fuzzy analysis of nodal heads will be $2 \times [nf - 1) \times N + 1$.

Gupta et al. [17] extended the application of an impact table based methodology to the fuzzy analysis of pressure-deficient networks. Herein, nodal flows are the additional dependent parameters which are assumed to change monotonically with independent fuzzy parameters in the given fuzzy interval.

Moosavian [10] suggested obtaining a Jacobian matrix for the dependent parameters with respect to fuzzy independent parameters. This Jacobian matrix is used the same as an impact table. Gupta and Bhave [4] developed an impact table using the signs of gradient values obtained numerically. Moosavian [10] suggested obtaining gradient values through a chain rule of differentiation by using a first-order partial derivative of pipe head loss equations (Eq. 10.1) and node flow continuity equations (Eq. 10.2) with respect to dependent parameters and fuzzy independent parameters.

10.4.2.2 Solution Using Complete Enumeration

A fuzzy analysis solution can also be obtained through complete enumeration. In this case, the uncertain parameters take one of the two extreme values at any α-cut for optimal value of any dependent parameter. Thus, considering X pipe roughness coefficient as uncertain, there will be 2^X combinations of uncertain variables. Karmakar [18] suggested a hydraulic analysis with different possible combinations. From the solutions of different analyses, maximum and minimum values of dependent parameters can be selected to obtain a membership function. Sivakumar et al. [19] also suggested the same method and called it the vertex method. The number of possible combinations goes on increasing with the number of uncertain parameters. Hence, the method is feasible for only a small network. The number of combinations of uncertain parameters considering uncertainty in nodal demands and pipe roughness coefficients for the medium size networks of the city of Khorramshahr and the Gittikhadan area of Nagpur in India will be 2^{45+65} ($= 1.3 \times 10^{33}$) and $2^{141+179}$ ($= 2.1 \times 10^{96}$), respectively. Considering approximately 20 seconds for 1000 EPANET runs for the network of the Gittikhadan area on Intel Core 2Duo central processing unit (CPU) P8700 (Intel, Santa Clara, California) at 2.53 GHz and 4.00 GB RAM [9], the time required to check all combinations will be several years.

10.4.2.3 Solution Using Newton–Raphson Method

Spiliotis and Tsakiris [20] suggested an approach based on the monotony of node flow continuity equations to develop a membership function for dependent parameters considering fuzzy nodal demands. They converted fuzzy analysis problems to crisp analysis problems and applied the Newton–Raphson method [21] to estimate the fuzzy nodal pressures. The main limitation of this approach is that it can be used only for uncertain nodal demands and not for uncertain pipe roughness coefficient. Also, the membership function for discharge (or velocity) cannot be obtained by using this methodology.

In fact, it was found that the membership function of nodal heads can be obtained directly by selecting lower values of demands at all nodes for maximum values of dependent parameters and higher values of demands at all nodes for minimum values of dependent parameters.

10.4.2.4 Solution Using Partial Gradient Values

Using the co-content model [22], Moosavian and Lence [9] showed that the gradient of nodal demand with respect to the nodal head is always less than zero. Thus, both of them are inversely proportional. That is, as the nodal demands increase, the pressure head at that node decreases. Not only at that node, but at all other nodes also, the head decreases as the required flows and corresponding friction losses increase. Therefore, the maximum/minimum value of the nodal head occurs at the minimum/maximum value of nodal demands.

Further, considering the head loss using Eq. (10.3) in terms of pipe resistance and evaluating signs of gradients of pipe resistance with respect to heads at upstream and downstream nodes of that pipe, they showed that the gradient is positive with respect to the head at the upstream node and negative with respect to the head at the downstream node. Therefore, it was suggested that, to find the maximum value of the nodal head at a node, resistance for all pipes with flow directions entering the node should be set to their maximum values, and resistance for all pipes with flow directions exiting the node should be set to their minimum values. Now, the question is about pipe resistance constants to be taken for pipes not connected at that node. They suggested selecting minimum values of pipe resistances for pipes not connected at that node if the nodal head was to be maximized and selecting maximum values of pipe resistances for pipes not connected at that node if the nodal head was to be minimized.

As compared to methods based on the impact table [4] that determine gradient numerically and its direction for all the dependent parameters with respect to all the uncertain independent parameters, the method of Moosavian and Lence [9] uses partial information. The membership function of the nodal heads is therefore approximate. Further, the method cannot provide membership for pipe discharges (or velocity).

10.5 Comparison of Fuzzy Analysis Methods and Discussion

Fuzzy analysis methods, as briefed in Section 10.4, are compared on various aspects and shown in Table 10.1. The following general observations can be made from Table 10.1.

1) The optimization based fuzzy analysis methods are mainly useful when output parameters change non-monotonically with input parameters. The number of hydraulic simulations required in case of optimization methods are fewer than required by analysis methods based on complete enumeration [18, 19]. However, these are many times more than that required by other methods based on analysis [4, 9]. With consideration of monotonous relationship, the nonlinear optimization problem changes to a discrete optimization problem which can be easily solved using the cross-entropy method [14].
2) The impact table method of Gupta and Bhave [4] and gradient-based method of Moosavian [10] differ only in the way the gradients of dependent parameters with respect to independent fuzzy parameters are obtained. Both provide the same solution.

Table 10.1 Comparison of various fuzzy analysis methods.

Methods	Suggested by	Category	Uncertain parameters considered	Dependent parameters considered	Relationship between parameters	Size of networks analyzed	Remarks
Sequential quadratic programming (SQP)	Revelli and Ridolfi [5]	Optimization	Nodal demands and pipe roughness coefficients	Nodal heads and pipe discharges	Non-monotonous	Two networks of small size (≤ 25 nodes)	Fuzzy analysis of networks cannot be carried out using HW head loss formula Moosavian and Lence [9] observed that the method did not converge to the global optimum, and a suboptimal, narrower range of nodal pressures were obtained for all nodes of medium-size networks
Genetic algorithm (GA)	Branisavljevic and Ivetic [3]	Optimization	Pipe roughness coefficients	Nodal heads and pipe discharges	Non-monotonous	Large size (453 nodes and 616 pipes)	
	Sivakumar et al. [13]	Optimization	Pipe roughness coefficients	Nodal heads and pipe discharges	Non-monotonous	Four networks (maximum 31 nodes)	As the uncertain pipe roughness variables are generated as a continuous function, it caused slight inaccuracy in the results, particularly in the lower range of roughness values
	Haghighi and Asl [6]	Optimization	Nodal demands and pipe roughness coefficients	Nodal heads, pipe velocities, and discharges	Non-monotonous	Two networks (maximum 46 nodes)	Several optimization problems are solved simultaneously by modifying multi-objective GA Requires number of hydraulic simulations, thereby restricting its use for large networks

Cross-entropy	Shibu and Reddy [14]	Optimization	Pipe roughness coefficients	Nodal heads and pipe discharges	Monotonous	A small network of four nodes	
Many-objective particle swarm optimization (MO-PSO)	Sabzkouhi and Haghighi [7, 8]	Optimization	Nodal demands and pipe roughness coefficients	Nodal heads, pipe velocity, and pipe discharges	Non-monotonous	Two networks (maximum 46 nodes)	Several optimization problems are solved simultaneously Computations efforts are less compared to methods based on modified NSGA-II but still huge [9]
Impact table	Gupta and Bhave [4], Bhave and Gupta [22], Gupta and Dongre [24]	Analysis	Nodal demands, pipe roughness coefficients	Nodal heads, pipe velocity, and pipe discharges	Monotonous	Three networks (maximum 141 nodes and 179 pipes)	Computational burden is less Moosavian and Lence [9] showed its application on a network with 1892 nodes and 2467 pipes
	Gupta et al. [17]	Analysis	Nodal demands and pipe roughness coefficients	Nodal outflows and heads, and pipe discharges	Monotonous	One small network	Assumes a monotonous change of nodal outflows with change in nodal demands and pipe roughness coefficients

(Continued)

Table 10.1 (Continued)

Methods	Suggested by	Category	Uncertain parameters considered	Dependent parameters considered	Relationship between parameters	Size of networks analyzed	Remarks
Newton–Raphson	Spiliotis and Tsakiris [20]	Analysis	Nodal demands	Nodal heads	Non-monotonous	Small network with 11 nodes	Uses monotony of continuity equations Membership functions for only nodal heads can be obtained
Complete enumeration	Karmakar [18]	Analysis	Pipe roughness coefficients	Nodal heads and pipe discharges	Monotonous	One small network with four nodes	Considered uncertainty in head loss formula and residual chlorine as dependent parameters Applicable for only small networks as computation burden increases manifold with size of problem
	Sivakumar et al. [19]	Analysis	Pipe roughness coefficients	Nodal heads and pipe discharges	Monotonous	Three small networks (maximum nine nodes network).	Applicable for only small networks
Based on partial gradient values	Moosavian and Lence [9]	Analysis	Nodal demands and pipe roughness coefficients	Nodal heads	Monotonous	Two large networks	Provides approximate membership functions and that too for only nodal heads

3) All the methods based on analysis use the monotony of output parameters with input parameters, except that based on the Newton–Raphson method [20], which uses the monotony of continuity equations. An assumption of monotony may not be true in the case of pipe discharges. The membership function needs to be modified in such cases as discussed elsewhere in this chapter with the results of the network used for practical application.
4) Some of the methods are capable of considering both nodal demands and pipe roughness coefficients as uncertain, some consider only one of them. Methods that consider only pipe roughness coefficients as uncertain can be easily extended to include uncertainty in nodal demands.
5) Fuzzy analysis can be used: (i) to verify whether uncertainties in some hydraulic quantities are compatible with the working requirements of the network so that changes to the system can be made during the planning stage of a design [5, 23], (ii) in predicting accurate water quality parameters through modeling that requires correct information of pipe velocities, and (iii) to limit the extreme values of dependent parameters within permissible limits during operation stages by monitoring valves in different pipes and thereby increasing or decreasing discharges in pipes and head values at demand nodes [4]. Thus, both nodal heads and pipe discharges are useful in making decisions. Therefore, fuzzy methods that provide membership functions of nodal heads only have a limited use.
6) The method of Gupta and Bhave [4] based on the impact table can consider both nodal demands and pipe roughness as uncertain and provide membership functions of both nodal heads and pipe discharges. In fact, any combination of parameters from pipe resistance constants, pipe discharges, nodal demands, and nodal head values can be considered dependent parameters, as long as network solvability rules are satisfied. It provides accurate membership functions, and computational efforts are much less in terms of computational time and cost than any optimization-based methods.

10.6 Illustration of Fuzzy Analysis Using Impact Table

The network shown in Figure 10.4 taken from Bhave and Gupta [22] is used for the illustration of methodology. It has two source nodes (1 and 2), nine demand nodes (3, ..., 11), and 14 pipes (1, ..., 14).

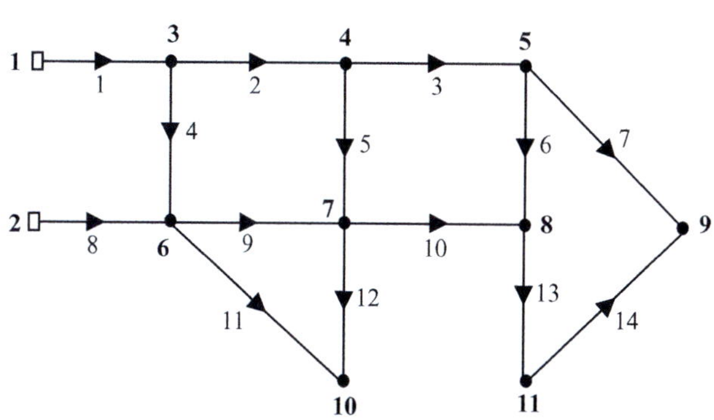

Figure 10.4 Two-source network.

Nodes 1 and 2 are elevated reservoirs with normal HGL values of 100 and 95 m, respectively. Maximum uncertainty of ±0.5 m exists in HGL. The normal demands at nodes 3, ..., 11 are 5, 6, 4, 3, 8, 5, 4, 6, and 3 m^3/min. Uncertainty in nodal demands are considered as ±5%. HW coefficients are also considered uncertain. Pipe 6 has been recently installed. Therefore, its C value is assumed to be 130, a precise value. The pipe data are given in Table 10.2. The maximum, normal and minimum C values of other pipes are as given in the table. The EPANET 2.0 [12] is used for simulation of the network.

Table 10.2 Pipe data.

Pipe	Length (m)	Diameter (mm)	HW coefficient		
			Maximum	Normal	Minimum
(1)	(2)	(3)	(4)	(5)	(6)
1	200	500	105	100	95
2	1000	450	105	100	90
3	800	300	100	95	90
4	900	400	100	95	85
5	1000	400	100	95	85
6[a]	1200	350	—	130	—
7	1500	200	100	95	90
8	300	400	100	95	90
9	900	300	100	95	90
10	1000	300	100	95	90
11	1500	250	100	95	90
12	1000	300	100	95	90
13	1100	250	95	90	85
14	1300	200	90	85	80

[a] Recently installed pipe.

Fuzzy analysis is carried out to obtain membership functions of all pipe discharges and nodal heads. The flow directions in pipes are fixed such that the pipe discharges are positive for analysis with normal values for all independent parameters. The pipe network is analyzed with all independent parameters at their normal values. The obtained pipe discharges are shown in row 1, Table 10.3. The supplies at source nodes and HGL values at demand nodes are shown in row 1, Table 10.4. The steps for further analysis are:

Stepwise Analysis and Results

1) No analysis is necessary to know the impact of increase in any reservoir level or any nodal demand on the nodal heads. However, herein analysis is required as impacts on different pipe discharges are required which may increase or decrease. Therefore, the impact table is prepared considering changes in all the independent parameters.
2) Independent parameters are selected one by one to increase the value to maximum by keeping others at normal. First of all, H_1 is selected and is set to its maximum value of 100.50 m.
3) By comparing the obtained value of each dependent parameter with its corresponding value in row 1, (see Tables 10.3 and 10.4), it is found whether a dependent parameter has

Table 10.3 Normal and extreme values of pipe discharges.

		Pipe discharges (m³/s)													
Row	**Value**	Q_1	Q_2	Q_3	Q_4	Q_5	Q_6	Q_7	Q_8	Q_9	Q_{10}	Q_{11}	Q_{12}	Q_{13}	Q_{14}
(1)	(2)	(3)	(4)	(5)	(6)	(7)	(8)	(9)	(10)	(11)	(12)	(13)	(14)	(15)	(16)
1	Nor	0.573	0.387	0.150	0.102	0.137	0.039	0.044	0.161	0.143	0.116	0.070	0.030	0.072	0.022
2	Max	0.626	0.422	0.168	0.133	0.165	0.056	0.049	0.211	0.170	0.133	0.083	0.043	0.079	0.027
3	Min	0.514	0.347	0.134	0.070	0.104	0.024	0.040	0.117	0.123	0.100	0.059	0.014	0.066	0.018

Table 10.4 Normal and extreme values of supplies at source nodes and HGL values at demand nodes.

		Supply at source nodes (m³/s)		HGL at demand nodes (m)								
Row	**Value**	q_1	q_2	H_3	H_4	H_5	H_6	H_7	H_8	H_9	H_{10}	H_{11}
(1)	(2)	(3)	(4)	(5)	(6)	(7)	(8)	(9)	(10)	(11)	(12)	(13)
1	Normal	−0.573	−0.161	95.606	77.839	58.266	92.954	72.800	57.617	30.742	71.535	38.980
2	Max	−0.626	−0.211	96.971	82.346	65.708	94.172	77.789	65.114	42.964	76.670	50.054
3	Min	−0.514	−0.117	94.199	71.071	47.863	91.305	65.246	47.146	14.511	63.958	24.107

increased or decreased. The increased dependent parameter is indicated by I; while the decreased dependent parameter is indicated by D, as shown in row 1 of Tables 10.5 and 10.6. In Table 10.5, the impact of an increase in the value of an independent parameter on the value of nodal supplies at source nodes and nodal heads at demand nodes are shown. In Table 10.6, the impact of an increase in the value of an independent parameter on pipe discharges is shown. These tables are defined as impact tables.

4) Steps 2 and 3 are repeated for other independent parameters (rows 2, ..., 24 in the impact table, Tables 10.5 and 10.6). Note that since the HW coefficient of pipe 6 is precisely known, it is not included in column 2, which shows independent fuzzy parameters.
5) The maximum value of q_1 is obtained from network analysis by changing the independent parameters to their maximum values or minimum values as indicated in column 3 of Table 10.5.

Table 10.5 Impact of increase in head at source node and nodal demand on HGL at demand node.

Row	Increase in input parameter	Increase/decrease in node Supply		Increase/decrease in nodal HGL								
		q_1	q_2	H_3	H_4	H_5	H_6	H_7	H_8	H_9	H_{10}	H_{11}
(1)	(2)	(3)	(4)	(5)	(6)	(7)	(8)	(9)	(10)	(11)	(12)	(13)
1	H_1	I	D	I	I	I	I	I	I	I	I	I
2	H_2	D	I	I	I	I	I	I	I	I	I	I
3	q_3	I	I	D	D	D	D	D	D	D	D	D
4	q_4	I	I	D	D	D	D	D	D	D	D	D
5	q_5	I	I	D	D	D	D	D	D	D	D	D
6	q_6	I	I	D	D	D	D	D	D	D	D	D
7	q_7	I	I	D	D	D	D	D	D	D	D	D
8	q_8	I	I	D	D	D	D	D	D	D	D	D
9	q_9	I	I	D	D	D	D	D	D	D	D	D
10	q_{10}	I	I	D	D	D	D	D	D	D	D	D
11	q_{11}	I	I	D	D	D	D	D	D	D	D	D
12	C_1	I	D	I	I	I	I	I	I	I	I	I
13	C_2	I	D	D	I	I	I	I	I	I	I	I
14	C_3	I	D	D	D	I	I	I	I	I	I	I
15	C_4	I	D	D	D	D	I	I	D	D	I	D
16	C_5	I	D	D	D	I	I	I	I	I	I	I
17	C_7	I	D	D	D	D	I	I	I	I	I	I
18	C_8	D	I	I	I	I	I	I	I	I	I	I
19	C_9	D	I	I	I	I	D	I	I	I	I	I
20	C_{10}	D	I	I	I	I	D	D	I	I	D	I
21	C_{11}	D	I	I	I	I	D	I	I	I	I	I
22	C_{12}	I	D	D	D	D	I	D	D	D	I	D
23	C_{13}	D	I	I	I	I	D	D	D	I	D	I
24	C_{14}	D	I	I	I	I	D	D	D	I	D	D

Table 10.6 Impact of increase in head at source node and nodal demand on pipe discharge.

Row	Increase in input parameter	Increase/decrease in pipe discharge													
		Q_1	Q_2	Q_3	Q_4	Q_5	Q_6	Q_7	Q_8	Q_9	Q_{10}	Q_{11}	Q_{12}	Q_{13}	Q_{14}
(1)	(2)	(3)	(4)	(5)	(6)	(7)	(8)	(9)	(10)	(11)	(12)	(13)	(14)	(15)	(16)
1	H_1	I	I	I	I	I	I	I	D	D	D	D	I	D	D
2	H_2	D	D	D	D	D	D	D	I	I	I	I	D	I	I
3	q_3	I	D	D	D	D	D	D	I	I	I	I	D	I	I
4	q_4	I	I	D	D	D	D	D	I	I	I	I	D	I	I
5	q_5	I	I	I	D	I	D	D	I	I	I	I	D	I	I
6	q_6	I	I	I	I	I	I	I	I	D	D	D	I	D	D
7	q_7	I	I	I	I	I	I	I	I	I	D	I	D	D	D
8	q_8	I	I	I	D	I	I	I	I	I	I	I	D	D	D
9	q_9	I	I	I	D	I	I	I	I	I	I	I	D	I	I
10	q_{10}	I	I	I	I	I	I	I	I	I	D	I	I	D	D
11	q_{11}	I	I	I	D	I	I	I	I	I	I	I	D	I	D
12	C_1	I	I	I	I	I	I	I	D	D	D	D	I	D	D
13	C_2	I	I	I	D	I	I	I	D	D	D	D	I	D	D
14	C_3	I	I	I	D	D	I	I	D	D	D	D	I	D	D
15	C_4	I	D	D	I	D	D	D	D	I	I	I	D	I	I
16	C_5	I	I	D	D	I	D	D	D	D	I	D	I	I	I
17	C_7	I	I	I	D	D	D	I	D	D	D	D	I	D	D
18	C_8	D	D	D	D	D	D	D	I	I	I	I	D	I	I
19	C_9	D	D	D	I	D	D	D	I	I	I	D	I	I	I
20	C_{10}	D	D	D	I	I	D	D	I	I	I	I	D	I	I
21	C_{11}	D	D	D	I	D	D	D	I	D	I	I	D	I	I
22	C_{12}	I	I	I	D	I	I	I	D	I	D	D	I	D	D
23	C_{13}	D	D	D	I	I	I	D	I	I	I	I	D	I	I
24	C_{14}	D	D	D	I	I	I	D	I	I	I	I	D	I	I

6) Independent parameters are changed, column-wise, as indicated in columns 4–13 of the impact Table 10.5 and columns 3–16 of Table 10.6, and the network is re-analyzed. The maximum values of other dependent parameters q_2, Q_1, ..., Q_{14} and H_3, ..., H_{11} are successively obtained. The obtained maximum values of dependent parameters are shown in row 2, Tables 10.3 and 10.4.
7) To obtain the minimum values of the dependent parameters, the independent parameters are appropriately changed to their extreme values, and the network is consequently re-analyzed. The obtained minimum values of the dependent parameters are shown in row 3, Tables 10.3 and 10.4.

It can be observed that:

1) As the HGL values H_1 and H_2 increase at the source nodes, the HGL values H_3, ..., H_{11} at the demand nodes also increase, as seen in the impact table (Table 10.5). This is as it should logically be. Hence, the impact table can be filled directly without additional simulations for this and computational efforts can be reduced.

2) As the nodal demands $q_3, ..., q_{11}$ increase, the source node supplies q_1 and q_2 increase, while the HGL values at demand nodes $H_3, ..., H_{11}$ decrease. This is also as it should logically be. Hence, additional simulations for an impact table can be avoided, if the nodal demands are uncertain and membership functions of only nodal heads are required. The proper extreme values of nodal demands can be chosen directly to obtain the extreme values of nodal heads.
3) Since most likely values of independent parameters occur at the same point of time, node flow continuity equations are satisfied using values of dependent parameters in the solution. However, extreme values of dependent parameters occur at different points of time. Therefore, node flow continuity equations at demand nodes are not satisfied for the pipe discharges in Table 10.3. This will also be true for source nodes with multiple off-taking pipes.
4) The impact of uncertainty on nodal heads is seen more on the nodes away from the source. It can be seen that node 9, farthest from the source nodes, is most affected (Table 10.4, column 11).
5) The impact of uncertainty on pipe discharges is more than compared to nodal heads. The minimum velocity in all pipes is observed to be more than 0. Hence, flow reversal is not observed in any case.

10.7 Practical Application

The network used by Sabzkouhi and Haghighi [7, 8] to evaluate the performance of a MO-PSO for fuzzy analysis is considered for obtaining a solution using the impact table method. This example pipe network belongs to the industrial park of Khorramshahr located in southwest Iran in the vicinity of the Persian Gulf. The network has been constructed in a flat area with about a 2 m elevation above mean sea level. The network consists of 45 demand nodes, 65 polyethylene pipes, and an elevated tank with a 27 m head, as shown in Figure 10.5. Other information of the network is given in Tables 10.7 and 10.8.

To study the extent to which the uncertainties spread out over the system and influence the responses, an uncertainty of ±15% in network demands and HW coefficient was assumed. The input membership functions are discretized in five levels of uncertainty with $\alpha = 0, 0.25, 0.5, 0.75$, and 1.

The fuzzy analysis results for the network are obtained using the impact table method of Gupta and Bhave [4] considering the values of n, r, and K as used in EPANET in the head loss equation (Eq. 10.1). The minimum nodal head for any supply is 18 m. The nodal pressures at different α-cuts are shown in Table 10.9 and the pipe discharges at different α-cuts are shown in Table 10.10. The pipe velocities can be obtained with the known pipe discharges and their membership functions can also be obtained. The following can be observed from Tables 10.9 and 10.10.

1) There are several nodes at which residual pressure is less than 18 m. All values less than 18 m are emphasized with bold typeface. The deficiency in pressure at nodes 24, 33, 38, and 44 is observed at α-cut level of 0 only. Nodes 39, 40, and 45 are observed pressure deficient at α-cut levels of 0 and 0.25. Nodes 28, 29, 30, 36, and 37 are observed pressure deficient at α-cut levels of 0, 0.25, and 0.5. Nodes 41, 42, 43, and 46 are observed pressure deficient at α-cut levels of 0, 0.25, 0.5, and 0.75. There is no deficiency in pressure at other nodes. This indicates relative criticality of different nodes in the network. Nodes away from the source are more critical and with slight uncertainty in dependent parameters show deficiency in residual pressure.

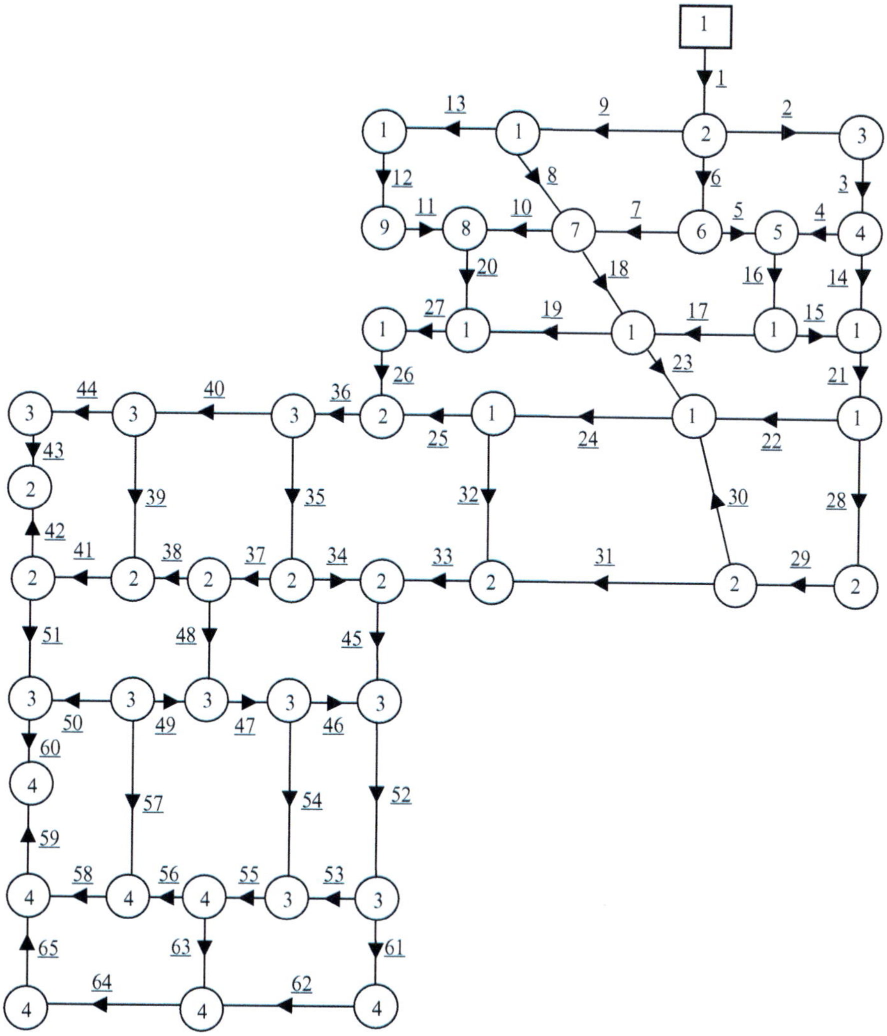

Figure 10.5 Pipe network of Khorramshahr city (Iran).

2) There are few pipes in which minimum discharge is observed to be negative at different α-cut levels. This indicates flow reversal. For example, in pipes 42, 43, 50, and 51 flow reversal is noticed at α-cut level of 0 only, and in pipes 11, 30, 39, 46, and 53 flow reversal is noticed at α-cut levels of 0 and 0.25. Obviously, at certain α-cut levels the velocity shall be 0. The point to be noted here is that the minus signs show the flow reversal. Therefore, the minimum velocity at this α-cut level should be zero as there will be some combination of uncertain parameters that will make the pipe discharge zero in the pipe having flow reversal.
3) The method required a total of 991 hydraulic simulations (111 to generate impact table and 220 for each α-cuts of 0, 0.25, 0.50, and 0.75). Earlier, Haghighi and Asl [6] carried out fuzzy analysis of this network using NSGA-II, and Sabzkouhi and Haghighi [7, 8] used MO-PSO. The number of hydraulic simulations required with NSGA-II is quite large, about 2 620 000 for α-cuts of 0 and 0.5. Similarly, the MO-PSO requires a total of 590 040 simulations at α-cuts of 0 and 0.5. Thus, fuzzy analysis using the impact table approach is easier as no optimization algorithm is needed and the results can be obtained with fewer simulations. The number of hydraulic simulations for two α-cuts at 0 and 0.5 is only 551.

Table 10.7 Pipe lengths and diameter.

Pipe No	Length (m)	Diameter (mm)	Pipe No	Length (m)	Diameter (mm)	Pipe No	Length (m)	Diameter (mm)	Pipe No	Length (m)	Diameter (mm)
(1)	(2)	(3)	(1)	(2)	(3)	(1)	(2)	(3)	(1)	(2)	(3)
1	25	572.8	18	231.2	322.6	35	141	409	52	150	254.4
2	210.1	322.6	19	223.1	322.6	36	200	454.4	53	200.1	145.4
3	100	254.4	20	200	227.2	37	110	322.6	54	150	227.2
4	158.8	227.2	21	100	227.2	38	90	322.6	55	110	322.6
5	50	363.6	22	223.7	100	39	141	113.6	56	90	100
6	100	454.4	23	115.6	322.6	40	200	227.2	57	150	113.6
7	192.9	254.4	24	281.8	409	41	200	181.8	58	200	181.8
8	115.5	363.6	25	100	322.6	42	91	181.8	59	100	363.6
9	250.8	409	26	100	409	43	50	145.4	60	50	113.6
10	107.1	181.8	27	100	454.4	44	200	145.4	61	175.1	227.2
11	100	100	28	141	181.8	45	150	227.2	62	308.4	145.4
12	100	145.4	29	142	181.8	46	202.1	100	63	156.2	181.8
13	149.2	363.6	30	163	100	47	100	322.6	64	290.6	254.4
14	200	145.4	31	362.9	100	48	150	322.6	65	138.4	322.6
15	156.1	181.8	32	141	113.6	49	90	113.6	—	—	—
16	100	454.4	33	95.91	113.6	50	200	145.4	—	—	—
17	126.9	322.6	34	204.1	181.8	51	150	145.4	—	—	—

Table 10.8 Demand in L/s at each node.

Node	Demand (L/s)	Node	Demand (L/s)	Node	Demand (L/s)	Node	Demand L/s)	Node	Demand L/s)
(1)	(2)	(1)	(2)	(1)	(2)	(1)	(2)	(1)	(2)
2	2.29	11	1.95	20	3.95	29	17.46	38	4.03
3	1.09	12	2.41	21	1.52	30	11.69	39	4.93
4	2.55	13	4.32	22	4.27	31	9.97	40	3.01
5	3.69	14	5.83	23	3.67	32	11.40	41	4.04
6	0.00	15	4.09	24	4.69	33	3.56	42	4.90
7	5.21	16	4.34	25	5.02	34	4.12	43	43.89
8	3.15	17	2.67	26	3.10	35	2.70	44	2.51
9	7.01	18	6.39	27	4.74	36	3.91	45	5.54
10	67.13	19	4.59	28	11.26	37	4.05	46	9.93

Table 10.9 Normal and extreme values of nodal heads.

Node No.	Nodal heads (*m*) at α-cut								
	1	0.75		0.5		0.25		0	
	Normal	Min.	Max.	Min.	Max.	Min.	Max.	Min.	Max.
(1)	(2)	(3)	(4)	(5)	(6)	(7)	(8)	(9)	(10)
1	27.000	27.000	27.000	27.000	27.000	27.000	27.000	27.000	27.000
2	24.94	24.931	24.948	24.92	24.954	24.908	24.96	24.895	24.966
3	24.846	24.81	24.875	24.765	24.898	24.71	24.917	24.641	24.933
4	24.713	24.649	24.766	24.572	24.81	24.479	24.846	24.367	24.876
5	24.581	24.502	24.649	24.408	24.706	24.298	24.755	24.168	24.796
6	24.745	24.696	24.787	24.637	24.823	24.568	24.853	24.486	24.878
7	24.218	24.081	24.335	23.922	24.435	23.736	24.521	23.518	24.596
8	23.504	23.238	23.732	22.927	23.926	22.562	24.093	22.134	24.236
9	23.754	23.536	23.941	23.28	24.101	22.979	24.238	22.625	24.355
10	24.082	23.924	24.218	23.74	24.334	23.525	24.434	23.273	24.52
11	24.316	24.195	24.421	24.052	24.51	23.885	24.587	23.689	24.652
12	24.192	24.039	24.322	23.857	24.431	23.640	24.524	23.382	24.602
13	24.448	24.345	24.535	24.224	24.610	24.081	24.674	23.914	24.728
14	*23.758*	23.562	23.929	23.335	24.077	23.072	24.205	22.766	24.317
15	23.048	22.743	23.314	*22.389*	23.545	21.98	23.747	21.505	23.923
16	22.953	22.637	23.228	22.272	23.468	21.849	23.677	21.358	23.86
17	24.071	23.895	24.219	23.683	24.343	23.428	24.447	23.121	24.534
18	23.363	23.098	23.592	22.79	23.79	22.432	23.962	22.014	24.11

(Continued)

Table 10.9 (Continued)

Node No.	Nodal heads (*m*) at α-cut								
	1	0.75		0.5		0.25		0	
	Normal	Min.	Max.	Min.	Max.	Min.	Max.	Min.	Max.
(1)	(2)	(3)	(4)	(5)	(6)	(7)	(8)	(9)	(10)
19	23.059	22.753	23.325	22.399	23.557	21.989	23.758	21.513	23.934
20	22.806	22.476	23.094	22.094	23.345	21.653	23.564	21.14	23.756
21	23.831	23.607	24.019	23.34	24.177	23.019	24.31	22.636	24.421
22	23.639	23.383	23.855	23.08	24.038	22.721	24.192	22.296	24.322
23	22.017	21.484	22.469	20.856	22.853	20.114	23.181	19.234	23.46
24	21.160	20.535	21.698	19.807	22.163	18.957	22.564	**17.962**	22.911
25	22.034	21.581	22.428	21.058	22.772	20.452	23.071	19.747	23.333
26	21.534	20.998	21.999	20.378	22.404	19.659	22.756	18.822	23.063
27	21.482	20.937	21.954	20.307	22.365	19.576	22.722	18.725	23.034
28	19.688	18.853	20.411	17.886	21.038	16.763	21.583	15.453	22.057
29	19.592	18.749	20.323	17.772	20.956	16.638	21.507	15.315	21.986
30	19.68	18.844	20.404	17.874	21.032	16.748	21.577	15.434	22.052
31	21.715	21.188	22.17	20.576	22.563	19.865	22.904	19.035	23.200
32	22.346	21.945	22.695	21.482	23.000	20.946	23.266	20.325	23.499
33	20.849	20.183	21.425	19.409	21.922	18.508	22.352	17.456	22.724
34	21.125	20.520	21.650	19.82	22.105	19.007	22.501	18.060	22.846
35	21.236	20.649	21.744	19.972	22.186	19.186	22.571	18.272	22.906
36	19.335	18.437	20.113	17.399	20.789	16.196	21.377	14.795	21.889
37	*19.162*	18.252	19.952	17.200	20.638	15.981	21.236	14.563	21.758
38	20.685	19.995	21.281	*19.196*	21.798	18.268	22.245	17.186	22.634
39	20.433	19.713	21.055	18.879	21.595	17.909	22.063	16.777	22.471
40	20.336	19.598	20.973	18.744	21.526	17.75	22.006	16.591	22.423
41	18.768	17.812	19.599	16.709	20.323	15.433	20.955	13.950	21.508
42	18.507	17.536	19.351	16.418	20.088	15.124	20.733	13.622	21.297
43	18.476	17.502	19.323	16.379	20.063	15.081	20.71	13.574	21.276
44	20.559	19.841	21.178	19.009	21.713	18.039	22.177	16.907	22.578
45	19.175	18.272	19.960	17.228	20.643	16.020	21.239	14.615	21.759
46	18.557	17.591	19.398	16.477	20.132	15.188	20.774	13.693	21.336

Membership functions for nodal heads are shown in Figure 10.6. Similarly, membership functions for pipe velocities in a few typical pipes are shown in Figure 10.7. The membership functions for velocity need to be modified for those pipes in which flow direction changes. Suppose velocity changes its sign at certain α-cuts and becomes negative. For any α-cut lower than this α-cut, the minimum velocity will be zero. A typical membership function for velocity (can be considered similarly for discharge also) is shown in Figure 10.8(a) obtained by considering velocity to change monotonically with independent parameters. The

Table 10.10 Normal and extreme values of pipe discharges.

Pipe No	Pipe discharges (L/s) at α-cut								
	1	0.75		0.5		0.25		0	
	Normal	Min.	Max.	Min.	Max.	Min.	Max.	Min.	Max.
(1)	(2)	(3)	(4)	(5)	(6)	(7)	(8)	(9)	(10)
1	322.57	310.474	334.666	298.377	346.763	286.281	358.859	274.185	370.956
2	28.73	25.612	32.147	22.771	35.891	20.183	39.992	17.829	44.484
3	27.64	24.548	31.031	21.731	34.748	19.168	38.822	16.837	43.287
4	15.949	13.266	18.858	10.782	22.022	8.468	25.473	6.292	29.243
5	115.392	105.604	125.611	96.245	136.265	87.314	147.358	78.809	158.897
6	156.267	144.59	168.366	133.337	180.884	122.51	193.824	112.113	207.186
7	40.876	35.924	46.208	31.324	51.951	27.047	58.139	23.067	64.807
8	55.852	46.937	65.077	38.306	74.637	29.93	84.553	21.776	94.848
9	135.283	125.058	145.978	115.292	157.154	105.972	168.825	97.087	181.001
10	27.31	24.215	30.688	21.388	34.367	18.815	38.369	16.489	42.72
11	3.341	2.11	4.547	0.694	5.787	−1.087	7.093	−2.102	8.49
12	10.351	9.035	11.684	7.659	13.074	5.871	14.544	4.483	16.115
13	77.481	73.848	81.185	70.264	84.977	66.691	88.871	62.988	92.883
14	9.141	8.029	10.38	7.033	11.762	6.141	13.304	5.342	15.026
15	12.799	11.067	14.684	9.474	16.737	8.006	18.978	6.652	21.427
16	127.651	117.872	137.81	108.471	148.352	99.448	159.28	90.803	170.601
17	110.532	101.322	120.119	92.487	130.088	84.024	140.445	75.931	151.196
18	64.208	56.037	72.65	48.099	81.394	40.346	90.468	32.713	99.896
19	82.829	75.436	90.561	68.38	98.636	61.654	107.061	55.256	115.839
20	27.501	23.746	31.491	20.172	35.751	16.615	40.314	13.278	45.21
21	19.53	17.647	21.571	15.904	23.787	14.287	26.199	12.778	28.829
22	3.793	3.154	4.488	2.569	5.245	2.051	6.073	1.605	6.981
23	86.08	78.707	93.806	71.681	101.89	64.996	110.339	58.648	119.161
24	86.189	78.926	93.753	71.954	101.627	65.257	109.815	58.788	118.326
25	73.211	65.847	80.873	58.774	88.842	51.983	97.122	45.453	105.721
26	101.9	93.691	110.341	85.71	119.018	77.936	127.929	70.31	137.074
27	106.24	97.994	114.729	89.986	123.464	82.194	132.444	74.563	141.668
28	13.067	11.717	14.528	10.464	16.115	9.289	17.844	8.167	19.735
29	11.547	10.222	12.979	8.988	14.535	7.823	16.231	6.684	18.087
30	2.706	1.745	3.648	0.646	4.613	−0.789	5.627	−1.574	6.71
31	4.571	3.93	5.287	3.36	6.085	2.861	6.972	2.452	7.958
32	8.388	7.323	9.553	6.346	10.827	5.447	12.223	4.619	13.753
33	9.29	7.958	10.74	6.734	12.321	5.609	14.047	4.575	15.935
34	21.518	19.077	24.122	16.786	26.903	14.635	29.874	12.611	33.052

(Continued)

Table 10.10 (Continued)

Pipe No	Pipe discharges (L/s) at α-cut								
	1	0.75		0.5		0.25		0	
	Normal	Min.	Max.	Min.	Max.	Min.	Max.	Min.	Max.
(1)	(2)	(3)	(4)	(5)	(6)	(7)	(8)	(9)	(10)
35	126.978	119.994	134.029	113.078	141.145	106.234	148.325	99.463	155.568
36	171.16	163.826	178.5	156.498	185.844	149.177	193.192	141.864	200.542
37	100.44	93.993	107.024	87.684	113.742	81.519	120.591	75.501	127.571
38	33.068	29.675	36.62	26.426	40.346	23.312	44.267	20.322	48.415
39	3.738	2.398	5.003	0.734	6.267	−1.446	7.567	−2.542	8.931
40	32.782	30.076	35.673	27.541	38.764	25.163	42.074	22.931	45.625
41	32.066	29.247	35.049	26.584	38.205	24.071	41.542	21.7	45.071
42	10.076	7.646	12.402	5.095	14.637	2.399	16.792	−0.472	18.87
43	7.384	5.352	9.537	3.429	11.827	1.603	14.271	−0.135	16.892
44	19.074	17.19	21.105	15.443	23.294	13.827	25.658	12.341	28.212
45	26.118	23.373	29.034	20.788	32.137	18.351	35.44	16.051	38.959
46	2.408	1.533	3.206	0.304	3.987	−1.199	4.776	−1.896	5.592
47	46.788	42.681	51.053	38.725	55.483	34.912	60.083	31.227	64.862
48	64.272	59.475	69.225	54.831	74.34	50.334	79.623	45.983	85.084
49	14.784	13.233	16.461	11.799	18.273	10.476	20.232	9.256	22.35
50	5.035	3.71	6.417	2.422	7.874	1.138	9.419	−0.203	11.064
51	10.73	8.111	13.352	5.439	16.014	2.617	18.748	−0.573	21.581
52	24.966	22.289	27.787	19.743	30.767	17.311	33.917	14.972	37.253
53	6.17	4.237	8.103	2.222	10.073	−0.164	12.084	−2.248	14.038
54	40.26	36.594	44.097	33.098	48.11	29.774	52.302	26.632	56.681
55	41.5	38.213	44.909	35.05	48.444	32.011	52.106	29.103	55.9
56	9.527	8.399	10.749	7.357	12.071	6.393	13.505	5.501	15.06
57	5.839	4.679	7.068	3.574	8.375	2.511	9.773	1.472	11.276
58	11.327	9.777	12.991	8.336	14.778	6.998	16.696	5.765	18.755
59	32.175	29.187	35.176	26.204	38.201	23.219	41.266	20.225	44.399
60	11.715	9.507	13.963	7.296	16.277	5.004	18.679	2.445	21.185
61	14.766	13.187	16.479	11.737	18.331	10.415	20.328	9.223	22.476
62	12.256	10.743	13.902	9.361	15.686	8.111	17.614	6.997	19.688
63	28.963	26.185	31.873	23.538	34.92	21.022	38.108	18.644	41.442
64	35.678	32.451	39.01	29.324	42.451	26.291	46.007	23.352	49.684
65	25.748	22.532	29.047	19.392	32.434	16.323	35.92	13.321	39.512

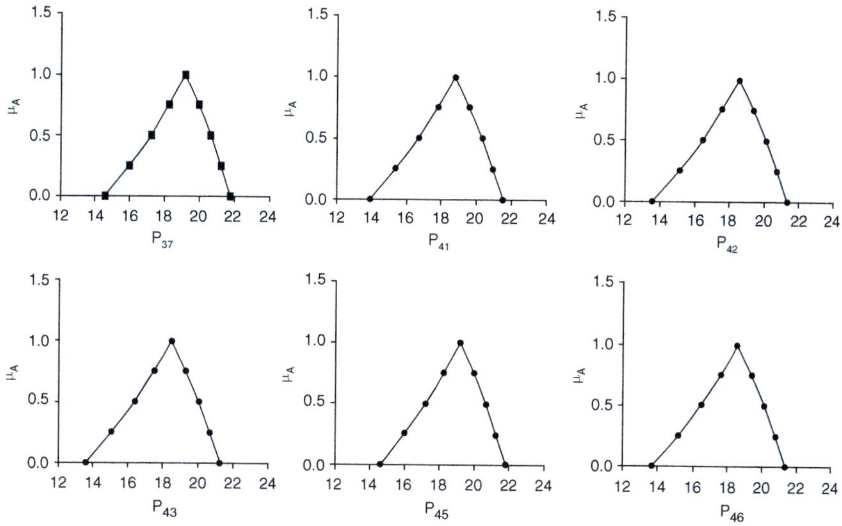

Figure 10.6 Membership functions of nodal heads by impact table method.

membership function can be modified for velocity, as shown in Figure 10.8(b). The accuracy of the membership function will depend on the number of α-cut levels.

10.8 General Comments and Approximate Membership Functions

The membership functions of the dependent parameters are triangular, when the membership functions of the independent parameters are triangular. However, if the membership function of an independent parameter is trapezoidal, the membership functions of all dependent parameters, influenced by this independent parameter, will be trapezoidal. Even though the membership functions of the independent parameters have straight-line boundaries, the membership functions of the independent parameters have curvilinear boundaries. In practice, however, the difference between curvilinear boundaries and straight-line boundaries of dependent parameters is small. A large number of simulations are required for fuzzy analysis and computational work increases manifold as the size of the network increases due to the increase in the number of unknowns for which the membership function is to be obtained and also due to the increased number of uncertain parameters to be considered with an increase in the size of network. Some methodologies have been suggested to reduce computational efforts and approximate the membership function.

Xu [25] suggested the use of a linearized model, based on the first-order Taylor series expansion to reduce the computational work for large networks. The linear model determines the gradient at the central values of the fuzzy parameters describing the imprecision in model parameters and extends it to obtain the membership function. A typical membership function of nodal heads with curvilinear boundaries AB and AC is shown in Figure 10.9. The linearized model of Xu [25] will thus approximate the membership function by triangle ADE.

Moosavian and Lence [9] used the monotonic behavior of a continuity equation to obtain an approximate gradient to reduce the computational wok. Their approximate membership function will be as shown by triangle AFG. To reduce the computational effort for large networks, Gupta and Bhave [4] suggested obtaining extreme values of dependent parameters correctly at 0 and 1 α-cuts only. Thus, points A, B, and C are determined correctly. Then AB and AC can be joined by a straight line. This approximation for dependent parameters is better as it fixes extreme values at the base correctly.

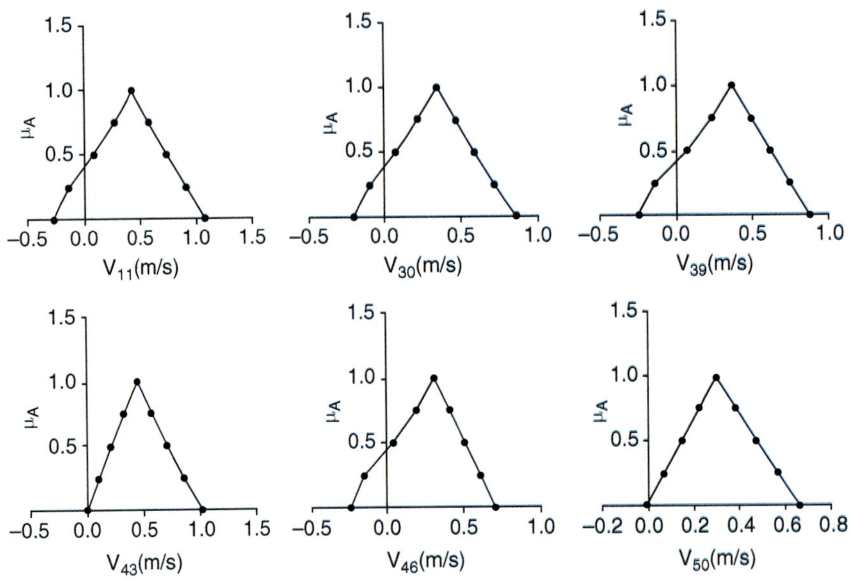

Figure 10.7 Membership functions of pipe velocities by impact table method.

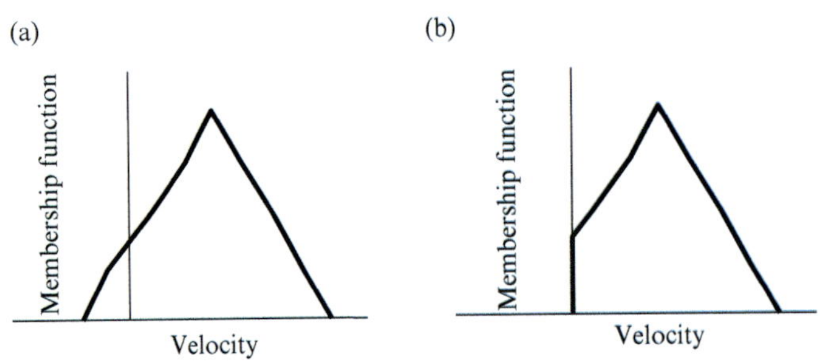

Figure 10.8 Membership functions for pipe velocity: (a) considering monotonous change with independent parameters; (b) considering non-monotonous change with independent parameters.

Dongre and Gupta [23] used a fuzzy analysis approach in a reliability-based optimal design of WDNs with uncertainty in nodal demands and pipe roughness values. Membership of unknown nodal heads was considered by defining desirable and less acceptable nodal heads [26, 27]. The GA methodology was used for pipe size optimization under uncertainty of nodal demands and pipe roughness coefficients to ensure fuzzy approximated nodal heads within the acceptable nodal head membership functioned at each node.

10.9 Summary

Fuzzy analysis involves obtaining the membership functions of dependent parameters considering uncertainty in independent parameters. This is achieved either through optimization-based methods or analysis-based methods. Optimization-based methods are general and can be used in both monotonous and non-monotonous change of dependent parameters with uncertain parameters. However, they are more useful when dependent parameters change non-monotonically with change in independent parameters. They require more computational work.

Analysis-based methods use monotonic relationships between dependent and independent parameters. Due to the monotonic behavior, the appropriate value of the independent parameter is either its maximum value or its minimum value, and this is so in combination with other independent parameters also. Thus, the boundary-value constraint in which the independent parameter can take any value between the two extremes and thus have infinite possibilities is reduced to an alternate constraint in which the independent parameter takes only one of the two extreme values. Complete enumeration of all the combinations is possible for small networks but not practicable for large networks. Methods based on impact tables determine the impact of the change of independent parameters and use the same to obtain extreme values at different α-cuts. Alternatively, a Jacobian matrix can be used to select appropriate extreme values of independent fuzzy parameters. This method can be applied to any type of network. Also, for pipes having flow reversal, the membership functions can be obtained using the method of an impact table by changing the negative values of pipe discharges to zero. Gupta et al. [17] used this method for fuzzy analysis of pressure-deficient networks in outflow at demand nodes for small networks. However, its applicability for large networks with pressure deficiency is to be further studied. Dongre and Gupta [23] used it for optimal pipe sizing under uncertainty of nodal demands and pipe roughness coefficients

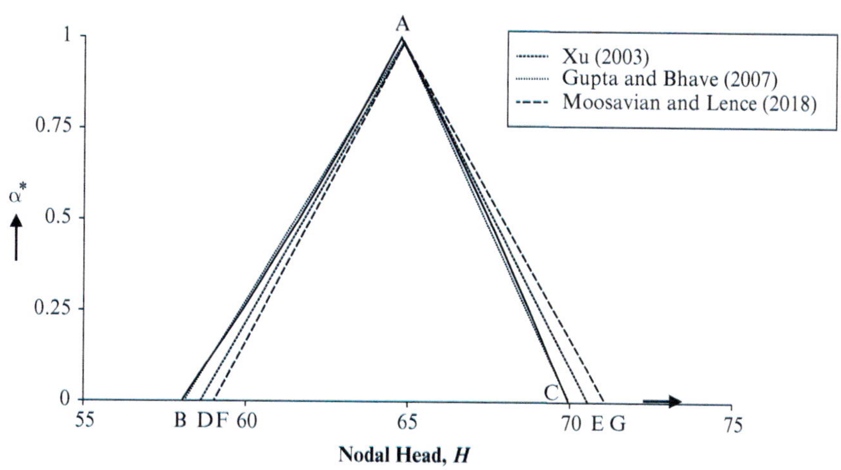

Figure 10.9 Approximate membership functions for nodal head H.

Acknowledgment

The authors greatly acknowledge the assistance provided by Amruta C.V., a past graduate student of VNIT, for analyzing the network of the city of Khorramshahr, Iran.

References

1 Kadu, M.S., Gupta, R., and Bhave, P.R. (2008). Optimal design of water networks using a modified genetic algorithm with reduction in search space. *Journal of Water Resources Planning and Management* 134 (2): 147–160.

2 Savic, D. and Walters, G. (1997). Genetic algorithms for least-cost design of water distribution networks. *Journal of Water Resources Planning and Management* 123 (2): 67–77.

3 Branisavljevic, N. and Ivetic, M. (2006). Fuzzy approach in the uncertainty analysis of the water distribution network of Becej. *Civil Engineering and Environmental Systems* 23 (3): 221–236.

4 Gupta, R. and Bhave, P.R. (2007). Fuzzy parameters in pipe network analysis. *Civil Engineering and Environmental Systems* 24 (1): 33–54.

5 Revelli, R. and Ridolfi, L. (2002). Fuzzy approach for analysis of pipe networks. *Journal of Hydraulic Engineering* 128 (1): 93–101.

6 Haghighi, A. and Asl, A.Z. (2014). Uncertainty analysis of water supply networks using the fuzzy set theory and NSGA-II. *Engineering Applications of Artificial Intelligence* 32: 270–282.

7 Sabzkouhi, A.M. and Haghighi, A. (2016). Uncertainty analysis of pipe-network hydraulics using a many objective particle swarm optimization. *Journal of Hydraulic Engineering* 142 (9): 1–12.

8 Sabzkouhi, A.M. and Haghighi, A. (2018). Closure to "uncertainty analysis of pipe-network hydraulics using a many objective particle swarm optimization." *Journal of Hydraulic Engineering* 144 (4). doi: 10.1061/(ASCE)HY.1943-7900.0001148.

9 Moosavian, N. and Lence, B.J. (2018). Approximation of fuzzy membership functions in water distribution network analysis. *Journal of Hydraulic Engineering* 144 (7). doi: 10.1061/(ASCE)HY.1943-7900.0001483.

10 Moosavian, N. (2020). A gradient method for fuzzy analysis of water distribution networks. *Journal of Hydraulic Research*. doi: 10.1080/00221686.2020.1744748.

11 Wall, M. (1996). *GAlib: A C++* Library of Genetic Algorithm Components, Documentation. https://github.com/pemryan/GAlib accessed 26 September 2022).

12 Rossman, L.A. (2000). *EPANET 2 Users Manual*. Cincinnati, OH: Water Supply and Water Resources Division, National Risk Management Research Laboratory, US Environmental Protection Agency.

13 Sivakumar, P., Prasad, R.K., and Chandramouli, S. (2016). Uncertainty analysis of looped water distribution networks using linked EPANET-GA method. *Journal of Water Resources Planning and Management* 30 (1): 331–358.

14 Shibu, A. and Janga Reddy, M. (2011). Uncertainty analysis of water distribution networks by fuzzy-cross entropy approach. *International Conference on Fuzzy Systems and Neural Computing (ICFSNC 2011)*. Paris, France. 14–16 November, 724–731.

15 Gupta, R. and Sreetu, S. (2020). Fuzzy analysis of water distribution networks using Jaya algorithm. *Journal of Indian Water Works Association* 52 (4): 258–263.

16 Rao, R. (2016). Jaya: a simple and new optimization algorithm for solving constrained and unconstrained optimization problems. *International Journal of Industrial Engineering Computations* 7 (1): 19–34.

17 Gupta, R., Harkutiya, J., Dongre, S., and Ormsbee, L. (2014). Fuzzy analysis of pressure-deficient water distribution networks. In: *EWRI World Environment and Water Resource Congress 2014*, organized by EWRI, Cincinnati, 1–5 June 2014.

18 Karmakar, S. (2011). Propagation of uncertainties in water distribution systems modelling. *Desalination and Water Treatment* 33: 107–117.

19 Sivakumar, P., Prasad, R.K., Chandramouli, S., and Majumdar, S. (2014). Uncertainty analysis of water distribution networks using linked EPANET-Vertex method. *Indian Journal of Innovative Research in Science Engineering and Technology* 3 (12): 17900–17911.

20 Spiliotis, M. and Tsakiris, G. (2012). Water distribution network analysis under fuzzy demands. *Civil Engineering and Environmental Systems* 29 (2): 107–122.

21 Abbasbandy, S. and Asady, B. (2004). Newton's method for solving fuzzy nonlinear equations. *Applied Mathematics and Computation* 159: 349–356.

22 Bhave, P.R. and Gupta, R. (2006). *Analysis of Water Distribution Networks*. New Delhi, India: Narosa Publishing House Pvt. Ltd.

23 Dongre, S.R. and Gupta, R. (2017). Optimal design of water distribution network under hydraulic uncertainties. *Journal of Risk Uncertainty Engineering System, ASCE-ASME* 3 (3): 1–11.

24 Gupta, R. and Dongre, S.R. (2018). Discussion of "Uncertainty analysis of pipe-network hydraulics using a many-objective particle swarm optimization" by Abdell Moradi Sabzkouhi and Ali Haghighi. *Journal of Hydraulic Engineering* 144 (4). doi: 10.1061/(ASCE)HY.1943-7900.0001148.

25 Xu, C. (2003). Discussion of "Fuzzy approach for analysis of pipe networks" by R. Rivelli and L. Ridolfi. *Journal of Hydraulic Engineering* 129 (7): 549–550.

26 Bhave, P.R. and Gupta, R. (2004). Optimal design of water distribution networks for fuzzy demands. *Journal of Civil Engineering and Environmental Systems* 21 (4): 229–245.

27 Xu, C. and Goulter, I.C. (1999). Optimal design of water distribution networks using fuzzy optimization. *Civil Engineering and Environmental Systems* 16 (4): 243–266.

11

Selection of Wastewater Treatment for Small Canadian Communities

An Integrated Fuzzy AHP and Grey Relational Analysis Approach

Guangji Hu, Haroon R. Mian, Manjot Kaur, James Hager, Kasun Hewage, and Rehan Sadiq

School of Engineering, University of British Columbia (Okanagan), Kelowna, BC, Canada

11.1 Introduction

Selecting an appropriate municipal wastewater treatment (WWT) technology has long been a challenge to small communities due to financial constraints, limited access to advanced WWT technologies, and a lack of technical expertise [1]. In Canada, 80% of municipalities are small communities with a population of greater than 5000, and many of them have a pressing need to improve their municipal WWT services to meet increasingly stringent sanitation regulations [2, 3].

To select the optimal WWT technology from a set of alternatives, economic, technological, and social issues need to be considered to ensure a community obtains the maximum benefit from WWT services at a reasonable cost [4–6]. Hence, the selection of WWT technologies can be seen as a multicriteria decision-making (MCDM) problem. MCDM techniques, such as the analytic hierarchy process (AHP), the technique for order of preference by similarity to ideal solution (TOPSIS), grey relational analysis (GRA), and data envelopment analysis (DEA) have been widely used to deal with complex optimal selection problems for different systems based on conflicting criteria [7–9]. These techniques use a structured and logical reasoning approach to model the decision-making process [5]. The above-mentioned MCDM techniques were proven effective in facilitating decision-making on the selection of optimal WWT technology in different countries [9–15].

Although MCDM techniques can identify an optimal selection by comparing alternatives based on selected criteria, their utility is often constrained when quantitative information provided to compare criteria is incomplete or imprecise [10, 13, 16]. In such cases, decision-makers often use qualitative descriptors (e.g. linguistic terms) to evaluate criteria. These terms are inherently reflective of the vagueness and ambiguity of human feeling and recognition and could result in undesirable, nonprobabilistic uncertainty during the decision-making process [15–17].

Nonprobabilistic uncertainty can be addressed using FL-based techniques (e.g. fuzzy set theory, fuzzy inference system, and fuzzy synthetic technique) [18–20]. The combined use of fuzzy-logic-based techniques and MCDM techniques, such as fuzzy AHP (F-AHP) and fuzzy TOPSIS (F-TOPSIS), have gained popularity in solving MCDM problems in the diverse fields of environmental engineering, where criteria information is often incomplete or imprecise [10, 13, 16, 17, 21–23]. In this study, an integrated F-AHP and GRA approach was proposed to facilitate decision-making on the selection of appropriate WWT alternatives for small communities in Canada.

Fuzzy Systems Modeling in Environmental and Health Risk Assessment, First Edition. Edited by Boris Faybishenko, Rehan Sadiq, and Ashok Deshpande.

GRA solves MCDM problems by aggregating incommensurate attributes for each alternative into a single composite value, while the weight of each attribute is subject to the decision-maker's judgment. Thus, how to properly determine attribute weights becomes a challenge in the application of this technique. To overcome this challenge, F-AHP is used to facilitate the selection of attribute weights by incorporating fuzzy logic (FL) for handling epistemic uncertainties in experts' judgment. In the proposed approach, suitable WWT alternatives for small Canadian communities and evaluation criteria were identified first. An integrated F-AHP and GRA approach was then used to rank the alternatives based on the identified criteria for the best selection.

11.2 Methodology

In this section, WWT technologies suitable for small communities were shortlisted as alternatives, and relevant criteria were identified for performance evaluation of the selected technologies. The fundamental concepts of the integrated F-AHP and GRA approach were also introduced.

11.2.1 Wastewater Treatment Alternatives and Criteria

The selection of an appropriate WWT alternative for a community is always a scenario-based problem, defined by the population size of the community and the characteristics of the influent and effluent [24]. As this study aims to identify an optimal WWT alternative for small communities in Canada, a hypothetical small WWT plant with a treatment capacity of 3000 population equivalent (p.e.) was assumed. The hypothetical WWT plant is capable of treating municipal wastewater influent at an average flow rate of 400 m^3 per day. Influent pollutants and their concentrations are standardized according to the reported characteristics of municipal wastewater from small communities [5, 14]. The effluent is assumed to be discharged into a nonsensitive environment without reuse, which is a common effluent disposal practice for small communities.

Seven conventional secondary WWT technologies were evaluated, including constructed wetlands (CW), stabilization ponds (SP), extended aeration lagoons (EALs), membrane bioreactor (MBR), rotating biological contactor (RBC), trickling filter (TF), and sequencing batch reactor (SBR) [5]. Due to the economic, stability, and required expertise considerations, advanced and tertiary WWT technologies were not shortlisted as alternatives. The selected WWT technologies were evaluated based on their performance on technical, economic, environmental, and social criteria. In this study, the goal of WWT alternative selection was defined to identify the most appropriate WWT alternative capable of generating maximum benefits (e.g. high contaminants removal, high operation stability, and low environmental and social impacts) at the lowest cost to the community based on the holistic evaluation of the four criteria. Each criterion consists of several subindices indicating the performance of a WWT alternative with respect to a specific parameter. For instance, organic matter removal is a subindex under the criterion of technical performance.

Based on a comprehensive literature review and extensive expert consultation, Molinos-Senante et al. developed 17 indicators for the assessment of the sustainability of WWT technologies for small communities [5]. These indicators provide an excellent reference to the development of subindices for the criteria evaluated in this study. In addition to these indicators, cold climate adaptability was also considered an important subindex under the technical performance criterion due to the fact that vast areas in Canada have a long and low-temperature winter season. Table 11.1 lists the criteria and subindices used to assess the overall

Table 11.1 Criteria and subindices to assess overall performance.

Criterion	Subindex	Description	Units	CW	SP	EA	MBR	RBC	TF	SBR
C1 (Economic)	EC1	Investment cost	€[a]/p.e.	219	199	239	355	355	347	391
	EC2	O&M costs	€/m^3	0.119	0.179	0.203	0.303	0.173	0.18	0.185
	EC3	Land area required	m^2/p.e.	4.5	4	0.52	0.25	0.4	0.35	0.35
C2 (Technical)	TP1	Organic matter removal	—	0.7	0.75	0.85	0.89	0.8	0.85	0.8
	TP2	Suspended solids removal	—	0.8	0.7	0.9	0.99	0.85	0.7	0.9
	TP3	Nitrogen removal	—	0.5	0.3	0.7	0.7	0.5	0.45	0.65
	TP4	Phosphorus removal	—	0.4	0.55	0.45	0.5	0.2	0.45	0.45
	TP5	Cold climate adaptability	—	0.1	0.1	0.5	0.5	0.9	0.9	0.5
	TP6	Reliability	—	0.5	0.5	0.7	0.5	0.7	0.3	0.7
	TP7	Ease to operation	—	0.7	0.7	0.3	0.1	0.3	0.3	0.5
C3 (Environmental impact)	EI1	Sewage sludge generation	kg/m^3	0.06	0.06	0.5	0.7	0.3	0.3	0.41
	EI2	Energy consumption	kWh/m^3	0.06	0.06	0.5	0.7	0.3	0.3	0.41
	EI3	Water reuse potential	—	1.73	0.85	1.22	0.85	0.85	0.85	0.85
	EI4	Resource recovery potential	—	0.3	0.3	0.5	0.7	0.5	0.5	0.7
C4 (Social impact)	SI1	Odors	—	0.7	0.7	0.3	0.3	0.5	0.5	0.3
	SI2	Noise	—	0.3	0.3	0.7	0.7	0.5	0.5	0.7
	SI3	Visual impact	—	0.1	0.1	0.7	0.7	0.7	0.9	0.5
	SI4	Public acceptance	—	0.7	0.7	0.5	0.5	0.5	0.3	0.5

EC: economic; TP: technical performance; EI: environmental impact; SI: social impact; CW: constructed wetland; SP: stabilization pond; EAL: extended aeration lagoon; MBR: membrane bioreactor; RBC: rotating biological contactor; TF: trickling filter; SBR: sequencing batch reactor; O&M: operation & maintenance; p.e.: population equivalent.

[a] The monetary cost is measured in euros (€). Converting the monetary cost into Canadian dollars would not affect the final ranking results.

performance of different WWT alternatives. Most of the subindex values were reported in the study by Molinos-Senante et al. [5]. For subindices such as cold climate adaptability and visual impact, which do not have quantitative values, linguistic terms such as *excellent*, *good*, *moderate*, *poor*, and *very poor* were used to describe their performance. Subjection grades such as 0.9, 0.7, 0.5, 0.3, and 0.1 were used to represent the linguistic terms, respectively [6].

11.2.2 Fuzzy Analytic Hierarchy Analysis

Fuzzy set theory combined with MCDM methods has been used extensively to manage uncertainty resulting from ambiguities of linguistic variables and vagueness of human judgments. The integration of fuzzy set theory with AHP, known as F-AHP, is one of the most commonly used fuzzy MCDM techniques for system optimization in different engineering fields [7, 17, 21]. F-AHP is a systematic reasoning approach to the alternative selection and justification problem using the concepts of fuzzy set theory and hierarchical structure analysis. Traditional AHP involves the principles of decomposition, pairwise comparisons, and priority vector generation and synthesis. In traditional AHP, decision-makers can specify preferences over the importance of each criterion in a judgment matrix using natural language or numerical values [5]. In F-AHP, pairwise comparisons in the judgment matrix are represented by fuzzy numbers (e.g. fuzzy triangular numbers). F-AHP uses fuzzy arithmetic and fuzzy aggregation operators to process the pairwise comparisons and calculate a sequence of weight vectors, which can then be used to derive the priority weights of criteria. The F-AHP procedure is explained in the following steps.

Step 1 involves developing a hierarchy structure for the studied system to decompose the complex decision problem into smaller, manageable elements of different hierarchical levels. A four-level hierarchical tree is illustrated in Figure 11.1. The top level of the hierarchy corresponds to the objective or goal, the bottom level corresponds to the evaluation alternatives or options, and the intermediate levels refer to criteria and their subindices. The nomenclature $S_{k,n}$ (Figure 11.1) represents the kth subindex of the nth criterion.

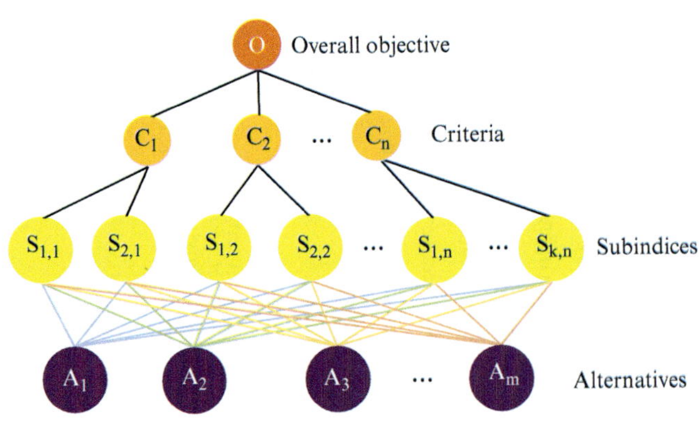

Figure 11.1 The hierarchical structure for F-AHP.

Step 2 entails establishing a fuzzy judgment matrix using pairwise comparisons. The elements of a particular subindex level are compared with other specific elements at the same level. A fuzzy judgment matrix $(\tilde{J})$ is developed using pairwise comparison $(\tilde{j}_{k,l})$ for each level [25]. For example, criteria C1 and C2 can form a pairwise comparison for evaluating their relative importance; the relative importance of subindices $S_{1,1}$ and $S_{2,1}$ can also be compared. A relative importance value of the pairwise comparison is assigned on a scale of 1–9, which can be fuzzified to address the fuzzy uncertainty of perception and meaning (Table 11.2) [25]. In this study, the fuzzification factor θ was set at 1 (e.g. $\tilde{4}$ = (3, 4, 5)). For n comparison items, the fuzzy judgment matrix is

$$\tilde{J} = \begin{vmatrix} \tilde{J}_{1,1} & \tilde{J}_{1,2} & \dots & \tilde{J}_{1,n} \\ \tilde{J}_{2,1} & \ddots & \dots & \tilde{J}_{2,n} \\ \vdots & \dots & \ddots & \vdots \\ \tilde{J}_{n,1} & \tilde{J}_{n,2} & \dots & \tilde{J}_{n,n} \end{vmatrix} \quad (11.1)$$

The importance value of the same criterion (or subindex) always equals 1 (i.e. $(\tilde{J}_{k,l} = 1, \forall k = l)$). The upper triangular entries ($\tilde{J}_{k,l}$) of the matrix need to be defined by the decision-maker, and the lower triangular entries are the inverted values of the upper entries (i.e. $\tilde{J}_{l,k} = 1/\tilde{J}_{k,l}$).

Table 11.2 Fuzzy numbers used for making pairwise comparisons.

Relative importance	Fuzzy number[a]	Definition[b]
$\tilde{1}$	$(1, 1, 1)$	Equal importance
$\tilde{3}$	$(3-\theta, 3, 3+\theta^{c})$	Weak importance
$\tilde{5}$	$(5-\theta, 5, 5+\theta)$	Strong importance
$\tilde{7}$	$(7-\theta, 7, 7+\theta)$	Very strong importance
$\tilde{9}$	$(8, 9, 9)$	Extreme importance
$\tilde{2}, \tilde{4}, \tilde{6}, \tilde{8}$	$(x-\theta, x, x+\theta)$	Intermediate values between two adjacent judgments
$1/\tilde{x}$	$(1/(x+\theta), 1/x, 1/(x-\theta))$	

[a] Fuzzy numbers are presented in minimum, most likely, and maximum values.
[b] The importance definition is in accordance with the description proposed by Saaty [26].
[c] θ: fuzzification factor.

In Step 3, the consistency of the pairwise comparison matrix is checked. The pairwise comparisons in matrix $\tilde{J}$ are often inconsistent in the preference responses of different decision-makers [27]; to avoid this problem, a consistency index (CI) is calculated as a measurement of the relative weight in the presence of inconsistency in responses. The CI is defined as

$$CI = \frac{\lambda_{max} - n}{n-1} \quad (11.2)$$

where λ_{max} is the maximum eigenvalue and n is the dimension of judgment matrix $\tilde{J}$. The final inconsistency in pairwise comparisons can be determined by calculating the consistency ratio CR = CI/RI, where RI is the random index as described in Table 11.3. The resulting CR should be less than 0.1 to ensure adequate consistency of pairwise comparisons.

The last step is to calculate the fuzzy weights. In this study, the geometric mean was adopted to estimate the weights. Fuzzy arithmetic operations were utilized over the matrix $\tilde{J}$ to calculate the fuzzy weights. The geometric mean was computed for each row

$$\tilde{J}_k = (\tilde{J}_{k1} \otimes \dots \otimes \tilde{J}_{kn})^{1/n} \quad (11.3)$$

$$\tilde{w}_k = \tilde{J}_k \otimes (\tilde{J}_1 \oplus \dots \oplus \tilde{J}_n)^{-1} \quad (11.4)$$

Table 11.3 Random index used to compute consistency ratio (CR).

n	1	2	3	4	5	6	7	8	9	10
RI[a]	0	0	0.52	0.89	1.11	1.25	1.35	1.40	1.45	1.49

[a] Random index.

where $\tilde{w}_k$ $(k=1,2,\ldots,n)$ are the fuzzy weights, and the sum of the most likely weights equals 1 (i.e. $\sum_k \tilde{w}_k = 1$).

11.2.3 Grey Relational Analysis

GRA was derived from the grey system theory established by Deng et al. in 1987 [28]. It measures the degree of similarity or difference between two sequences based on the grade of relation. GRA has the merit of point-set topology, which is capable of making a global comparison between two sets of data instead of local comparison between two data points [6].

The main procedure of GRA is converting the performance of all alternatives into a comparable sequence. This step is also known as grey relational generating. According to the values of criteria in all comparable sequences, a preference sequence is defined as the ideal option. After this, grey relational coefficients between all comparable sequences and the preference sequence are calculated. Based on the grey relational coefficients, the grey relational degree between the preference sequence and all comparable sequences is obtained. The most desirable alternative should have the highest grey relation degree as compared to the preference sequence.

In GRA, a decision matrix D with m alternatives and n criteria is defined as

$$D = \begin{vmatrix} x_{1,1} & \cdots & x_{1,j} & \cdots & x_{1,n} \\ \vdots & \cdots & \vdots & \cdots & \vdots \\ x_{i,1} & \cdots & x_{i,j} & \cdots & x_{i,n} \\ \vdots & \cdots & \vdots & \cdots & \vdots \\ x_{m,1} & \cdots & x_{m,j} & \cdots & x_{m,n} \end{vmatrix} \tag{11.5}$$

The decision matrix D is normalized to D' shown as

$$D' = \begin{vmatrix} x'_{1,1} & \cdots & x'_{1,j} & \cdots & x'_{1,n} \\ \vdots & \cdots & \vdots & \cdots & \vdots \\ x'_{i,1} & \cdots & x'_{i,j} & \cdots & x'_{i,n} \\ \vdots & \cdots & \vdots & \cdots & \vdots \\ x'_{m,1} & \cdots & x'_{m,j} & \cdots & x'_{m,n} \end{vmatrix} \tag{11.6}$$

Two equations are used to normalize criteria. For beneficial criteria, a higher value indicates better performance of the alternative and vice versa. Conversely, a lower value of the nonbeneficial (cost) criteria is preferable for the alternative. The beneficial and nonbeneficial criteria are normalized using Eqs. (11.7) and (11.8), respectively.

$$x'_{i,j} = \frac{x_{i,j}}{\max\{x_{i,j} \mid i=1,2,\ldots,m\}} \tag{11.7}$$

$$x'_{i,j} = \frac{\min\{x_{i,j} \mid i = 1, 2, \ldots, m\}}{x_{i,j}} \tag{11.8}$$

After normalization, the larger the value of $x'_{i,j}$, the better the performance is. The optimal normalized $x'_{0,j}$ should have a value of 1. Thus, a preference sequence $X_0 = \left\{x'_{0,j}; j = 1, 2, 3, \ldots, n\right\}$ can be identified as (1, 1, 1, ..., 1). The grey relational coefficient $\xi_{0i}(j)$ between the compared sequence $X_i = \left\{x'_{i,j} \mid j = 1, 2, 3, \ldots, n\right\}$ and the preference sequence X_0, for the jth criterion can be calculated as

$$\xi_{0i}(j) = \frac{\Delta_{min} + \zeta\Delta_{max}}{\Delta_{0i}(j) + \zeta\Delta_{max}}; i = 1, 2, \ldots, m; j = 1, 2, \ldots, n \tag{11.9}$$

where $\Delta_{0i}(j) = \mid x'_{0,j} - x'_{i,j} \mid$ is called the *relative grey relational space* denoting the absolute difference between the preference sequence and the compared sequence; $\Delta_{min} = min_i \{min_j \mid x'_{0,j} - x'_{i,j} \mid; j = 1, 2, ..., n\}; i = 1, 2, ..., m\}$ and $\Delta_{max} = max_i \{max_j \mid x'_{0,j} - x'_{i,j} \mid; j = 1, 2, ..., n\}; i = 1, 2, ..., m\}$. The ζ is the distinguishing coefficient, which has a value range of $0 \leq \zeta \leq 1$ and is normally set at 0.5. The grey relation grade ε is calculated as the sum of grey relational coefficients $\xi_{0i}(j)$

$$\varepsilon = \sum_{j=1}^{n} \xi_{0i}(j) \tag{11.10}$$

To consider the different weights of criteria, Eq. (11.10) is modified to

$$\varepsilon = w_j \sum_{j=1}^{n} \xi_{0i}(j) \tag{11.11}$$

In this study, w_j is the fuzzy weight of the jth criterion derived from the F-AHP. Thus, the resulting ε is also a fuzzy number, which needs to be defuzzified for ranking the alternatives. Various techniques can be used to defuzzify the fuzzy number, such as Yager's centroid method [28], Chen's ranking method [29], and the ordered-weighted average method [30]. In this study, Yager's centroid method and Chen's ranking method were both used to defuzzify the final fuzzy number, and the defuzzified results were compared. Chen's ranking method is performed by calculating the total utility $U_T(x)$ to rank the alternatives. For m alternatives, there will be m fuzzy numbers A_i, $i = 1, 2, ..., m$. For a triangular fuzzy number (a, b, c), the ranking is carried out by maximizing $c_{max} = \max(c_1, c_2, ..., c_m)$ and minimizing $a_{min} = \min(a_1, a_2, ..., a_m)$ a set of alternatives. For a triangular fuzzy number (a_1, b_1, c_1), the $U_T(x)$ is calculated as

$$U_T(x) = \frac{1}{2}\left[\frac{(c_1 - a_{min})}{(c_{max} - a_{min}) - (b_1 - c_1)} + 1 - \frac{(c_{max} - a_1)}{(c_{max} - a_{min}) - (b_1 - a_1)}\right], x = 1, \ldots, m \tag{11.12}$$

Finally, the larger the defuzzified grey relational grade, the closer the relationship between the reference sequence and the given optional alternative. Thus, the largest grey relational grade is desired as the optimal result.

11.3 Results and Discussion

The fuzzy weights of criteria and subindices were calculated using F-AHP, and the results are presented in Section 11.3.1. The fuzzy weights were then incorporated into the GRA to determine the grey relational grade for ranking the alternatives (Section 11.3.2).

11.3.1 Fuzzy Weights of Criteria and Subindices

Since the optimal selection problem was constructed as a hierarchy, the relations between criteria on succeeding subindices are obtained by making pairwise comparisons. The weights representing the relative importance of each criterion and subindex can be calculated using the eigenvalue technique. To calculate the weights of criteria and subindices, decision-makers need to base their judgments on the relative importance (i.e. importance value) of the contribution of each criterion (or subindex) toward achieving the overall goal. In this study, the importance value of each pairwise comparison was assigned based on the authors' knowledge, and the values can be adjusted to accommodate different expert judgments. The pairwise comparison matrix can be generated for the criteria using Eq. (11.1) as

$$\tilde{J}_C = \begin{matrix} \\ C1 \\ C2 \\ C3 \\ C4 \end{matrix} \begin{matrix} C1 & C2 & C3 & C4 \\ \left| \begin{matrix} \tilde{1} & \tilde{2} & \tilde{4} & \tilde{5} \\ 1/\tilde{2} & \tilde{1} & \tilde{4} & \tilde{4} \\ 1/\tilde{4} & 1/\tilde{4} & \tilde{1} & \tilde{2} \\ 1/\tilde{5} & 1/\tilde{4} & 1/\tilde{2} & \tilde{1} \end{matrix} \right| \end{matrix} \tag{11.13}$$

The CR was calculated as 0.032 (< 0.1), indicating good consistency between the pairwise comparisons. The fuzzy weights of criteria can be calculated using Eqs. (11.2) to (11.4)

$$J_{C1} = (\tilde{1} \otimes \tilde{2} \otimes \tilde{4} \otimes \tilde{5})^{1/4} = [(1,1,1) \otimes (1,2,3) \otimes (3,4,5) \otimes (4,5,6)]^{1/4} \\ = (1.87, 2.52, 3.09) \tag{11.14}$$

$$\tilde{J}_{C2} = [(1/\tilde{2}) \otimes \tilde{1} \otimes \tilde{4} \otimes \tilde{4}]^{1/4} = [(1/3, 1/2, 1/1) \otimes (1,1,1) \otimes (3,4,5) \otimes (3,4,5)]^{1/4} \\ = (1.32, 1.69, 2.24) \tag{11.15}$$

$$\tilde{J}_{C3} = [(1/\tilde{4}) \otimes (1/\tilde{4}) \otimes \tilde{1} \otimes \tilde{2}]^{1/4} \\ = [(1/5, 1/4, 1/3) \otimes (1/5, 1/4, 1/3) \otimes (1,1,1) \otimes (1,2,3)]^{1/4} \\ = (0.45, 0.60, 0.76) \tag{11.16}$$

$$\tilde{J}_{C4} = [(1/\tilde{5}) \otimes (1/\tilde{4}) \otimes (1/\tilde{2}) \otimes \tilde{1}]^{1/4} \\ = [(1/6, 1/5, 1/4) \otimes (1/5, 1/4, 1/3) \otimes (1/3, 1/2, 1/1) \otimes (1,1,1)]^{1/4} \\ = (0.33, 0.40, 0.54) \tag{11.17}$$

$$\tilde{w}_{C1} = \tilde{J}_{C1} \otimes (\tilde{J}_{C1} \oplus \tilde{J}_{C2} \oplus \tilde{J}_{C3} \oplus \tilde{J}_{C4})^{-1} = (0.28, 0.48, 0.78) \tag{11.18}$$

Similarly, $\tilde{w}_{C2}$, $\tilde{w}_{C3}$, and $\tilde{w}_{C4}$ were calculated as (0.20, 0.32, 0.56), (0.07, 0.12, 0.19), and (0.05, 0.08, 0.14), respectively. The same F-AHP algorithm was applied to calculate the fuzzy weights of subindices under each criterion. The pairwise comparison matrices of subindices under the four criteria were shown as

$$\tilde{J}_{EC} = \begin{matrix} & EC_1 & EC_2 & EC_3 \\ EC_1 & \tilde{1} & \tilde{3} & \tilde{5} \\ EC_2 & 1/\tilde{3} & \tilde{1} & \tilde{3} \\ EC_3 & 1/\tilde{5} & 1/\tilde{3} & \tilde{1} \end{matrix} \tag{11.19}$$

$$\tilde{J}_{TP} = \begin{matrix} & TP_1 & TP_2 & TP_3 & TP_4 & TP_5 & TP_6 & TP_7 \\ TP_1 & \tilde{1} & \tilde{1} & \tilde{3} & \tilde{3} & \tilde{1} & 1/\tilde{5} & 1/\tilde{2} \\ TP_2 & \tilde{1} & \tilde{1} & \tilde{3} & \tilde{7} & 1/\tilde{2} & 1/\tilde{7} & 1/\tilde{5} \\ TP_3 & 1/\tilde{3} & 1/\tilde{3} & \tilde{1} & \tilde{1} & 1/\tilde{5} & 1/\tilde{5} & 1/\tilde{5} \\ TP_4 & 1/\tilde{3} & 1/\tilde{7} & \tilde{1} & \tilde{1} & 1/\tilde{7} & 1/\tilde{7} & 1/\tilde{5} \\ TP_5 & \tilde{1} & \tilde{2} & \tilde{5} & \tilde{7} & \tilde{1} & 1/\tilde{3} & 1/\tilde{3} \\ TP_6 & \tilde{5} & \tilde{7} & \tilde{5} & \tilde{7} & \tilde{3} & \tilde{1} & \tilde{1} \\ TP_7 & \tilde{2} & \tilde{5} & \tilde{5} & \tilde{5} & \tilde{3} & \tilde{1} & \tilde{1} \end{matrix} \tag{11.20}$$

$$\tilde{J}_{EI} = \begin{matrix} & EI_1 & EI_2 & EI_3 & EI_4 \\ EI_1 & \tilde{1} & 1/\tilde{4} & 1/\tilde{5} & \tilde{1} \\ EI_2 & \tilde{4} & \tilde{1} & 1/\tilde{2} & \tilde{3} \\ EI_3 & \tilde{5} & \tilde{2} & \tilde{1} & \tilde{5} \\ EI_4 & \tilde{1} & 1/\tilde{3} & 1/\tilde{5} & \tilde{1} \end{matrix} \tag{11.21}$$

$$\tilde{J}_{SI} = \begin{matrix} & SI_1 & SI_2 & SI_3 & SI_4 \\ SI_1 & \tilde{1} & \tilde{1} & 1/\tilde{3} & 1/\tilde{7} \\ SI_2 & \tilde{1} & \tilde{1} & 1/\tilde{3} & 1/\tilde{7} \\ SI_3 & \tilde{3} & \tilde{3} & \tilde{1} & 1/\tilde{5} \\ SI_4 & \tilde{7} & \tilde{7} & \tilde{5} & \tilde{1} \end{matrix} \tag{11.22}$$

The CRs were computed as 0.042, 0.065, 0.018, and 0.027 for the pairwise comparison matri ces $\tilde{J}_{EC}$, $\tilde{J}_{TP}$, $\tilde{J}_{EI}$, and $\tilde{J}_{SI}$, respectively. All of the CRs indicate acceptable consistency of the pairwise comparisons of subindices. The fuzzy weights of subindices were calculated accordingly, and the results are shown in Table 11.4.

Table 11.4 Fuzzy weights of subindices under each criterion.

Criteria	Subindices	Fuzzy weights
C_1	EC_1	(0.431, 0.637, 0.919)
	EC_2	(0.171, 0.258, 0.401)
	EC_3	(0.075, 0.105, 0.159)
C_2	TP_1	(0.070, 0.102, 0.155)
	TP_2	(0.061, 0.087, 0.132)
	TP_3	(0.027, 0.038, 0.057)
	TP_4	(0.023, 0.031, 0.044)
	TP_5	(0.092, 0.139, 0.212)
	TP_6	(0.238, 0.335, 0.468)
	TP_7	(0.174, 0.267, 0.390)
C_3	EI_1	(0.195, 0.204, 0.217)
	EI_2	(0.249, 0.279, 0.311)
	EI_3	(0.279,0.310, 0.342)
	EI_4	(0.196, 0.206, 0.220)
C_4	SI_1	(0.191, 0.200, 0.212)
	SI_2	(0.191, 0.200, 0.212)
	SI_3	(0.232, 0.262, 0.290)
	SI_4	(0.313, 0.338, 0.364)

C: criterion; EC: economic; TP: technical performance; EI: environmental impact; SI: social impact.

11.3.2 Grey Relational Analysis Using Fuzzy Weights

After obtaining the fuzzy weights of the criteria and subindices, the GRA was performed by converting all the alternatives into the comparable sequence. Table 11.5 shows the normalized values of subindices using Eqs. (11.7) and (11.8).

The normalization values at the subindex level were used to calculate the grey relational space. The relative grey relational space $\Delta_{0i}(j)$ between each normalized value $x'_{i,j}$ and the preference sequence $x'_{0,j}$ was computed, and the resulting minimum ($\Delta_{\min}$) and maximum ($\Delta_{\max}$) values of $\Delta_{0i}(j)$ were 0 and 0.944, respectively. Based on the $\Delta_{0i}(j)$ values, the grey relation coefficient of each subindex can be calculated using Eq. (11.9). Table 11.6 shows the calculated grey relation coefficients.

The defuzzified grey relation grades of criteria given in Table 11.7 were further aggregated to calculate the grey relation grades of WWT alternatives using the same approach. As shown in Table 11.8, the computed grey relation grades of alternatives are also fuzzy numbers, which need to be defuzzified for ranking. The centroid method and Chen's ranking method were both used for the final defuzzification. Based on the defuzzified value, the grey relation grades of alternatives were ranked from the largest to the lowest. The largest grey relation grade represents the best WWT alternative for the community. The ranking results are shown in Table 11.9. The ranking results from the two defuzzification approaches were compared with that from the calculation using the most likely value, which represents the combination of the nonfuzzy-based AHP and GRA approaches.

Table 11.5 Normalized values of subindices of WWT alternatives.

	CW	SP	EA	MBR	RBC	TF	SBR
EC1	0.908	1.000	0.832	0.560	0.560	0.573	0.508
EC2	1.000	0.665	0.586	0.393	0.688	0.661	0.643
EC3	0.056	0.062	0.481	1.000	0.625	0.714	0.714
TP1	0.787	0.843	0.955	1.000	0.899	0.955	0.899
TP2	0.808	0.707	0.909	1.000	0.859	0.707	0.909
TP3	0.714	0.429	1.000	1.000	0.714	0.643	0.929
TP4	0.727	1.000	0.818	0.909	0.364	0.818	0.818
TP5	0.111	0.111	0.556	0.556	1.000	1.000	0.556
TP6	0.714	0.714	1.000	0.714	1.000	0.429	1.000
TP7	1.000	1.000	0.429	0.143	0.429	0.429	0.714
EI1	1.000	1.000	0.120	0.086	0.200	0.200	0.146
EI2	1.000	1.000	0.120	0.086	0.200	0.200	0.146
EI3	1.000	0.491	0.705	0.491	0.491	0.491	0.491
EI4	0.429	0.429	0.714	1.000	0.714	0.714	1.000
SI1	0.429	0.429	1.000	1.000	0.600	0.600	1.000
SI2	1.000	1.000	0.429	0.429	0.600	0.600	0.429
SI3	1.000	1.000	0.143	0.143	0.143	0.111	0.200
SI4	1.000	1.000	0.714	0.714	0.714	0.429	0.714

EC: economic; TP: technical performance; EI: environmental impact; SI: social impact; CW: constructed wetland; SP: stabilization pond; EAL: extended aeration lagoon; MBR: membrane bioreactor; RBC: rotating biological contactor; TF: trickling filter; SBR: sequencing batch reactor.

Table 11.6 Grey relation coefficients of subindices.

	CW	SP	EA	MBR	RBC	TF	SBR
EC_1	0.838	1.000	0.738	0.518	0.518	0.525	0.490
EC_2	1.000	0.585	0.533	0.437	0.602	0.582	0.570
EC_3	0.333	0.335	0.476	1.000	0.557	0.623	0.623
TP_1	0.689	0.750	0.913	1.000	0.824	0.913	0.824
TP_2	0.711	0.617	0.839	1.000	0.770	0.617	0.839
TP_3	0.623	0.452	1.000	1.000	0.623	0.569	0.869
TP_4	0.634	1.000	0.722	0.839	0.426	0.722	0.722
TP_5	0.347	0.347	0.515	0.515	1.000	1.000	0.515
TP_6	0.623	0.623	1.000	0.623	1.000	0.452	1.000
TP_7	1.000	1.000	0.452	0.355	0.452	0.452	0.623
EI_1	1.000	1.000	0.349	0.341	0.371	0.371	0.356
EI_2	1.000	1.000	0.349	0.341	0.371	0.371	0.356
EI_3	1.000	0.481	0.616	0.481	0.481	0.481	0.481
EI_4	0.452	0.452	0.623	1.000	0.623	0.623	1.000
SI_1	0.452	0.452	1.000	1.000	0.541	0.541	1.000
SI_2	1.000	1.000	0.452	0.452	0.541	0.541	0.452
SI_3	1.000	1.000	0.355	0.355	0.355	0.347	0.371
SI_4	1.000	1.000	0.623	0.623	0.623	0.452	0.623

EC: economic; TP: technical performance; EI: environmental impact; SI: social impact; CW: constructed wetland; SP: stabilization pond; EAL: extended aeration lagoon; MBR: membrane bioreactor; RBC: rotating biological contactor; TF: trickling filter; SBR: sequencing batch reactor.

Table 11.7 Defuzzified grey relation grades of the criteria.

	CW	SP	EA	MBR	RBC	TF	SBR
C_1	0.869	0.862	0.690	0.577	0.572	0.579	0.553
C_2	0.732	0.735	0.791	0.662	0.821	0.634	0.823
C_3	0.889	0.728	0.490	0.522	0.459	0.459	0.529
C_4	0.892	0.892	0.596	0.596	0.521	0.462	0.600

C: criterion; CW: constructed wetland; SP: stabilization pond; EAL: extended aeration lagoon; MBR: membrane bioreactor; RBC: rotating biological contactor; TF: trickling filter; SBR: sequencing batch reactor.

Table 11.8 Fuzzy grey relation grades of WWT alternatives.

Alternative	Fuzzy grey relation grade
CW	(0.536, 0.898, 1.493)
SP	(0.499, 0.835, 1.389)
EA	(0.369, 0.613, 1.031)
MBR	(0.274, 0.457, 0.766)
RBC	(0.355, 0.589, 0.996)
TF	(0.260, 0.434, 0.726)
SBR	(0.358, 0.593, 1.005)

CW: constructed wetland; SP: stabilization pond; EAL: extended aeration lagoon; MBR: membrane bioreactor; RBC: rotating biological contactor; TF: trickling filter; SBR: sequencing batch reactor.

As indicated in Table 11.9, both the centroid method and Chen's ranking method generated the same ranking results of WWT alternatives. Based on the ranking results from the integrated F-AHP and GRA approach, CW is the best WWT selection, followed by SP, EA, SBR, RBC, MBR, and TF. However, the results generated based on the most likely values, which represent the nonfuzzy AHP, indicate a different ranking list. CW and SP retain the

Table 11.9 Final grey relation grades of WWT alternatives and ranking.

Fuzzy centroid		Most likely value		Fuzzy Chen's ranking	
CW	0.976	CW	0.898	CW	0.537
SP	0.908	SP	0.835	SP	0.499
EA	0.671	RBC	0.613	EA	0.353
SBR	0.652	TF	0.593	SBR	0.340
RBC	0.647	SBR	0.589	RBC	0.336
MBR	0.499	EA	0.457	MBR	0.233
TF	0.473	MBR	0.434	TF	0.215

CW: constructed wetland; SP: stabilization pond; EAL: extended aeration lagoon; SBR: sequencing batch reactor; RBC: rotating biological contactor; TF: trickling filter; MBR: membrane bioreactor.

same top positions in the rank, but the remaining alternatives are ranked differently. The difference in rankings between the fuzzy-based and nonfuzzy-based approaches suggests that uncertainty resulting from human thought and recognition can have an influence on the decision-making process. Incorporating FL into traditional AHP can minimize uncertainty and generate more realistic results for informed decision-making.

Both CW and SP scored relatively high grey relation grades compared to others. The top four WWT alternatives have a common characteristic: they are all suspended-growth WWT technologies in which the microorganisms responsible for treating wastewater are suspended in the water. RBC, MBR, and TF are attached-growth WWT technologies in which microorganisms are attached to the media in the reactor, such as rotary discs, membranes, and filter media. Suspended growth WWT technologies are more commonly used than attached growth WWT technologies across the globe [31]. Although the attached growth WWT technologies have the advantages of low land area requirements and high contaminant removal efficiency, they require more complex pretreatment processes than suspended growth WWT technologies, significantly increasing overall treatment complexity and cost [31]. Both CW and SP are considered well-adapted WWT technologies for small communities due to their simplicity of operations, low maintenance costs, robustness, low impact, and excellent cold climate adaptability, as compared to the two suspended-growth WWT technologies (i.e. SBR and EA) [32]. The ranking generated by the integrated F-AHP and GRA approach provided useful information for the selection of WWT technologies for small Canadian communities. The developed approach also represents another application of grey theory in solving MCDM problems [33].

11.4 Conclusions

The selection of the most appropriate WWT alternative is a complicated, multiple-objective decision-making process involving uncertainty. This uncertainty primarily originates from the ambiguous thoughts of decision-makers concerning the weighting of evaluation criteria and subindices. An integrated F-AHP and GRA approach was proposed to compare WWT alternatives regarding their performance for selected criteria. F-AHP was applied to address uncertainty resulting from the ambiguity and vagueness of human thoughts, and GRA was introduced to aggregate the performance of WWT alternatives on individual subindices into a final grade for ranking. It was found that CW, SP, and EA are the top three appropriate WWT technologies for small communities in Canada. The ranking of WWT alternatives by the fuzzy-based approach differs from the ranking results generated by the nonfuzzy approach, suggesting that fuzzy uncertainty could affect the decision-making of alternative selection. The fuzzy-based approach can generate a more realistic ranking than the nonfuzzy approach. The proposed approach can also be applied to solve MCDM problems in other environmental engineering fields. This study is also associated with several limitations, including (i) the case study was hypothetical, (ii) the data used for alternatives evaluation were sourced from a European study, and (iii) the sensitivity of the proposed approach was not assessed. Future studies can be conducted by using localized real-world WWT data and performing a sensitivity analysis.

References

1 USEPA. (1992). *Wastewater treatment/disposal for small communities*. Cincinnati, OH.

2 Brubaker, E. (2007). Water and wastewater treatment in Canada tapping into private-sector capital, expertise, and efficiencies. In: *A Breath of Fresh Air: Market Solutions for Improving*

Canada's Environment (ed. N. Schneider), 48–61. Vancouver, British Columbia, Canada: The Fraser Institute.

3 Sancton, A. and Yound, R. (2009). *Foundations of Governance: Municipal Government in Canada's Provinces*. Toronto, Ontario, Canada: University of Toronto Press for the Institute of Public Administration of Canada.

4 Kalbar, P.P., Karmakar, S., and Asolekar, S.R. (2012). Selection of an appropriate wastewater treatment technology: a scenario-based multiple-attribute decision-making approach. *Journal of Environmental Manage* 113: 158–169.

5 Molinos-Senante, M., Gómez, T., Garrido-Baserba, M., Caballero, R., and Sala-Garrido, R. (2014). Assessing the sustainability of small wastewater treatment systems: a composite indicator approach. *Science of the Total Environment* 497–498C: 607–617.

6 Zeng, G., Jiang, R., Huang, G., Xu, M., and Li, J. (2007). Optimization of wastewater treatment alternative selection by hierarchy grey relational analysis. *Journal of Environmental Management* 82 (2): 250–259.

7 Karimi, A.R., Mehrdadi, N., Hashemian, S.J., and Nabi-bidhendi, G.R. (2011). Using of the fuzzy TOPSIS and fuzzy AHP methods for wastewater treatment process selection. *International Journal of Academic Research* 3 (1): 737–745.

8 Pophali, G.R., Chelani, A.B., and Dhodapkar, R.S. (2011). Optimal selection of full scale tannery effluent treatment alternative using integrated AHP and GRA approach. *Expert Systems with Applications* 38 (9): 10889–10895.

9 Sala-Garrido, R., Molinos-Senante, M., and Hernández-Sancho, F. (2011). Comparing the efficiency of wastewater treatment technologies through a DEA metafrontier model. *Chemical Engineering Journal* 173 (3): 766–772.

10 Christian, R.A., Lad, R.K., Deshpande, A.W., Desai, N.G. (2008). Fuzzy MCDM approach for addressing composite index of water and air pollution potential of industries. *International Journal of Digital Content Technology and Its Application* 2 (2): 64–71.

11 Egilmez, G., Gumus, S., Kucukvar, M., and Tatari, O. (2016). A fuzzy data envelopment analysis framework for dealing with uncertainty impacts of input-output life cycle assessment models on eco-efficiency assessment. *Journal of Cleaner Production* 129: 622–636.

12 Kalbar, P.P., Karmakar, S., and Asolekar, S.R. (2012). Technology assessment for wastewater treatment using multiple-attribute decision-making. *Technology and Society* 34 (4): 295–302.

13 Karimi, A.R., Mehrdadi, N., Hashemian, S.J., Nabi Bidhendi, G.R., and Moghaddam, R.T. (2011). Selection of wastewater treatment process based on the analytical hierarchy process and fuzzy analytical hierarchy process methods. *International Journal of Environmental Science and Technology* 8 (2): 267–280.

14 Molinos-Senante, M., Gómez, T., Caballero, R., Hernández-Sancho, F., and Sala-Garrido, R. (2015). Assessment of wastewater treatment alternatives for small communities: an analytic network process approach. *Science of the Total Environment* 532: 676–687.

15 Zyoud, S.H., Kaufmann, L.G., Shaheen, H., Samhan, S., and Fuchs-Hanusch, D. (2016). A framework for water loss management in developing countries under fuzzy environment: integration of fuzzy AHP with fuzzy TOPSIS. *Expert System with Applications* 61: 86–105.

16 Kamble, S.J., Singh, A., and Kharat, M.G. (2017). A hybrid life cycle assessment based fuzzy multi-criteria decision making approach for evaluation and selection of an appropriate municipal wastewater treatment technology. *Euro Mediterranean Journal Environmental Integration* 2 (1): 9.

17 Taylan, O., Bafail, A.O., Abdulaal, R.M.S., and Kabli, M.R. (2014). Construction projects selection and risk assessment by fuzzy AHP and fuzzy TOPSIS methodologies. *Applied Soft Computing Journal* 17: 105–116.

18 Sadiq, R. and Tesfamariam, S. (2009). Environmental decision-making under uncertainty using intuitionistic fuzzy analytic hierarchy process (IF-AHP). *Stochastic Environmental Research and Risk Assessment* 23 (1): 75–91.

19 Li, J., Liu, L., Huang, G., and Zeng, G. (2006). A fuzzy-set approach for addressing uncertainties in risk assessment of hydrocarbon-contaminated site. *Water Air and Soil Pollution* 171 (1–4): 5–18.

20 Sadiq, R. and Rodriguez, M.J. (2004). Fuzzy synthetic evaluation of disinfection by-products: a risk-based indexing system. *Journal of Environmental Management* 73 (1): 1–13.

21 Lima, F., Jr., Osiro, L., and Carpinetti, L. (2014). A comparison between fuzzy AHP and fuzzy TOPSIS methods to supplier selection. *Applied Soft Computing Journal* 21: 194–209.

22 Anagnostopoulos, K.P., Gratziou, M., and Vavatsikos, A.P. (2007). Using the fuzzy analytic hierarchy process for selecting wastewater facilities at prefecture level. *Euro Water* 19/20: 15–24.

23 Ouyang, X., Guo, F., Shan, D., Yu, H., and Wang, J. (2015). Development of the integrated fuzzy analytical hierarchy process with multidimensional scaling in selection of natural wastewater treatment alternatives. *Ecological Engineering* 74: 438–447.

24 Bottero, M., Comino, E., and Riggio, V. (2011). Application of the analytic hierarchy process and the analytic network process for the assessment of different wastewater treatment systems. *Environmental Modelling and Software* 26 (10): 1211–1224.

25 Tesfamariam, S. and Sadiq, R. (2006). Risk-based environmental decision-making using fuzzy analytic hierarchy process (F-AHP). *Stochastic Environmental Research and Risk Assessment* 21 (1): 35–50.

26 Saaty, T.L. (1977). A scaling method for priorities in hierarchical structure. *Journal of Mathematical Psychology* 15: 234–281.

27 Zahedi, F. (1986). Group consensus function estimation when preferences uncertain. *Operational Research* 34 (6): 883–894.

28 Deng, J.L. (1987). *Essential Topics on Grey System: Theory and Applications*. Wuhan, China: Huazhong University of Science and Technology Press.

29 Yager, R.R. (1980). On a general class of fuzzy connectives. *Fuzzy Sets and Systems* 4: 235–242.

30 Chen, S.H. (1985). Ranking fuzzy numbers with maximizing set and minimizing set. *Fuzzy Sets and Systems* 17: 113–129.

31 Sadiq, R., Rodriguez, M.J., and Tesfamariam, S. (2010). Integrating indicators for performance assessment of small water utilities using ordered weighted averaging (OWA) operators. *Expert System with Applications* 37 (7): 4881–4891.

32 Lienard, A., Boutin, C., Molle, P., Racault, Y., Brissaud, F., and Picot, B. (2004). Constructed wetlands and waste stabilization ponds for municipal wastewater treatment in France: comparison of performance and maintenance operations in terms of durability and reliability. *P. 11 in 9ème Conférence Sur Les Marais Artificiels et 6ème Conférence Sur Le Lagunage Naturel*, Avignon, France.

33 Bakhtavar, E. and Lotfian, R. (2017). Applying an integrated fuzzy gray MCDM approach: a case study on mineral processing plant site selection. *International Journal of Mining and Geo-Engineering* 51 (2): 177–183.

12

Fuzzy Logic Applications for Water Pipeline Risk Analysis

Anmol Vishwakarma and Sunil K. Sinha

Virginia Tech, Blacksburg, VA, United States of America

12.1 Introduction

Data are crucial to any analysis, but providing valuable information and knowledge from data requires an understanding of the system being studied to determine the framework and potential value in the analysis. This is also reinforced by the current data science frameworks that propose hybrid modeling as the new approach to scientific discovery. Every analysis is based on knowledge developed through an extensive review of literature and practice, which is the foundation of all data analyses in this report. Over the past 100 years, many technological advancements in engineering models, manufacturing, pipeline appurtenances, installation techniques, and improvement of national and international standards have affected the performance of water pipelines. Before performing any analysis, including data analysis to understand historical changes in break rates or installation trends, these technological advancements should be considered to identify any drastic changes in the trends for pipeline performance. The research team is engaging experts in water utilities, consulting engineers, and pipeline manufacturing associations, and is leveraging the Sustainable Water Infrastructure Management Center (SWIM) knowledgebase to build a strong foundation for data analyses for this study.

12.1.1 Water Transmission and Distribution Pipeline Infrastructure System Elements

The purpose of water pipeline systems is to supply water to customers and provide fire protection to communities with minimum disruptions and water losses. Three criteria guide this:

- **Structural reliability**: To maintain structural strength throughout the lifecycle with minimum costs.
- **Functional reliability**: To deliver water at sufficient pressures to customers with adequate hydraulic capability and minimize utility pumping costs.
- **Safety requirement**: To deliver water free of pathogens, contaminants from infiltration, and internal leaching from water pipelines.

Water distribution systems generally consist of pipelines, pumps, valves, hydrants, storage tanks, reservoirs, water wells, treatment plants, meters, fittings, and other hydraulic appurtenances that help in connecting and managing water flow from the treatment plants or sources to the customers. It is essential to understand the characteristics of these system elements and their interaction with water pipelines and other system elements.

Fuzzy Systems Modeling in Environmental and Health Risk Assessment, First Edition. Edited by Boris Faybishenko, Rehan Sadiq, and Ashok Deshpande.

12.1.2 Water Pipelines

Water pipelines are installed to transport safe and sufficient water to customers with reliable service. Based on function, water pipelines are categorized into transmission mains (to transport water from a source such as treatment plants or reservoirs), distribution mains (to transport treated water from transmission mains to curb stops), service lines (to transport treated water from curb stops to individual customers), and premise plumbing (to transport water within a customer facility). Different materials are usually preferred for each category, and the diameters get progressively smaller from transmission mains to premise plumbing, owing to the amount of water transported. For this research, transmission and distribution pipelines are included in this study, and service pipelines are not included.

12.2 Risk Models Literature Review

Many risk models have been developed to support decision-making for water pipe renewal prioritization. These include models developed to predict the likelihood of failure (LoF) and CoF separately, or combined together as a risk index. Aspects related to LoF [1, 2], failure [3, 4], deterioration, and remaining useful life [5–7, 10] and performance [8, 9] have been investigated and modeled based on different techniques. Most of the studies in the LoF domain have focused on predicting these metrics based on the structural integrity of the water pipe. Functional and safety aspects like water quality [10, 11] and hydraulic reliability [12, 13] have been explored in some studies but less often compared to structural integrity prediction models. The modeling techniques used include but are not limited to the support vector machine (SVM), fuzzy inference system (FIS), neurofuzzy system (NFS), artificial neural network (ANN), and genetic algorithm (GA).

A review of the relevant literature was performed with the aim of understanding the parameters required, assumptions for associated costs, and modeling techniques to gather the background knowledge behind developing the consequence of failure (CoF) model. Previous studies have focused on a range of issues for assessing the CoF, such as the types of costs that should be associated with water pipe failures [14], the typical number of water utilities that actually collect these parameters [15], water customer perceptions [16], approximate costs [17] and tools to estimate the consequence of water pipe failure costs [14]. The data gathered for these studies are based on surveys with consumers and water utility experts as participants to provide an approximate cost for different types of costs. Models determined from practice reviews are mostly weighted models with weights and scales assessed by an internal expert committee. Models in literature are either cost estimation tools like the Grand Central model [14] or weighted models built with weights assigned from experts [18]. One study from an oil and gas pipeline used an advanced mathematical neurofuzzy technique to predict CoF [19].

Studies have also introduced important concepts like hard and soft costs [17] to differentiate between the costs that are paid by the agency and the costs paid by the users or the impact to the environment. It is essential to differentiate between the two as many of the soft costs are not found in the water utility database and are otherwise difficult to estimate. The study also provided the distribution of direct agency and societal costs (with direct costs ~25–50% of total costs), distribution within direct costs (majority 52% going into legal costs and damage claims), and distribution within societal costs (majority 57% going into property damage not paid by water utilities). It was also found that the location of the pipe failure was the most significant factor, followed by the volume of the leak (correlated with pipe diameter, leak duration, and static pressure). The study also mentioned that there was no direct correlation of consequence costs with diameter of water pipes, as usually expected. In terms of consumer experience, a majority (60%) were willing to accept service disruptions for up to two hours, and a majority (86%) said that they were mostly or totally satisfied with their local water utilities [16]. It was also mentioned that the number of households and cars affected by the main water failure are the main drivers of societal costs. The study also quantified the justification of renewal with respect to failure rates. For example, service disruption to 20 households justified all renewal technologies, whereas, based on direct water utility costs, repair was favored over other renewal methods when the break rate was two per year. With service disruption to 50 households and 150 cars affected in traffic, all the renewal technologies were found to be justified.

Most of the studies have identified difficulties in the data collection phase and have identified that water utilities do not routinely collect the full triple bottom line (TBL) costs, limiting the overall risk management program benefits [20]. However, many parameters identified in the literature are difficult to estimate but play a crucial role in the overall CoF, like renewal complexity. Also, many findings in the literature and practice are based on data collected from large diameter water pipe failures. However, findings from other literature have shown diameter is not a correlated factor with CoF, and data collection for full TBL factors should be conducted for the entire network. The other area for improvement would be using raw data instead of surveyed data for CoF assessment because survey data could be biased and present errors in the findings. Practice review has shown that factors change with respect to priorities of water utilities, and as a result the model developed based on limited datasets or survey information would not be robust enough to be used by other water utilities. This problem can be solved by using a bigger dataset and a comprehensive list of parameters for the CoF assessment.

12.2.1 Summary of Literature Review

A literature review shows that many previous studies have focused on the estimation of parameters essential to CoF assessment based on surveys from water utilities and experts. These studies help in the input parameter assumptions when the data are not available. The models developed in the previous studies are mostly deterministic and use weighted techniques for CoF estimation. The main issues with weighted techniques are that they are not able to capture the relationships between the parameters, which does not reflect the natural environment accurately. Also, the parameters used in the literature are mostly focused on including hard and soft costs and have not considered the operational impacts and renewal complexity impacts, which also contribute to an increase in the CoFs. Also, the survey-based dataset may be biased, which can skew the results based on heuristic knowledge, which may have errors. Also, as the datasets are based on 3–5 participating water utilities, the models are not verified and validated with enough datasets to have enough reliability to be used in real-world situations. Still, these studies have helped improve the understanding of the factors important to consider while assessing CoF and have also provided a valuable correlation between the most correlated and least correlated factors. The assumptions for many TBL parameters, especially societal and environmental, can be used from these studies with a certain degree of confidence while indirectly deriving input parameters for the model development. However, the studies from literature are not comprehensive in capturing all the parameters for assessing the consequence.

12.3 Risk Modeling Practice Review with Participating Water Utilities

A comprehensive state-of-the-practice review was conducted to utilize the knowledge of experts from over 500 water utilities. The authors selected 25 large water utilities, geographically distributed across the US to understand critical issues and scenarios frequently experienced in the real world. Selecting large water utilities helped the research team capture diverse and critical scenarios with good quality information, as many of these water utilities have established asset management programs. During this five-year study, multiple online and in-person meetings with these experts helped the research team understand the diverse operational risk scenarios for water pipes. The water utility teams for these meetings comprised experts from different teams like infrastructure asset managers, water distribution and transmission system managers, chief water distribution operators, data scientists, GIS analysts and database managers. The participation of representatives from different departments helped the research team solicit additional information, like recent and historical case studies, internal risk modeling reports, capital and operational budget information, laboratory testing results, and risk models.

Based on the comprehensive practice review, it was found that CoF estimation has been performed at varying levels of detail. Most of the smaller utilities do not have a risk management program and, by extension, no CoF assessment method. In contrast, many larger water utilities have a risk management program and assess CoF for their water pipes. For example, the Washington Suburban Sanitary Commission (WSSC) has business risk evaluation models prepared to identify renewal needs for water pipes. WSSC estimates LoF based on remaining useful life, which are heuristically estimated for different management strategy groups (MSGs). These groups are essentially categories to stratify pipes based on historical performance. An example for stratifications used by WSSC for their distribution pipes is shown in Figure 12.1. WSSC also estimates the CoF of water pipes based on the weighted estimation of various economic and social impacts like replacement costs, service disruption, and public health impacts. These estimations are based on a combination of empirical knowledge and assumptions sourced from various agencies, such as the Food and Drug Administration (FDA) and the Environmental Protection Agency (EPA), and local agencies and relating it in terms of percentage to the revenue.

Although the authors found that many large water utilities have some form of CoF assessment technique, it is important to note that most of these methodologies are limited. Many large water utilities score different parameters and use weighted models to determine CoF or criticality on a subset of the water distribution system. For example, some utilities only developed CoF models for their large diameter prestressed concrete cylinder pipes (PCCPs), ignoring the distribution pipes. The modeling frameworks also vary among these water utilities. For example, the city of Prince George assesses CoF in terms of combinations of scenarios, EnWin water utility assesses criticality through internal expert discussion and priorities, the City of Burnaby assesses CoF at a strategic level to identify cohorts of critical assets for prioritization, the Region of Peel water utility assesses critical cohorts for developing a risk profile, Tarrant Regional Water District uses scoring on different factors to assess CoF for its predominantly large diameter pipe network, San Diego County Water Authority assesses risk for its large diameter PCCP for prioritizing rehabilitation [21], and DC Water has comprehensive CoF weightage and scaling criteria for different parameters for large diameter pipes to identify renewal needs.

12.3.1 Summary of Practice Review

Practice review suggests that parameters used in models developed by water utilities are very different. This large variance can be attributed to the difference in priorities for the water utilities. The models are mostly developed with the help of consultants who use heuristic knowledge, which can be biased. Also, model testing, verification, and validation require large datasets to be effective, which these models lack because these are only tested with one water utility's dataset. Therefore, the models developed by water utilities lack robustness and cannot be applied to datasets outside their water utilities. The factors influencing the CoF of water pipes, as summarized from literature and practice review, are summarized in Table 12.1. Practice review suggests a lot of variances in the parameters. This can be attributed to the difference in priorities and impact effects for different water utilities.

ATTRIBUTES					ATTRIBUTES TO BE SET					
Material	**Make and Construction year**	**Size**	**Cathodically protected**	**Lining**	**Physical Effective Life**	**Additional Life Gained through Rehab**	**Maximum Potential Life (3)**	**Management Strategy Option to Start**	**Management Strategy for replacement**	**Notes**
CIP	Pit cast (always pre-1995)	3″–4″	No CP	Unlined	40	0	40	Immediate Replace with 4″	Replace	(1)
CIP	Pit cast (always pre-1995)	6″–8″	No CP	Unlined	40	0	40	Immediate Replace with 8″	Replace	(2)
CIP	Pit cast (always pre-1995)	10″–14″	No CP	Unlined	75	15	86	Clean and Line	Replace	
CIP	Pit cast (always pre-1995)	6″–8″	No CP	Lined	85	0	90	Replace		
CIP	Pit cast (always pre-1995)	10″–14″	No CP	Lined	95	30	120	Add Cathodic Protection	Replace	
CIP	Spun cast (always post-1955 pre 1966)	3″–4″	No CP	Unlined	40	0	40	Immediate Replace with 4″	Replace	(1)
CIP	Spun cast (always post-1955 pre 1966)	6″–8″	No CP	Unlined	40	0	40	Immediate Replace with 8″	Replace	(2)
CIP	Spun cast (always post-1955 pre 1966)	10″–14″	No CP	Unlined	60	15	72	Clean and Line	Replace	
CIP	Spun cast (always post-1955 pre 1966)	6″–8″	No CP	Lined	65	0	65	Replace		
CIP	Spun cast (always post-1955 pre 1966)	10″–14″	No CP	Lined	70	30	97	Add Cathodic Protection	Replace	
CIP	Spun cast (always post-1955 pre 1966)	3″–14″	No CP	Lined	45	0	45	Replace with 4″	Replace	
CIP	Spun cast (always post-1955 pre 1966)	6″–8″	No CP	Lined	45	0	45	Replace		

Figure 12.1 Water distribution pipe stratifications by WSSC for performing risk analyses.

(Continued)

ATTRIBUTES					ATTRIBUTES TO BE SET					
Material	Make and Construction year	Size	Cathodically protected	Lining	Physical Effective Life	Additional Life Gained through Rehab	Maximum Potential Life (3)	Management Strategy Option to Start	Management Strategy for replacement	Notes
CIP	Spun cast (always post-1955 pre 1966)	10″–14″	No CP	Lined	50	30	78	Add Cathodic Protection	Replace	
DIP	Class 51 pre-2002(non-polywrapped)	4″–6″	No CP	Lined	50	0	40	Replace with 4″ to 8″	Replace	
DIP	Class 50 pre-2002(non-polywrapped)	8″–14″	No CP	Lined	50	30	78	Add Cathodic Protection	Replace	
DIP	Class 51 pre-2002(non-polywrapped)	4″–6″	No CP	Lined	75	0	75	Replace with 4″ to 8″	Replace	
DIP	Class 50 pre-2002(non-polywrapped)	8″–14″	No CP	Lined	75	30	101	Add Cathodic Protection	Replace	
DIP	Class 54 Post- 2006(polywrapped)	4″–6″	No CP	Lined	100	0	100	Replace with 4″ to 8″	Replace	
DIP	Class 54 Post- 2006(polywrapped)	8″–14″	No CP	Lined	100	30	125	Add Cathodic Protection	Replace	
PVC	DR 14	4″–12″	No CP		75	0	75	Replace		
PVC	DR 18	4″–12″	No CP		75	0	75	Replace		
AC	Asbestos cement	4″–12″	No CP		65	0	65	Replace		
Steel	Steel	4″–12″	No CP		75	0	75	Replace		
Other	Copper or Galvanized <3″	<3″	No CP		40	0	40	Replace with 4″		

(1) Existing unlined ≤ 4″ CIP experiences high break rates, recommended immediate replacement
(2) Existing unlined 6″ CIP has significant tuberculation, hydraulically insufficient, recommended immediate replacement
(3) Rehab is assumed to occur at 5% remaining in the PEL.

Figure 2.1 (Cont'd)

Table 12.1 Parameters used in the literature.

		Literature review				Practice review					
	Influencing factors	**[22]**	**[23]**	**[15]**	**[14]**	**Austin Water**	**WSSC**	**City of Burnaby**	**City of Prince George**	**TRWD**	**DC Water**
	Pipe type	WW	20–90 in. W	W	W	W	W	W	W	W	W
Categories	**Modeling technique**	**Hierarchical weighted model**	**Used GCM [14] Deterministic**	**Survey (5 utilities) based estimation**	**Cost estimation methods for different factors**		**Delphi process based decision-making**		**108 scenarios based CoF evaluation**		
Pipe characteristics and renewal complexity	Pipe characteristics (material, length, depth, diameter)	✓ (age, length, diameter, depth, soil)				✓ (diameter)					✓ (material, diameter)
	Access	✓								✓	
	Distance to nearby utilities	✓	✓							✓	
Direct costs	Legal costs			✓			✓				
	Non-revenue water/ extra pumping costs			✓							
	Labor/contracting/ other infrastructure repair costs		✓	✓							
	Emergency purchases		✓	✓			✓				

(Continued)

Table 12.1 (Continued)

		Literature review				Practice review					
	Influencing factors	**[22]**	**[23]**	**[15]**	**[14]**	**Austin Water**	**WSSC**	**City of Burnaby**	**City of Prince George**	**TRWD**	**DC Water**
	Pipe type	**WW**	**20–90 in. W**	**W**	**W**	**W**	**W**	**W**	**W**	**W**	**W**
Categories	**Modeling technique**	**Hierarchical weighted model**	**Used GCM [14] Deterministic**	**Survey (5 utilities) based estimation**	**Cost estimation methods for different factors**		**Delphi process based decision-making**		**108 scenarios based CoF evaluation**		
Social impact	Disruption to nearby critical services	✓	✓	✓		✓				✓	✓
	Rail/road disruption	✓	✓	✓	✓	✓				✓	✓
	Supply disruption/ outage hours		✓	✓		✓	✓				
	Customer confidence			✓			✓				
	Likelihood of injury/ public health threat/ water quality impact		✓	✓			✓				
	Flooding costs				✓						
Environmental impact	Distance to water bodies	✓				✓					
	Land use	✓									
Operational impact	Flow rate/pressure										✓
	Redundancy										✓
	Seismic zone	✓			✓						✓

12.4 Methodology of Likelihood of Failure Assessment and CoF Model Development

12.4.1 Hypothesis

The authors propose that a CoF model developed for water pipes using a comprehensive list of parameters and employing complex mathematical techniques will be able to predict the CoF of water pipes more accurately.

12.4.2 Likelihood of Failure

Modeling the LoF is not within the scope of this study. The LoF scores are determined from the performance curve developed as the other part of this study [24]. This was based on a list of comprehensive performance indicators determined from previous studies [24–26]. The performance curve is translated into LoF values [24, 27–30], which can be combined with CoF values to show the overall pipe risk. The translation of performance attributes to LoF values is determined based on the methodology illustrated in Table 12.2.

Table 12.2 Converting performance/survival probability to LoF values.

Performance index (0–5 scale)	Performance index (other scale) *	Survival probability	LoF index rating	Category	Description
0–1	$X = \frac{x}{Y} \times 5$	0.00–0.15	0–1	Very low	New pipe installed correctly
1–2		0.15–0.35	1–2	Low	Little deterioration to the pipe
2–3		0.35–0.65	2–3	Medium	Fair level of deterioration without any previous repairs
3–4		0.65–0.85	3–4	High	Old pipe with 1–2 repairs previously done
4–5		0.85–1.00	4–5	Very high	Heavy deterioration of the pipe with more than 2 repairs done previously or major rehabilitation

12.4.3 CoF Model Development

We propose the inclusion of a comprehensive list of parameters from extensive literature and practice reviews and interviews with experts from water utilities. This approach will be more comprehensive in capturing the TBL parameters, along with operational and renewal parameters [31]. Models are developed based on the fuzzy logic (FL) approach to counter the inherent data uncertainty with knowledge in water pipeline risk modeling. The method of model development is process-driven (i.e. it converges the information obtained from data and real-world knowledge into the model, which will help the model to mimic the real-world scenarios). The models are also piloted with real-world data by extensively engaging participating water utilities.

This study provides a methodology for developing a CoF model based on FL to capture and model the essential parameters needed to accurately model the consequence of the failure of a water pipeline. The performance of the models is assessed based on the granularity and accuracy of the results. The parameters have been defined based on extensive literature and practice reviews and previous surveys conducted to develop CoF parameters. The contribution of this study in the CoF assessment methodology are as follows:

- The data structure is more comprehensive than the previous studies and includes parameters capturing the operational conditions and renewal complexity along with the triple bottom line (TBL) parameters. These have been verified by discussions with multiple water utilities.
- The inherent ambiguity in determining the input parameters for a CoF assessment can be accounted for by employing soft computing techniques. Soft computing is the use of approximate calculations to provide imprecise but usable solutions to complex computational problems. The principal constituents of soft computing techniques include neural networks, FL, and GAs [32]. Minor components include probabilistic programming, chaotic programming, and machine learning, among others [33].
- The models are also verified using sensitivity analysis (to check the model performance with respect to parameters) and piloted based on data collected from multiple utilities.

12.4.3.1 Data Structure for CoF Model

The development of the data structure was based on the findings from a comprehensive literature and practice review. The subparameters identified from the literature are categorized to improve global understanding. These are presented in Figure 12.2. The modules

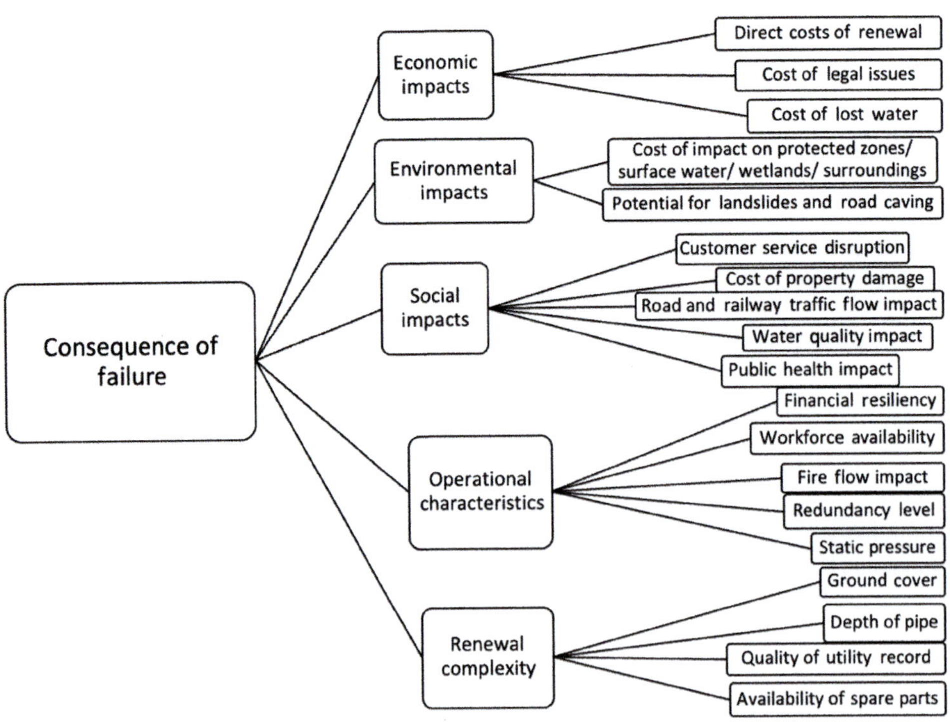

Figure 12.2 Categories considered for CoF estimation for water pipelines.

cover different aspects of the CoF assessment, including the TBL parameters as well as the parameters capturing the impact of operational parameters and the complexity of renewal activities.

12.4.3.2 CoF Model Parameters List and Membership Functions

Based on a comprehensive literature review and practice review, 12 subparameters were identified in the economic impact, social impact, and environmental impact modules. These were updated into 18 parameters based on operational impacts and renewal complexity related to CoF assessment and were discussed with water utilities. The proposed 18 subparameters are categorized into five modules in Table 12.3. All the membership functions for the parameters are triangular membership functions except some trapezoidal membership functions used to capture values above a certain limit in the *very high* category range.

Table 12.3 Consequence of failure model parameters and assumptions.

Parameters	Units	Membership function ranges	Membership function	Definition
Economic impact module (0–5)				
Direct cost of renewal	$	Very low ($0–$10 000)	Gaussian bell	Direct costs include direct and indirect labor costs, materials and equipment purchases, construction, engineering, repaving, landscaping, other contract repair costs, and laboratory services
		Low ($5000–$45 000)	Gaussian bell	From literature and practice reviews, it can be observed that water utilities consider this as one of the most important parameters
		Medium ($25 000–$125 000)	Gaussian bell	
		High ($75 000–$450 000)	Gaussian bell	
		Very high (>$225 000)	Trapezoidal	
Cost of lost water	$	Low ($0)	Gaussian bell	It includes the cost of water lost due to failure, which depends on the time taken to fix the break/leak
		Medium ($25 000)	Gaussian bell	Includes cost of additional supply or bottled water
		High ($50 000)	Trapezoidal	
Cost of legal issues	$	Low ($0–$20 000)	Gaussian bell	This is the cost of any legal fines and proceedings due to a failure event
		Medium ($10,00–$110 000)	Gaussian bell	This can occur due to flooding in residential, commercial, or industrial areas, and the property owners could sue the water utility
		High ($60 000–$3 000 000)	Trapezoidal	
Economic impact	Scale	Very low (0–1.25)	Gaussian bell	This module can also estimate results in $ values as well as in 1–5 scale
		Low (0–2.5)	Gaussian bell	However, a 1–5 scale is preferred for easy compatibility with the other modules
		Medium (1.25–3.75)	Gaussian bell	
		High (2.5–5)	Gaussian bell	
		Very high (3.75–5)	Gaussian bell	Centroid defuzzification method is used

(Continued)

Table 12.3 (Continued)

Parameters	Units	Membership function ranges	Membership function	Definition
Environmental impact module (0–5)				
Cost of impact on surface water, wetlands, and streams	\$	Very low (\$0–\$5000)	Gaussian bell	Chlorinated water contamination into nearby streams, water bodies, and wetlands can contaminate source water, causing a threat to aquatic life, in turn affecting the fish industry and recreational activities
		Low (\$2500–\$22 500)	Gaussian bell	
		Medium (\$12 500–\$150 000)	Gaussian bell	
		High (\$80 000–\$750 000)	Gaussian bell	The input value is determined based on proximity to wetlands, surface water bodies, and streams
		Very high (\$415 625-\$1 500 000)	Trapezoidal	
Potential for landslides	Scale	Low (0–2.5)	Gaussian bell	*Low*, *medium*, and *high* soil slip potential values are determined from NRCS SSURGO database
		Medium (0–5)	Gaussian bell	
		High (2.5–5)	Gaussian bell	This can indicate the potential for landslides High soil slip potential means high potential for landslides
Environmental impact	Scale	Very low (0–1.25)	Gaussian bell	Environmental impact module estimates the total impact of water pipeline failure on the natural environment
		Low (0–2.5)	Gaussian bell	
		Medium (1.25–3.75)	Gaussian bell	
		High (2.5–5)	Gaussian bell	Centroid defuzzification method is used
		Very high (3.75–5)	Gaussian bell	
Social impact module (0–5)				
Customer service disruption	Scale	No customers (0–1.25)	Gaussian bell	The input values are determined based on land use of the land above the pipe
		Residential (0–2.5)	Gaussian bell	Land use data are obtained from the National Land Cover Database (NLCD) or the municipal/state website
		Dry (retailers) (1.25–3.75)	Gaussian bell	
		Wet (using water as input like restaurants, food processor, beverage maker, computer chip manufacturers)* (2.5–5)	Gaussian bell	Open space is indicative of no customers, and as a result the customer service disruption is minimum Critical customers face the highest impact due to service disruption
		Critical customers (hospitals, schools, etc.) (3.75–5)	Gaussian bell	
Road and railway traffic flow impact	\$	Low (\$0–\$150 000)	Gaussian bell	The input values are determined based on proximity to roads and railway tracks
		Medium (\$75 000–\$875 000)	Gaussian bell	
		High (\$750 000–\$5000 000)	Trapezoidal	These data are obtained from the state DOT website Water pipe failures can disrupt traffic costing human-hours and delays in commercial and business activities

Table 12.3 (Continued)

Parameters	Units	Membership function ranges	Membership function	Definition
Water quality impact	Scale	Low (0–2) Medium (0.5–4.5) High (3–5)	Gaussian bell Gaussian bell Gaussian bell	Infiltration into the water distribution system after a failure event can affect water quality Input values are determined using proximity to superfund sites and wastewater pipelines
Cost of property damage	$	Low (< $5000) Medium ($2500–$102 500) High ($52 500–$500 000)	Gaussian bell Gaussian bell Trapezoidal	Flooding due to water pipeline failures can cause property damage in residential, commercial, and industrial buildings Input values are determined based on land use and proximity to residential areas and buildings obtained from NLCD and municipal/state website
Social impact	Scale	Very low (0–1.25) Low (0–2.5) Medium (1.25–3.75) High (2.5–5) Very high (3.75–5)	Gaussian bell Gaussian bell Gaussian bell Gaussian bell Gaussian bell	Social impacts include the parameters which indicate consequences to the customers and the public after a water failure event Centroid defuzzification method is used
Operational impact module (0–5)				
Financial consequence	%	Low (0%–1%) Medium (1%–3%) High (>3%)	Gaussian bell Gaussian bell Trapezoidal	Failure events can impose stress on the operating budget of the water utility This parameter is estimated as a ratio of $\frac{\textit{Annual Expenditure on Reactive Renewal}}{\textit{Annual O \& M Budget}}$ High financial consequence means less operational impact
Workforce availability	Scale	Not available (0–2) Manageable (0.5–4.5) Easily available (3–5)	Gaussian bell Gaussian bell Gaussian bell	Workforce ability is essential to isolate a failed section of the water pipe by operating valves and renewing the pipe High workforce availability means less operational impact High workforce availability is assumed for larger water utilities and low for smaller water utilities Manageable class includes availability through external contractors
Fire flow impact	Scale	Low (0–2) Medium (0.5–4.5) High (3–5)	Gaussian bell Gaussian bell Gaussian bell	The input values are based on proximity to fire hydrants High fire flow impact means a high operational impact and proximity to multiple fire hydrants, which could be disrupted in case of a failure event
Static pressure	psi	Low (0–25) Medium (20–80) High (50–200)	Gaussian bell Gaussian bell Trapezoidal	High static pressure translates into high operational impact due to loss of greater volume of water and difficulty in controlling the flow Static pressure indicates the amount of water loss The input value is determined from the water utility database

(Continued)

Table 12.3 (Continued)

Parameters	Units	Membership function ranges	Membership function	Definition
Redundancy level	%	No backup (0–100)	Gaussian bell	Extra redundant systems translate into low operational impact as water flow can be directed into redundant pipes using valves and keep the service uninterrupted
		One redundant system (80–120)	Gaussian bell	Redundancy is based on the pipe function and is assumed 200 for distribution pipes and 0 for transmission pipes
		Extra redundant systems (100–200)	Gaussian bell	
Operational impact	Scale	Very low (0–1.25)	Gaussian bell	Operational impacts consider the financial and O&M conditions of the water utility and the pipe
		Low (0–2.5)	Gaussian bell	Centroid defuzzification method is used
		Medium (1.25–3.75)	Gaussian bell	
		High (2.5–5)	Gaussian bell	
		Very high (3.75–5)	Gaussian bell	
Renewal complexity module (0–5)				
Ground cover	Scale	Open space (0–2)	Gaussian bell	Pipes under buildings and water surfaces are more complicated to perform pipe renewal activities
		Roads and railways (1–4)	Gaussian bell	The input value is obtained from NLCD or municipal/state database
		Buildings and/or water surface (3–5)	Gaussian bell	
Depth of pipe	ft	Low (0–6)	Gaussian bell	Pipes at lower depths are easier to renew due to ease of accessibility
		Medium (4–10)	Gaussian bell	Input value is obtained from the water utility database
		High (8–30)	Trapezoidal	
Quality of utility record	Scale	Comprehensive database (0–2)	Gaussian bell	Good utility records help in a timely and accurate location of failure, and information from utility records helps in preparedness
		Partial database (1–4)	Gaussian bell	Input value is based on the quality of obtained data from the water utility
		No database (3–5)	Gaussian bell	
Availability of spare parts	Scale	Easily available (0–2)	Gaussian bell	Availability of spare parts and repair materials translates into timely renewal
		Manageable (1–4)	Gaussian bell	
		Not available (3–5)	Gaussian bell	
Renewal complexity	Scale	Very low (0–1.25)	Gaussian bell	Renewal complexity module considers parameters essential to capture the relative difficulty in renewing the pipe, which can influence the time taken to isolate and fix a water pipeline failure
		Low (0–2.5)	Gaussian bell	Centroid defuzzification method is used
		Medium (1.25–3.75)	Gaussian bell	
		High (2.5–5)	Gaussian bell	
		Very high (3.75–5)	Gaussian bell	
Consequence of failure				
Consequence of failure	Scale	Very low (0–1.25)	Gaussian bell	Centroid defuzzification method is used
		Low (0–2.5)	Gaussian bell	
		Medium (1.25–3.75)	Gaussian bell	
		High (2.5–5)	Gaussian bell	
		Very high (3.75–5)	Gaussian bell	

12.4.3.3 CoF Model Parameters Description

Economic Impact

1) **Cost of renewal**: If a utility fails, it has to be necessarily fixed either by repair, rehabilitation, or replacement so that a particular level of service is achieved and the utility is in functioning mode again. This cost includes the costs of labor, materials, outside construction contractors, and land restoration. High cost indicates high consequences and vice versa. It is divided into three categories in the CoF model:
   a) Low ($0–$10 000)
   b) Medium ($5000–$100 000)
   c) High ($20 000–$2 000 000).
2) **Cost of legal issues**: In some cases, customers or affected individuals/firms may sue the utility for damage, injury, or loss due to failure in terms of claims, fines, insurances, noncompliance issues, attorney fees, etc. This cost is in addition to the cost of renewal and the cost of property damage. The higher the cost, the higher the consequences. It is divided into three categories in the CoF model:
   a) Low (< $100 000)
   b) Medium ($50 000–$500 000)
   c) High ($250 000–$1 000 000).
3) **Cost of lost water**: Due to failure, a large volume of treated clean water is lost. A huge investment is incurred in installing the treatment plant and in treating water. Failure results in the loss of this investment and the cost incurred to treat the water. This cost also must be accounted for when estimating consequences. It is divided into three categories in the CoF model:
   a) Low (< $4000)
   b) Medium ($2000–$100 000)
   c) High ($50 000–$750 000).

Environmental Impacts

4) **Quality impacts on surface water and/or surroundings**: Break out of wastewater pipes will lead to a large amount of sewer water flooding. This will lead to contamination of surface water bodies, which will affect aquatic life. Impacts on surroundings other than property (as described in parameter 2) should also be incorporated here. It is divided into five categories in the CoF model:
   a) Very low ($0–$5000)
   b) Low ($0–$15 000)
   c) Medium ($5000–$150 000)
   d) High ($75 000–$750 000)
   e) Very high ($300 000–$10 000 000).
5) **Potential landslide or road caving**: If the extent of failure is quite large and if a large amount of water is flooded, chances of landslides or road caving can cascade into other costs. It is divided into two categories in the CoF model:
   a) Desirable (0–3.5)
   b) Undesirable (1.5–5).

Social Impacts

6) **Cost to property damage**: The failure of any pipe results in loss not only to the pipe itself but also to other assets/infrastructure, mainly buildings, structures, and vehicles. These are basically the third-party damage cost that the utility might have to pay or is paid by the owner itself. The higher the cost, the higher the consequence. It is divided into three categories in the CoF model:
   a) Low (<$20 000)
   b) Medium ($10 000–$100 000)
   c) High ($50 000–$1 000 000).
7) **Customers service disruption**: Failure leads to service disruption as the water supply is interrupted. Different kinds of customers are affected in a variety of ways. It is divided into five categories in the CoF model:
   a) Open space (0–1.25)
   b) Residential (0–2.5)
   c) Dry (retailers) (1.25–3.75)
   d) Wet (using water as input like restaurants, food processor, beverage maker, computer chip manufacturers) (2.5–5)
   e) Critical customers (hospitals, schools, etc.) (3.75–5).
8) **Road and railway traffic flow impact**: Water pipeline failure can lead to disruption of traffic on roads and railway due to road closures and can translate into human-hours lost and potential business lost on freight corridors and can be accounted for as financial costs. It is divided into five categories in the CoF model:
   a) Very low ($0–$20 000)
   b) Low ($0–$150 000)
   c) Medium ($75 000–$1 500 000)
   d) High ($750 000–$7 500 000)
   e) Very high ($3 000 000–$15 000 000).
9) **Water quality impact**: Water pipe failures can lead to infiltration from nearby wastewater pipes or sediments and impact the quality of water supplied to people. This can also take the form of redness in water due to corroded pipes. It is divided into three categories in the CoF model:
   a) Low (0–2)
   b) Medium (0.5–4.5)
   c) High (3–5).

Operational Impacts

10) **Financial resilience**: Water utilities with large budgets (typically large utilities) can mitigate consequences due to water pipeline failures using the extra financial resiliency or using insurance programs. It is divided into three categories in the CoF model:
   a) Low (0–2)
   b) Medium (0.5–4.5)
   c) High (3–5).
11) **Workforce availability**: The presence of an adequate and skilled workforce can supplement a good database and can mitigate the consequences of water pipeline failures. Skilled operators can drastically reduce the consequences by identifying problems and implementing solutions faster. It is divided into three categories in the CoF model:
   a) Low (0–2)
   b) Medium (0.5–4.5)
   c) High (3–5).

12) **Fire flow impact**: Water pipes providing water to fire hydrants may pose the risk of unavailability of water during fire incidents for buildings, public spaces, hotels, etc., after a failure event:
    a) Low (0–2)
    b) Medium (0.5–4.5)
    c) High (3–5).
13) 13. **Static pressure**: Pipes working at higher operating pressures can have higher consequences due to a greater volume of water escaping in case of failure:
    a) Low (0–25 psi)
    b) Medium (20–80 psi)
    c) High (50–200 psi).
14) **Redundancy level**: A water distribution network with redundancies can ensure that the failure of water pipes with redundancy would not disrupt the service as the water flow can be redirected into the redundant pipes with valves. This can mitigate the consequences of a water pipeline failure. It is divided into three categories in the CoF model:
    a) Extra redundant systems (100–200)
    b) One redundant system (80–120)
    c) No backup (0–100).

Renewal Complexity Data Structure

15) **Ground cover**: This refers to the type of surface or structure under which the pipe has been buried. The complexity in renewing the failed pipe section depends on whether the pipe is buried under water or land or buildings, etc. It is divided into three categories in the CoF model:
    a) Open space (0–2)
    b) Roads and railways (1–4)
    c) Buildings and/or water surface (3–5).
16) **Depth**: Pipes at greater depths are more difficult to renew due to the extra effort needed for excavation and the greater difficulty in encountering a failure event before the effects magnify. It is divided into three categories in the CoF model:
    a) Low (0–6 ft)
    b) Medium (4–10 ft)
    c) High (8–30 ft).
17) **Quality of utility record**: It indicates the reliability of existing records on buried utilities. The availability of accurate historical utility records for the project can significantly reduce the potential for encountering unexpected underground utilities. It is divided into three categories in the CoF model:
    a) Good (0–2)
    b) Fair (0.5–4.5)
    c) Poor (3–5).
18) **Availability of spare parts**: A failed utility must be repaired, and for that the required materials must be available. If materials are not easily available, it will increase the time for which the utility remains out of service, causing annoying inconvenience to people, and thus consequences will be large. It is divided into three categories in the CoF model:
    a) Not available (0–2)
    b) Difficult (0.5–4.5)
    c) Easily available (3–5).

12.4.3.4 CoF Index Rating Scale

Determining the scale of model output was performed after discussion with water utilities and experts. This study utilizes the 1–5 scaling system for the CoF rating. The 1–5 scale is the preferred range, and it is also a balanced tradeoff between the granularity and statistical significance. Table 12.4 shows the description of the 1–5 index ratings.

Table 12.4 CoF index rating description.

CoF rating	Category	Description
0–1	Insignificant	Minimum to no renewal costs, environmental impacts, social disruption, easily renewable
1–2	Minor	Minor impact due to failure on renewal costs, environmental or social impact, and is easy to renew
2–3	Moderate	Some CoF parameters have moderate impacts on renewal costs, environmental or social impact, or are not very easy to renew
3–4	Major	May have a high impact on renewal costs, environmental or social impact, or is very hard to renew
4–5	Catastrophic	Critical and may have a very high impact on renewal costs, environmental impact, or social impact

12.4.3.5 Fuzzy Logic Model Details

FL was conceptualized from the fuzzy set theory introduced by Zadeh in his seminal work in 1965, which was introduced as an alternative to binary set theory. FL allowed graded memberships of inputs that were more intuitive and closer to real-world situations [34]. As shown in Eq. (12.1), the crisp sets based on binary logic are not intuitive and realistic in modeling input parameters, especially for risk analysis where data might be imprecise and uncertain.

$$\mu_s(X) = \begin{cases} 1 \; if \; X \in S \\ 0 \; if \; X \notin S \end{cases} \tag{12.1}$$

where $\mu_s(X)$ is the membership of the parameter X, and S refers to the crisp set.

Fuzzy sets, on the other hand, can have a grade of membership mapped on [0,1]. Formally, if we assume A to be the universe of discourse X and x be an element within X, the characteristic function $\mu_A(X)$ of fuzzy set A can be expressed as

$$\mu_A : X \rightarrow [0,1] \tag{12.2}$$

From Eq. (12.1), we can understand that the membership degree of any element x in X can be described as

$$\mu_A(x) \in [0,1] \tag{12.3}$$

where $\mu_A(x)$ is the membership function of fuzzy set A. Based on this understanding, the fuzzy set A can be defined as

$$A = \{(x, \mu_A(x)) \mid x \in X\} \tag{12.4}$$

The two most widely used fuzzy inference methods are Mamdani and Sugeno. This study uses the Mamdani fuzzy inference method [35] over the Sugeno method as the Mamdani method has widespread acceptance and provides a fuzzy set as an output instead of a mathematical expression, which has better interpretability into a single output using defuzzification [9].

Membership functions are used in FIS to model parameter ranges into a fuzzy space. A membership function is a line or curve that defines how each point in the input space is mapped to the degree of membership. The membership function can take linguistic values such as *low*/*high* or *small*/*large*, etc. There are several types of membership functions, which include triangular shaped, trapezoidal shaped, gaussian shaped, generalized bell shaped, sigmoidally shaped, s shaped, z shaped, and Π shaped. For this research, the authors used different membership functions, like triangular, trapezoidal, and gaussian bell membership functions. The membership functions were assigned based on secondary data and discussions with water utility experts. The membership functions were developed based on heuristic knowledge to capture the expert knowledge from industry professionals and the literature. The generated membership functions were then reviewed and evaluated by multiple professionals, including engineers from various water utilities, consultants, and pipe associations.

Input parameters projected into a fuzzy space using membership functions are modeled based on a fuzzy rule based (FRB) system. This system is built to represent relationships between input and the expected output using if–then statements. To model dependencies between parameters, FRBs can also use Boolean logic operators like AND (both parameters are included), OR (including one or both parameters), and NOT (exclude the parameter). An advantage of using different Boolean operators is that it helps reduce the number of rules in the model when all parameters are not required for specific outputs, especially when considering best and worst rating outputs. Similar to the development of membership functions, the FRB for this expert system model was developed based on the heuristic knowledge from literature and industry experts. Cumulatively, there are 381 rules in the proposed model.

Based on the membership functions, rules, and operators, the given crisp input can be mapped into an output fuzzy set. After that, the implication operation is used to compute the output fuzzy set of each operation. Then all the output fuzzy sets are combined into one fuzzy set based on aggregated operation utilizing summation (sum of the rule output sets) operator to represent all truncated values from the output membership function. Finally, the fuzzy output set is converted into a crisp CoF rating within [0,5] using defuzzification. Defuzzification can be performed based on different methods like centroid, bisector, middle of maximum (the average of the maximum value of the output set), largest of maximum, and smallest of maximum. For this research, the authors used the centroid method for defuzzification which returns the center of the area encompassed by the output fuzzy set. The centroid method is the most popular defuzzification method as it can find the central tendency of the aggregated output fuzzy set.

The authors chose a hierarchical framework for the FIS which can evaluate different categories as modules within the overall CoF model. The advantage of using a hierarchical framework is twofold. First, a hierarchical framework can help reduce the overall number of rules reducing the computational complexity and thus increasing the interpretability of the model. Second, a hierarchical framework helps in the model verification and validation process as the final outputs can be back-tracked to individual modules and the parameters included within the module. This helps verify discrepancies during validation and identify potential modifications needed. As a result, a total of six separate FL models are developed and arranged hierarchically, as shown in Figure 12.3.

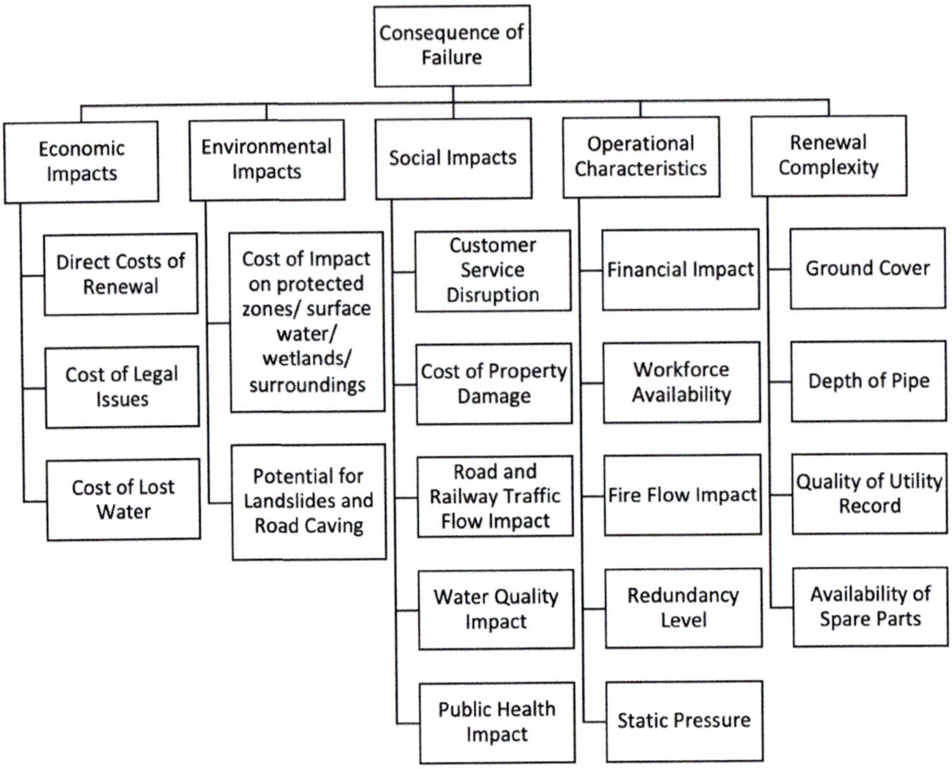

Figure 12.3 Model layout.

12.4.3.6 Model Verification and Validation Protocols

The authors developed a novel protocol to investigate the reliability of the results based on testing the qualitative and quantitative aspects of the model. Qualitative aspects relate to model performance in different risk scenarios faced in real-world situations. Quantitative aspects look at the model performance compared to historical data with respect to accuracy and precision metrics like root mean square error (RMSE), mean absolute error (MAE), etc. The tasks outlined in this section discuss protocols to test and validate CoF models in real-world scenarios.

- **Task 1**: Understanding the water pipeline system complexities and influential factors for modeling the CoFs to reduce epistemic uncertainties and improve risk modeling based on a comprehensive literature and practice review.
- **Task 2**: Identify critical parameters and necessary assumptions based on the findings of Task 1.
- **Task 3**: Collect available historical data from water utilities.
- **Task 4**: Perform data preprocessing and QA/QC and generate descriptive statistics to understand data characteristics and reduce aleatory uncertainty.
- **Task 5**: Understand available algorithms and choose the algorithm capable of
  a) Handling data uncertainty: This includes performing data preprocessing based on system understanding to reduce aleatory and epistemic uncertainties.
  b) Combining data with knowledge following a process-driven approach: One of the major issues with risk modeling for water pipeline systems is the lack of reliable data. As a result, it is imperative to use algorithms that can incorporate knowledge and system understanding.

- **Task 6**: Run the model based on available data and assumptions and get the CoF results for water pipe segments on a 0–5 scale.
- **Task 7**: Evaluation and verification of results
  a) Set up an evaluation team with water utilities to discuss model performance based on their dataets.
  b) Face validity to check model parameters, rules, and results.
  c) Sensitivity analysis to check how uncertainty propagates through the model and identify parameters with the highest influence over model results.
  d) Comparing results with a competing model already being used by water utilities.
- **Task 8**: Validate model
  a) Set up a validation team with a large water utility having enough resources to collect data and provide feedback on model performance.
  b) List critical parameters (based on Task 1 and Task 7), sources and measurement protocol (based on Task 8) to collect data. Data can come from three types of measurements:
     i) Directly recorded: This includes measurements made by water utility based on laboratory or field tests using instruments.
     ii) Indirectly obtained: These are second-hand data obtained from external sources and not necessarily recorded by the water utility.
     iii) Educated guess: These can come based on experience and heuristic knowledge of water utility experts and can be digitized for analysis purposes.
- As the result of the model is represented on a 0–5 scale, it is unlikely that the definition of each scale in the proposed research will match the validation team's definition. A viable option will be to use piecewise validation of the model modules. Ground truth metrics can be selected for each module within the CoF model to evaluate the accuracy of model predictions.
- Collect data based on a robust sampling methodology.
- Perform data pre-processing and QA/QC and generate descriptive statistics to understand data characteristics. Also, parameters which cannot be measured on the ground will be estimated based on Task 2 and Task 4.
- Run the model and generate results.
- Evaluate accuracy of model outputs with the selected metrics. Qualitative discussions of the model predictions versus the ground truth will also be provided to contextualize the results.

12.5 Results from the Application of the CoF Model to Data from 500 Water Utilities in the US

12.5.1 Overall Distribution (Miles) of Pipes in the US in Different CoF Categories

The results for the data collected across water utilities in the US from the CoF model are presented in Table 12.5.

The bar chart for Table 12.5 is presented in Figure 12.4 showing the proportion of pipes in different CoF ranges. The results showed that the majority of the larger diameter pipes (commonly falling into transmission pipe category) fall into higher consequence category, as failure of similar pipes can disrupt the service downstream. The overall distribution of pipe samples in different CoF categories is shown in Figures 12.5–12.7.

Table 12.5 Distribution of water pipes (as a percentage) falling in the different consequence ratings.

Type	Material	Diameter (in.)	Mileage	Distribution in different CoF ratings (% of samples)				
				Catastrophic	Major	Moderate	Minor	Insignificant
Metallic	DI	≤16	50 310	0.00	0.02	24.09	75.89	0.00
		16–36	8137	0.99	8.09	90.92	0.00	0.00
		>36	772	70.32	29.68	0.00	0.00	0.00
	Steel	≤16	7488	0.00	0.02	31.85	68.13	0.00
		16–36	1877	10.44	10.40	79.16	0.00	0.00
		>36	1903	89.72	10.28	0.00	0.00	0.00
	CI	≤16	45 267	0.00	0.01	13.80	86.19	0.00
		16–36	4133	4.52	2.65	92.82	0.00	0.00
		>36	264	86.63	13.37	0.00	0.00	0.00
Cementitious	RCCP	≤16	35	0.00	0.00	35.48	39.14	25.38
		16–36	816	8.05	8.09	52.06	31.80	0.00
		>36	458	85.33	14.67	0.00	0.00	0.00
	PCCP	≤16	576	0.00	0.04	38.58	55.02	6.37
		16–36	1943	12.57	13.54	41.51	32.38	0.00
		>36	1397	99.52	0.48	0.00	0.00	0.00
	AC	≤16	21 416	0.00	0.00	32.59	39.99	27.42
		16–36	2194	0.00	0.24	60.36	39.40	0.00
		>36	<1	96.55	3.45	0.00	0.00	0.00
Plastic	PVC	≤16	26 271	0.00	0.00	16.23	46.74	37.02
		16–36	3293	0.43	0.90	38.95	59.72	0.00
		>36	63	80.00	20.00	0.00	0.00	0.00
	HDPE	≤16	399	0.00	0.01	6.33	45.39	48.27
		16–36	379	0.00	9.32	10.25	80.43	0.00
		>36	1	70.21	29.79	0.00	0.00	0.00
Composite	Composite	≤16	2	0.00	0.00	33.06	25.81	41.13
		16–36	2	0.92	8.26	31.19	59.63	0.00
		>36	6	100	0.00	0.00	0.00	0.00

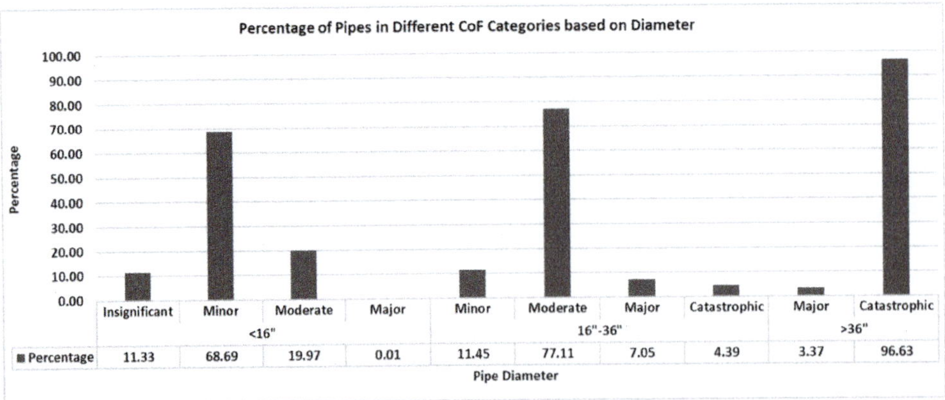

Figure 12.4 Percentage of pipes in different CoF categories based on diameter stratification.

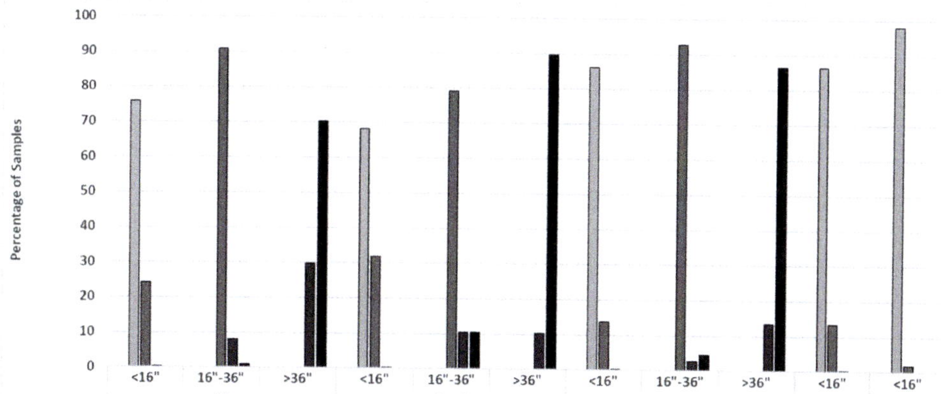

Category	DI <16"	DI 16"-36"	DI >36"	Steel <16"	Steel 16"-36"	Steel >36"	CI <16"	CI 16"-36"	CI >36"	Copper <16"	Br <16"
Insignificant	0	0	0	0	0	0	0	0	0	0	0
Minor	75.89	0	0	68.13	0	0	86.19	0	0	86.49	98.15
Moderate	24.09	90.92	0	31.85	79.16	0	13.8	92.82	0	13.49	1.85
Major	0.02	8.09	29.68	0.02	10.4	10.28	0.01	2.65	13.37	0.01	0
Catastrophic	0	0.99	70.32	0	10.44	89.72	0	4.52	86.63	0	0

Material and Diameter

□ Insignificant ■ Minor ■ Moderate ■ Major ■ Catastrophic

Figure 12.5 Percentage of metallic pipes in different CoF categories based on material and diameter.

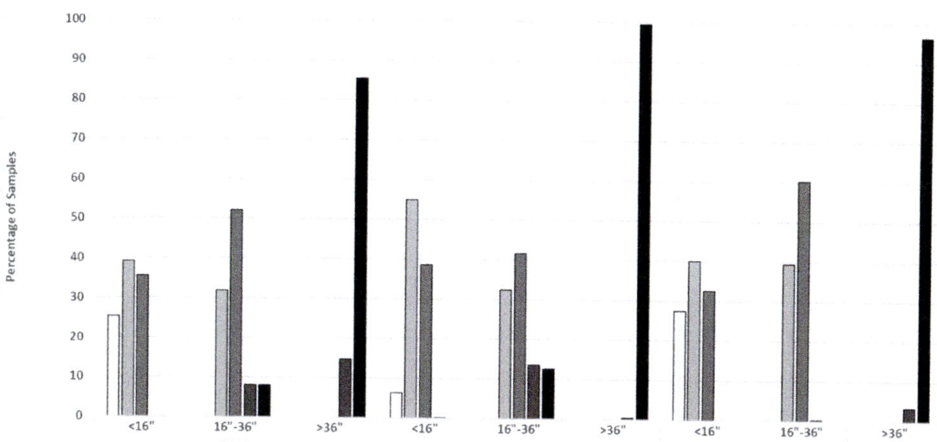

Category	RCCP <16"	RCCP 16"-36"	RCCP >36"	PCCP <16"	PCCP 16"-36"	PCCP >36"	AC <16"	AC 16"-36"	AC >36"
Insignificant	25.38	0	0	6.37	0	0	27.42	0	0
Minor	39.14	31.8	0	55.02	32.38	0	39.99	39.4	0
Moderate	35.48	52.06	0	38.58	41.51	0	32.59	60.36	0
Major	0	8.09	14.67	0.04	13.54	0.48	0	0.24	3.45
Catastrophic	0	8.05	85.33	0	12.57	99.52	0	0	96.55

Material and Diameter

□ Insignificant ■ Minor ■ Moderate ■ Major ■ Catastrophic

Figure 12.6 Percentage of cementitious pipes in different CoF categories based on material and diameter.

12.5.2 Key Findings from the Results

Based on the results, it could be summarized that CoF ratings are consistently higher for larger diameter pipes. However, there is a difference between the CoF ratings of different materials in the diameter categories which can be attributed to the pipe's operational function (distribution, transmission, and redundancy). The authors also found that CoF

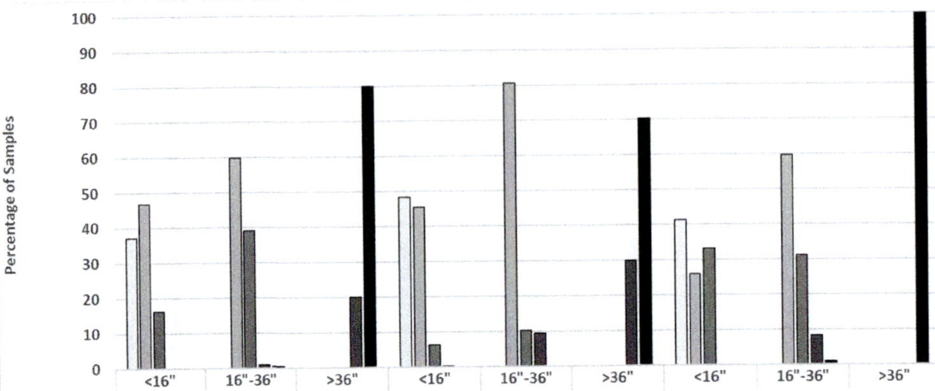

	<16"	16"-36"	>36"	<16"	16"-36"	>36"	<16"	16"-36"	>36"
		PVC			HDPE			Composite	
Insignificant	37.02	0	0	48.27	0	0	41.13	0	0
Minor	46.74	59.72	0	45.39	80.43	0	25.81	59.63	0
Moderate	16.23	38.95	0	6.33	10.25	0	33.06	31.19	0
Major	0	0.9	20	0.01	9.32	29.79	0	8.26	0
Catastrophic	0	0.43	80	0	0	70.21	0	0.92	100

Material and Diameter

Insignificant Minor Moderate Major Catastrophic

Figure 12.7 Percentage of plastic and composite pipes in different CoF categories based on material and diameter.

ratings increase from average values as the pipe location changes to more critical ones. As a result, location plays an important role in CoF modeling. Also, the model estimates higher CoF ratings for larger diameter pipes, which are mostly transmission pipes serving a larger network of distribution pipes. Results from correlation analysis showed that the model is most correlated to redundancy in the pipe system followed by pipe pressure and financial impact. One of the least correlated parameters are ground cover and road and railway traffic flow impact. The model results are most sensitive to results from economic impact and operational impact.

12.6 Conclusions

Based on this study, it was found that risk assessment is an important component of an advanced asset management program for a water utility and has multiple benefits, like the support for water pipeline renewal prioritization decisions, allocation of budgets, and moving from a reactive to a proactive asset management culture. This study highlighted the drawbacks in the previous methods of assessing the CoFs and application of risk matrix in the risk management framework for water utilities. These drawbacks ranged from the lack of comprehensive parameters, suboptimal methodologies, and limited verification and validation for CoFs assessment models. The proposed improvements are based on comprehensive literature and practice review from within and outside the water industry. The CoF model follows a process guided approach which integrates heuristic knowledge (from the literature review, practice review, and interviews with water utility experts) with data-driven machine learning methods. It was also found that CoF ratings are consistently higher for larger diameter pipes. The CoF ratings were found to increase from average values as the pipe location changes to more important ones. As a result, location plays an important role in CoF modeling. Also, the model was found to estimate higher CoF ratings for larger diameter pipes, which are mostly transmission pipes serving a larger network of distribution pipes. This study also found that larger diameter CI (cast iron), AC (asbestos cement), and PVC (polyvinyl chloride) pipes fall into the highest risk category ratings. Smaller-diameter AC, PVC, HDPE (high-density polyethylene) and steel pipes were found to have the lowest risk values. The research team also found that pressure is an important parameter to incorporate into CoF modeling frameworks as pressure with water pipes can explain the extent of damage to surroundings in the event of failure. However, the research team found that pressure data are usually not integrated well with the water pipe's geospatial information and are mostly a derived estimate using pressure zone information.

12.7 Recommendations

The authors found that risk assessment is critical to an advanced asset management program as it supports renewal prioritization decisions, helps in allocation of capital improvement and O&M budgets, and helps move water utilities from reactive to a more proactive form of asset management. It is recommended that water utilities need to collect better CoF data to help develop and support more comprehensive models which can determine CoF costs more accurately. Based on this study, it can also be suggested that probabilistic CoF modeling can capture uncertainty better than deterministic models, and modeling techniques for water pipes need to move from a deterministic to a probabilistic framework. However, one caveat with probabilistic analysis is that it requires higher volumes of good quality data. It is also recommended to include hard and soft costs of failure and not just the direct cost to the water utility for CoF modeling. The final recommendation from this study would be that decision support systems should provide maximum information to the decision-makers rather than being fully prescriptive. This is because it is complex to incorporate decision attitudes for different organizations into a single framework. In these situations, the more information the water utility decision-makers have, the more usable the decision support system will be for water utilities.

References

1 Higgins, M.S. et al. (2012). Numbers don't lie: PCCP performance and deterioration based on a statistical review of a decade of condition assessment data. *Pipelines 2012*. https://acppa.org/wp-content/uploads/2011/05/027-Numbers-Dont-Lie-PCCP-Performance-and-Deterioration.pdf (accessed 4 October 2022). 298–306.

2 Zarghamee, M.S. et al. (2011). Statistical analysis of condition assessment data and prediction of future performance of PCCP. *Pipelines 2011*. https://ascelibrary.org/doi/10.1061/41187%28420%2916 (accessed 4 October 2022). 160–169.

3 Bubtiena, A.M. et al. (2011). *Application of Artificial Neural Networks in Modeling Water Networks*. IEEE.

4 Kutyłowska, M. (2015). Neural network approach for failure rate prediction. *Engineering Failure Analysis* 47: 41–48.

5 Christodoulou, S. et al. (2010). Proactive risk-based integrity assessment of water distribution networks. *Water Resources Management* 24 (13): 3715–3730.

6 Marzouk, M. and Osama, A. (2017). Fuzzy-based methodology for integrated infrastructure asset management. *International Journal of Computational Intelligence Systems* 10 (1): 745.

7 Valis, D. et al. (2017). *Modelling Water Distribution Network Failures and Deterioration*. IEEE.

8 Bruaset, S. et al. (2018). Performance-based modelling of long-term deterioration to support rehabilitation and investment decisions in drinking water distribution systems. *Urban Water Journal* 15 (1): 46–52.

9 St. Clair, A.M. and Sinha, S.K. (2014). Development of a fuzzy inference performance index for ferrous drinking water pipelines. *Journal of Pipeline Systems Engineering and Practice* 5 (3): 04014003.

10 Dawood, T., Elwakil, E., Novoa, H.M.J., and Delgado, F.G. (2020). Water pipe failure prediction and risk models: state-of-the-art review. *Canadian Journal of Civil Engineering* 47 (10): 1117–1127.

11 Sadiq, R. et al. (2007). Water quality failures in distribution networks-risk analysis using fuzzy logic and evidential reasoning. *Risk Analysis* 27 (5): 1381–1394.

12 Tabesh, M. et al. (2011). Calibration of water distribution hydraulic models: a comparison between pressure dependent and demand driven analyses. *Urban Water Journal* 8 (2): 93–102.

13 Yannopoulos, S. and Spiliotis, M. (2013). Water distribution system reliability based on minimum cut-set approach and the hydraulic availability. *Water Resources Management* 27 (6): 1821–1836.

14 Cromwell, J.E. (2002). *Costs of Infrastructure Failure*. American Water Works Association.

15 Grigg, N. et al. (2007). *Main Break Prediction, Prevention, and Control*. Awwa Research Foundation.

16 Damodaran, N. et al. (2005). *Customer Acceptance of Water Main Structural Reliability*. American Water Works Association.

17 Gaewski, P.E. and Blaha, F.J. (2007). Analysis of total cost of large diameter pipe failures. In: *Proceedings of the AWWA Research Symposium Distribution Systems: The Next Frontier*, Reno, NV.

18 Vladeanu, G.J. and Matthews, J.C. (2019). Consequence-of-failure model for risk-based asset management of wastewater pipes using AHP. *Journal of Pipeline Systems Engineering and Practice* 10 (2): 04019005.

19 Parvizsedghy, L. and Zayed, T. (2015). Consequence of failure: neurofuzzy-based prediction model for gas pipelines. *Journal of Performance of Constructed Facilities* 30 (4): 04015073.

20 Raucher, R.S. (2017). *Managing Infrastructure Risk: The Consequence of Failure for Buried Assets*. Water Research Foundation.

21 Coghill, M.R. (2013). Prestressed concrete cylinder pipe management: communication methodologies and decision support tools for the San Diego County water authority: a case study. In: *Pipelines 2013: Pipelines and Trenchless Construction and Renewals: A Global Perspective* (ed. S. Arnaout and L. Slavin), 787–799, American Society of Civil Engineers.

22 Vladeanu, G.J. and Matthews, J.C. (2019). Consequence-of-failure model for risk-based asset management of wastewater pipes using AHP. *Journal of Pipeline Systems Engineering and Practice* 10 (2). doi: 10.1061/(ASCE)PS.1949-1204.0000370.

23 Gaewski, P.E. and Blaha, F.J. (2007). Analysis of total cost of large diameter pipe failures. *Proceedings of the AWWA Research Symposium Distribution Systems: The Next Frontier*. Reno, Nevada. http://infra-tect.com/wp-content/uploads/2014/11/Analysis-of-Total-Cost-of-Large-Diameter-Pipe-Failures.pdf (accessed 4 October 2022).

24 Xu, H. and Sinha, S.K. (2020). *Applying survival analysis to pipeline data: gaps and challenges*. Pipelines 2020. https://doi.org/10.1061/9780784483213.017 (accessed 4 October 2022). 148–158.

25 Vishwakarma, A. (2019). *Development of a Performance Analysis Framework for Water Pipeline Infrastructure Using Systems Understanding*. Virginia Tech.

26 Vishwakarma, A. and Sinha, S. (2019). *Water pipeline infrastructure systems understanding for performance analysis of cast iron pipes*. Pipelines 2019. https://ascelibrary.org/doi/book/10.1061/9780784482506 (accessed 4 October 2022). 398–407.

27 Xu, H. (2017). *Seismic Fragility Assessment of Bridge and Its Network Using Response Surface Models*. Harbin Institute of Technology.

28 Xu, H. and Sinha, S.K. (2019). *A framework for statistical analysis of water pipeline field performance data*. Pipelines 2019: 180–189.

29 Xu, H. et al. (2020a). Development of a fuzzy inference performance rating system for water pipelines using a comprehensive list of input variables. *Pipelines 2020*: 178–188.

30 Xu, H. et al. (2020b). *Development of a Fuzzy Inference Performance Rating System for Water Pipelines Using a Comprehensive List of Input Variables*. American Society of Civil Engineers.

31 Vishwakarma, A. and Sinha, S.K. (2020). *Development of a consequence of failure model and risk matrix for water pipelines infrastructure systems*. Pipelines 2020: 169–177.

32 Chen, C. et al. (2004). Fuzzy logic-based life-cycle costs analysis model for pavement and asset management. In: *6th International Conference on Managing Pavements*, Brisbane, Queensland, Australia.

33 Zadeh, L.A. (1997). The roles of fuzzy logic and soft computing in the conception, design and deployment of intelligent systems. In: *Software Agents and Soft Computing Towards Enhancing Machine Intelligence* (ed. H.S. Nwana and N. Azarmi), 181–190. Springer.

34 Zadeh, L.A. (1965). Fuzzy sets. *Information and Control* 8 (3): 338–353.

35 Mamdani, E.H. and Assilian, S. (1975). An experiment in linguistic synthesis with a fuzzy logic controller. *International Journal of Man-Machine Studies* 7 (1): 1–13.

13

Fuzzy Logic Applications for Water Pipeline Performance Analysis

Hao Xu and Sunil K. Sinha

Virginia Tech, Blacksburg, VA, United States of America

13.1 Introduction

Data are crucial to any analysis but providing valuable information and knowledge from data requires an understanding of the system being studied to determine the framework and potential value in the analysis. This is also reinforced by the current data science frameworks that propose hybrid modeling as the new approach to scientific discovery. Every analysis is based on knowledge developed through an extensive review of literature and practice, which is the foundation of all data analyses in this chapter. In the past century, many technological advancements in engineering models, manufacturing, pipeline appurtenances, installation techniques, and improvement of national and international standards have affected the performance of water pipelines. Before performing any analysis, including data analysis to understand historical changes in break rates, of installation trends, these technological advancements should be considered to identify any drastic changes in the trends for pipeline performance. The research team is engaging experts in water utilities, consulting engineers, and pipeline manufacturing associations, and is leveraging the Sustainable Water Infrastructure Management Center (SWIM) knowledgebase to build a strong foundation for data analyses for this research.

13.1.1 Water Transmission and Distribution Pipeline Infrastructure System Elements

The purpose of water transmission and distribution systems is to supply water to customers and provide fire protection to communities with minimum disruptions and water losses. Three criteria guide this:

- Structural reliability: To maintain structural strength throughout the lifecycle with minimum costs.
- Functional reliability: To deliver water at sufficient pressures to customers with adequate hydraulic capability and minimize utility pumping costs.
- Safety requirement: To deliver water free of pathogens, contaminants from infiltration, and internal leaching from water pipelines.

Water distribution systems generally consist of pipelines, pumps, valves, hydrants, storage tanks, reservoirs, water wells, treatment plants, meters, fittings, and other hydraulic appurtenances that help in connecting and managing water flow from the treatment plants or

Fuzzy Systems Modeling in Environmental and Health Risk Assessment, First Edition. Edited by Boris Faybishenko, Rehan Sadiq, and Ashok Deshpande.

sources to the customers. It is essential to understand the characteristics of these system elements and their interaction with water pipelines and other system elements.

13.1.2 Water Pipelines

Water pipelines are installed to transport safe and sufficient water to customers with reliable service. Based on function, water pipelines are categorized into transmission mains (to transport water from a source such as treatment plants or reservoirs), distribution mains (to transport treated water from transmission mains to curb stops), service lines (to transport treated water from curb stops to individual customers), and premise plumbing (to transport water within a customer facility). Different materials are usually preferred for each category, and the diameters get progressively smaller from transmission mains to premise plumbing, owing to the amount of water transported. The overall categories of the pipeline materials as shown in [1] are used in this chapter. For this research, transmission and distribution pipelines are included in this study, and service pipelines are not included.

13.2 Performance Models Literature Review

There has been extensive research on diverse statistical modeling techniques. This chapter is not aimed at distinguishing or judging the different ways of classifying these models. In this chapter, several major techniques are selected, and for each type of them, the definition, purpose, pros, and cons are discussed. This section provides an overview of a large body of work that quantifies water pipe infrastructure deterioration by analyzing historical performance data.

13.2.1 Deterministic Models

Deterministic models can be used in cases where the relationships between components and the precise measurements from the studied system are known. The two major approaches in this type of modeling can be categorized into empirical and mechanistic. An example of empirical approach for water pipes is finding the relation between failure rates and the influencing system parameters. The main limitation, apart from the availability of precise measurements, is that this approach can only be applied to a cohort of prespecified pipes. An example of the mechanistic approach related to water pipes is the prediction of the service lifetime of each pipe. The main limitation of this approach is that it assumes that the failure mode and mechanisms of the studied pipes are well understood.

Deterministic models can work well to predict corrosion rates, estimate the remaining wall thickness, and relate its expected pipe service lives given that enough precise measurements are available from the field. These models can be expressed in the form of equations with limited parameters and can be only applied at a cohort level. Cohorts can be developed based on the homogeneity of certain pipes with respect to material, diameter, internal or external protections, external environment, and other key common features. Regression models can provide a simplified method of learning and determining spatiotemporal relationships while developing failure rate curves. It relies on learning from historical patterns in the feature set to predict future failure rates. Previous studies have employed various types of regression techniques like linear, exponential, and gaussian to determine relationships and estimate future failure rates as a function of different influencing factors. Previous studies have shown that most of the existing methods in literature are regression-based algorithms, combining mechanistic empirical analysis, and opinions from experienced engineers [2].

13.2.2 Probabilistic Models

Statistical modeling is a conventional technique used for predicting the expected service life of infrastructure assets [3–7]. Specifically, the utilization of statistical modeling from observed data also helps in describing key data characteristics. Statistical methodology is typically applied to homogeneous cohorts based on common characteristics of the asset data. It requires historical failure or condition data for application. The main limitation of this technique stems from the requirement of large historical data as well as the inability to recognize patterns for future predictions.

Probabilistic modeling analysis in the pipeline domain relies on the relative frequency or instances of asset failure. Historical condition or performance data along with asset attributes can be used to predict the future probability of failure [8–10]. These have been consistently applied for pavements, bridges, and other critical infrastructure asset management for supporting renewal (repair, rehabilitation and replacement) decisions. The main limitation is that these models require extensive historical data [11–20].

13.2.3 Advanced Mathematical Techniques

Advanced mathematical models used for water infrastructure can be categorized into learning methods like ANNs (artificial neural networks) and expert systems like fuzzy logic (FL). These two types of advanced mathematical models present approaches that have been used frequently for infrastructure deterioration [21].

13.2.3.1 Artificial Neural Network Model

Machine learning algorithms are generally considered advanced algorithms. There have been several works of literature focusing on applying such algorithms into pipe condition assessment problems. In the problem of building the relations between output and input variables, the utilization of supervised learning algorithms is common since output variables can act like tags. Such algorithms include ANN, support vector machine (SVM), naive Bayes, decision trees, random forest, etc. There has been much research applying and comparing such algorithms in an assessment of a pipeline's condition, but there is no general conclusion that a specific algorithm outperforms the others. Some algorithms may outperform the others depending on the modeling framework and the feature set. ANN models have been explored for various applications in previous studies [7, 20, 22–24].

The machine learning based model is the most data-demanding type of model. It can outperform statistical models when the data are sufficient and accurate enough. However, the water industry's current situation cannot meet the enormous data needs in most cases. An ANN is used to model the deterioration or failure of pipe infrastructure [25]. A neural network comprises interconnected processing elements, often referred to as *neurons* that work together to provide a result. Each element is relatively simple but can become very complicated when multiple interconnecting networks are involved. When properly trained, the ANN model mimics the human brain's functioning through pattern recognition and generalization capabilities. Overall, this methodology provides a useful tool for modeling asset deterioration or failure due to its nonlinearity, adaptivity, and learning capabilities. ANN modeling entails the use of all variables that influence the service life of a pipe. Due to its highly complex, nonlinear, and parallel computing, the ANN is primarily inspired by how information in a biological nervous system is processed. The input information and the predicted outputs are linked with the use of functional relationships. This approach can be used to develop an individual asset or a network.

13.2.3.2 Fuzzy Logic Models

FL is a mathematical method used to deal with systems provided with inaccurate information or uncertainty. Variables are assigned a degree of membership on a continuous interval [0, 1] and subjected to approximate reasoning in a way that the human brain functions. Fuzzy methods are recognized to handle systems that are subject to uncertainties, ambiguities, and contradictions. In general, this nonlinear model provides a rules-based expert system capable of imitating the human thought process. Researchers have verified the advantages of fuzzy inference techniques over some other models such as the weighted summation ones. St. Clair [26] developed a novel performance index for metallic pipes using fuzzy inference methodology. Altogether, 27 parameters were used for this model. The fuzzy model can handle only several input and output variables easily as the model is made out of many membership functions, which accounts for many uncertainties in an event. The model testing is performed using the data from water utilities. It was found that the fuzzy model accurately predicted the performance of the pipelines utilizing several if–then statements.

FL modeling entails the use of FL-based techniques that possess the ability to incorporate engineering judgment to predict infrastructure deterioration. This type of model is often used where data are scarce, cause–effect knowledge is imprecise, and observations and model criteria are expressed in vague or fuzzy terms.

13.3 Water Utility Practice Review

To evaluate and predict pipe conditions, statistical models such as Weibull curves have been used to predict failure. The remaining service life is determined based on that. In current practice, most of the renewal practices are based on the risk analysis results. The business risk exposure (BRE) provides the criticality of the pipe, and priority is assigned to the pipe for renewal activities. Some case studies show prioritization based on economic analysis, such as the triple bottom line (TBL) or other models, such as the Nessie curve. Other cases use pipe failure prediction results from a statistical model, such as a Weibull curve, or physical analysis. Models should be validated with a large volume of data across utilities to make them more robust and reliable. In some case studies, the models that are used by utilities were not validated. When using the models, reasonable assumptions should be made if some data are not available. Risk analysis to determine likelihood of failure (LoF) and consequence of failure (CoF) is challenging. Currently, many utilities use their in-house practices to calculate LoF and CoF. There is a strong need for common standards for calculating LoF and CoF. Furthermore, more advanced models are needed to quantify LoF accurately. The renewal prioritization is highly dependent on the improvement of condition evaluation, prediction, and risk analysis. The case studies reviewed are summarized in Table 13.1.

Previous studies from practice can be summarized to have used mostly heuristic models with limited parameters and data. The issue with these models is that, although it might work on the particular utility dataset, it will have difficulty generalizing datasets for water pipelines in other scenarios.

Table 13.1 Summary of case studies.

Utility	Technique	Prerequisite	Comments
WSSC	Performance curve	FEM	Can be used for preliminary risk evaluation of PCCP but developed based on conservative assumptions and does not consider the coating's bonding effect
Sydney Water	Extreme value analysis	Pitting corrosion	Reliability and certainty of condition assessment could be improved by further monitoring of the pipeline condition over the entire length Predictions using structural and statistical modeling are currently based on the inspection results of six locations, and an assumption is made that these locations represent a homogeneous condition around the inspection site
PWB	Weibull curve	Pipe and environment data	Leak data per pipe segment of the pipeline with a similar installation date is needed to fit the distribution Segment assumed to be 340 ft
City of Edmonton	ANN	Pipe and environment data	They were unable to predict with sufficient advance notice for this to be suitable Simulation related to the number of breakages that would occur during a period not to the time of the breakage
WSSC	Risk model (BRE score)	LoF, CoF, risk mitigation	The PoF is based on the remaining asset life Accuracy of such estimation depends on the type of BRE being evaluated: basic or advanced CoF is estimated with the support of tables developed by WSSC using a Delphi method Risk mitigation options are included in a resiliency factor, which measures how many resources and what plans are in place in anticipation of asset failure BRE is estimated as a function of PoF, CoF, and risk mitigation options BRE scores and costs are used as decision-support tools
PWB	Risk model (BRE score)	PoF, CoF	The BRE is estimated as a function of CoF and LoF (estimated on a 1–5 scale, where 1 = minimum tier, and 5 = maximum) When considering CoF, some consequences are so high that they are considered in a separate category known as *above tier 5*. These are extremely high consequences such as multiple life losses and others
SPU	Strategic asset management plan (SAMP)	Criticality criteria	The criticality of the pipes is the basis for prioritizing renewal activities. The criticality (or risk) of the pipe is composed of CoF and LoF Criteria for defining critical assets are based on the cost of repair, the number of affected customers, and the time needed for the repair, among other factors
SPU	TBL analysis		Has developed organizational structures to deal with improved project review and decision-making TBL analysis considers economic, social, and environmental impacts of a project (including positive and negative effects)
Sydney Water	PARMS-PRIORITY	Long-term capital investment, KANEW model	Utilizes the program PARMS-PRIORITY to predict the condition of water pipelines The five core tasks of this model are: risk calculation, failure prediction, cost assessment, data exploration, and scenario evaluation Utilizes a ranking system that is based on the product of LoF and CoF

13.4 Performance Index Model Development

This section aims to develop a robust fuzzy inference system to predict performance indices for different types of water pipe types, such as including metallic, cementitious, and plastic pipes. Table 13.2 provides definitions of the proposed performance indices. Metallic includes pipe materials such as gray cast iron, ductile iron, galvanized iron, and steel pipes; cementitious includes pipe materials such as prestressed concrete cylinder pipes, bar-wrapped pipes, asbestos cement, and reinforced concrete cylinder pipes; and plastic pipes include pipe materials like polyvinyl chloride and high-density polyethylene. It should be noted that these materials can be categorized further based on design standards, also commonly referred to as vintage. This study makes use of a comprehensive list of variables, including structural, operational, and environmental factors that can capture the performance of these different materials based on their internal and external environments [22, 23, 27, 28]. The results from the proposed FL model can also be piloted with utilities for verification and validation purposes based on the protocols defined in the subsequent sections. The performance model and a management plan will help in the prioritization of future inspection and maintenance practices.

13.4.1 Performance Scale

A scale is important to develop in order to visualize results from the performance model. Based on the comments from a 25-member panel of experts from water utilities, consultants, and academics, it was decided that a 1–5 scaling system would work best in this case. The 1–5 scale is most widely used, consistent with rating systems of other assets managed by water utilities, and can also provide a balanced tradeoff between granularity and statistical significance [26, 29].

Table 13.2 Proposed performance index definition.

Grade	Representation	Definition	Color
1	Excellent	Very low probability of failure Excellent expected service life Almost no deterioration or corrosion	Dark green
2	Good	Low probability of failure Satisfactory expected service life A little deterioration or corrosion	Light green
3	Fair	Acceptable probability of failure Acceptable expected service life Acceptable deterioration or corrosion	Yellow
4	Poor	Hard-to-accept probability of failure Short expected service life Obvious deterioration or corrosion	Orange
5	Very poor	Unacceptable probability of failure Unacceptable expected service life Unacceptable deterioration or corrosion The pipe can no longer provide service (needs to be replaced)	Red

13.4.2 Performance Parameters Selection

For metallic pipelines, the performance rating system consists of four subdivided modules, as shown in Figure 13.1, which is based on extensive literature and practice reviews. According to the data source, the parameters can be marked as direct records, indirect records, or educated guesses. The reliability of direct records is the highest, while the reliability of the educated guesses is the lowest among the three categories of parameter sources.

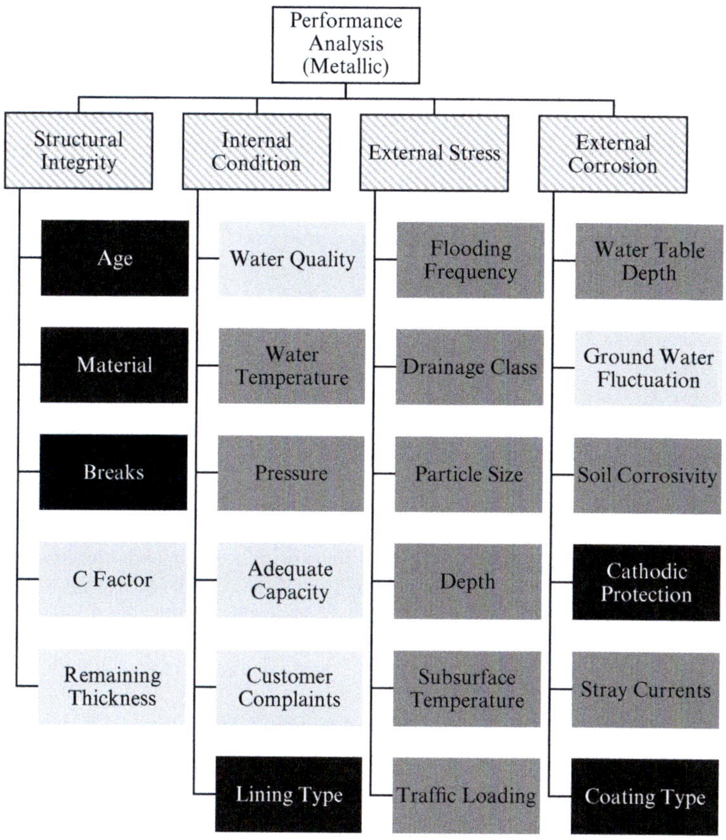

Figure 13.1 Data structure for performance analysis of metallic pipelines (diagonal patterns indicate performance model modules; black indicates parameters from direct record; dark gray indicates parameters derived indirectly; light gray indicates parameters requiring educated guesses).

The performance rating system for cementitious pipelines also consists of four modules. The development of the modules is performed similarly to metallic pipelines, with some differences in the variables selected based on the influencing factors. For example, the structural integrity module of cementitious pipelines does not have a C factor and remaining thickness. There are no stray currents or coating types in the external corrosion module, but there is a calcareous reaction that is identified as influential and relevant. The modules and parameters are shown in Figure 13.2.

The performance rating system for plastic pipelines consists of three modules. The corrosion module is eliminated as plastic pipes do not corrode. The external stress module is developed similarly for metallic, cementitious, and plastic pipelines, albeit with some

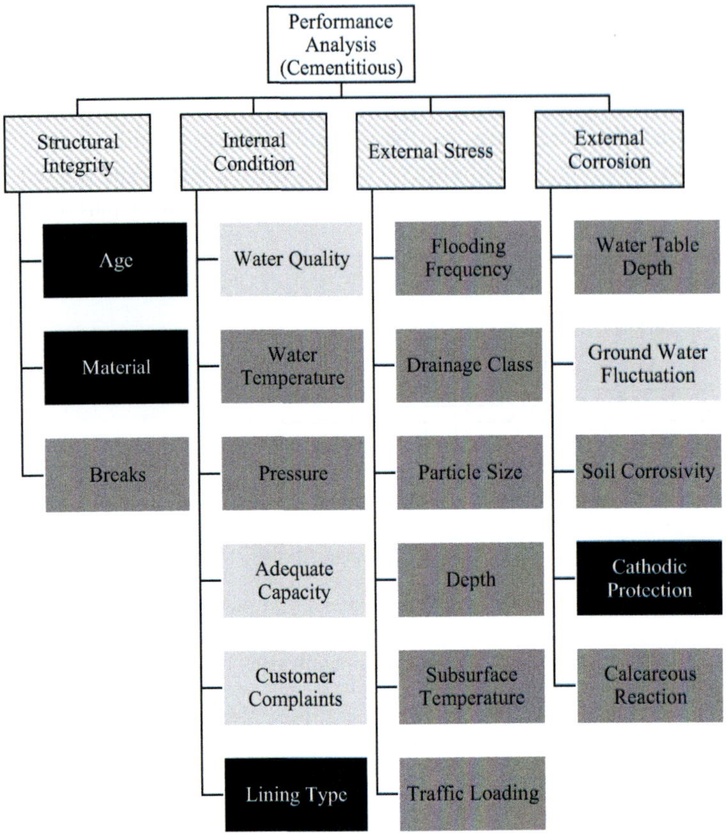

Figure 13.2 Data structure for performance analysis of cementitious pipelines (diagonal patterns indicate performance model modules; black indicates parameters from direct record; dark gray indicates parameters derived indirectly; light gray indicates parameters requiring educated guesses).

differences in the parameters based on the key influencing factors. For example, for plastic pipelines, there is a factor called ovality in the structural integrity module, and there is no lining type in the internal condition module. The modules and parameters are shown in Figure 13.3.

13.4.3 Parameter Membership Functions

In a FL model, the identified parameters can be configured to interact with other parameters using a fuzzy rule base. In order to correctly map the parameters into a fuzzy space, different membership functions can be used, as shown in Table 13.3. It shows the parameters and the membership function ranges used for the metallic pipes model. We have developed similar membership functions for the cementitious and plastic pipe models for applicability over an entire water pipe system. We have only used triangular membership functions as those are simple to understand and are more applicable to this problem. The nature of membership is such that these exist in overlapping ranges to map input parameters with degrees of membership for different membership functions within fuzzy space.

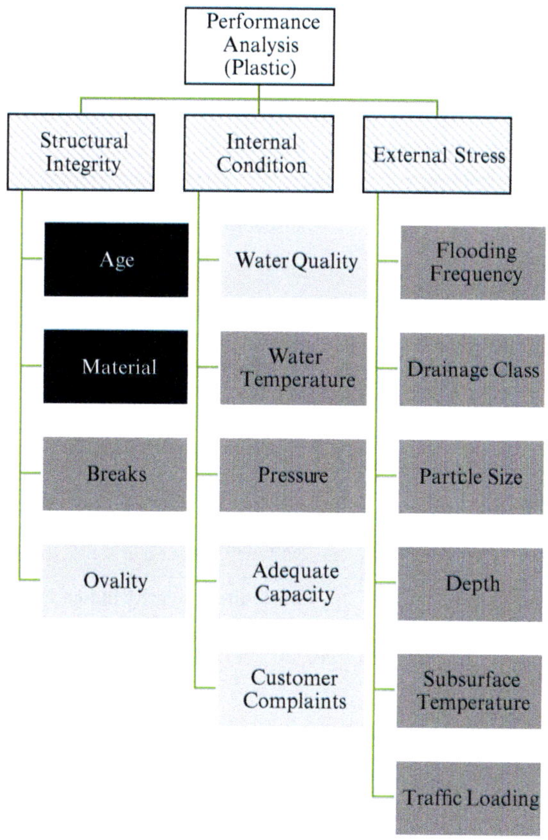

Figure 13.3 Data structure for performance analysis of plastic pipelines (diagonal patterns indicate performance model modules; black indicates parameters from direct record; dark gray indicates parameters derived indirectly; light gray indicates parameters requiring educated guesses).

Table 13.3 Parameters for metallic pipes model.

Structural integrity module			
Input variables	**Units**	**Membership function ranges**	**Explanation**
Pipe age	%	Young (<25) Medium (25–50) Old (>50)	Percentage of life completed based on suggested design life
Pipe material	Type	Good (CI <1930) Fair (CI >1930) Fair (DI asphaltic) Fair (DI others) Poor (ST <1930) Fair (ST 1930–1980) Good (ST >1980)	Pipe vintage based on design standards affects real-world performance

(Continued)

Table 13.3 (Continued)

Structural integrity module

Input variables	Units	Membership function ranges	Explanation
Pipe break	Number	Good (0–1) Fair (1–5) Poor (>5)	Historical breaks indicate susceptibility to future failures
C factor (tuberculation)	Value	High (90–160) Medium (70–110) Low (0–80)	Higher C-factor values indicate lower pipe tuberculation
Remaining wall thickness	%	Good (100–90%) Fair (89–75%) Poor (74–0%)	Wall thickness indicates susceptibility to hydraulic hoop stress
Output variables	**Units**	**Membership function ranges**	**Explanation**
Structural integrity	Scale	Excellent (1–2) Good (1–3) Fair (2–4) Poor (3–5) Very poor (4–5)	Structural integrity indicates the condition of pipe material

Internal condition module

Input variables	Units	Membership function ranges	Explanation
Water quality (aggressivity index)	Index	Good (<10) Fair (10–12) Poor (>12)	The aggressivity index can be estimated based on treated water chemical characteristics and can affect internal lining and pipe wall
Water temperature	Fahrenheit	Good (low) Fair (medium) Poor (high)	High water temperature facilitates pipe wall corrosion
Pressure	Psi	Good (low) Fair (medium) Poor (high)	High static and transient pressures in pipes can lead to pipe bursts
Pipe capacity	Linguistic	Good (yes) Poor (no)	Under-designed pipes put into service can create a weak link in pipe network
Customer complaints	Linguistic	Good (no) Poor (yes)	Customer feedback related to water pressure and aesthetics can indicate infiltration due to leaks or corrosion
Lining type	Linguistic	Good (lined) Poor (unlined)	Pipes can be lined during production/service to improve resistance to corrosion from treated water

(Continued)

Table 13.3 (Continued)

Internal condition module			
Output variables	**Units**	**Membership function ranges**	**Explanation**
Internal condition	Scale	Excellent (1–2) Good (1–3) Fair (2–4) Poor (3–5) Very Poor (4–5)	Internal condition of the pipe indicates vulnerability to internal corrosion and pressure-related failures
External stress module			
Input variables	**Units**	**Membership function ranges**	**Explanation**
Flooding frequency	Index	Excellent (none) Good (rare) Fair (occasional) Poor (frequent)	Flooding/standing water in the soils after wet-weather events can facilitate external corrosion
Drainage class	Index	Good (well drained) Fair (somewhat poorly drained) Poor (poorly drained)	Natural soil drainage properties affect duration of standing water and can influence external corrosion
Particle size	Index	Good (fine) Fair (medium) Poor (coarse)	Soil types with large particle sizes can create external pressure points for crack initiation
Buried depth	Feet	Low (<4 ft) Medium (4–10 ft) High (>10 ft)	Pipe depth indicates susceptibility to traffic loading
Subsurface temperature	Fahrenheit	Good (low) Fair (medium) Poor (high)	High temperature facilitates chemical reactions and external pipe corrosion
Traffic loading	Index	Good (low) Fair (medium) Poor (high)	Live loading from traffic causes compressive forces on the pipe wall Bending stresses can also be induced if the pipeline is not evenly supported
Output Variables	**Units**	**Membership Function Ranges**	**Explanation**
External stress	Scale	Excellent (1–2) Good (1–3) Fair (2–4) Poor (3–5) Very poor (4–5)	External stress indicates the influencing factors that are outside the pipe boundary

(Continued)

Table 13.3 (Continued)

External corrosion module			
Input variables	**Units**	**Membership function ranges**	**Explanation**
Water table depth	Feet	Good (>10) Fair (5–15) Poor (0–10)	Water table depth can indicate underground pipe flooding and corrosion risks
Ground water fluctuation	Index	Good (low) Fair (medium) Poor (high)	Groundwater fluctuation can destabilize pipe trench and also induce corrosion mechanisms
Soil corrosivity for steel	Ohm-cm	Low corrosivity (>3000) Medium corrosivity (2000–3000) High corrosivity (<2000)	Soil resistivity has a direct impact on the degree of corrosion in underground pipelines A decrease in resistivity relates to an increase in corrosion activity
Cathodic protection	Linguistic	Good (yes) Poor (no)	Cathodic protections using bonded anodes and impressed currents can change chemical pathways for corrosion from pipe to soil
Stray currents	Linguistic	Good (no) Poor (yes)	Stray currents are caused by a local direct current flowing through the earth from sources like electrified rail tracks and can lead to pipe corrosion
Coating type	Linguistic	Good (yes) Poor (no)	Coatings and polyethylene encasements can provide corrosion resistance in aggressive environments
Output variables	**Unit**	**Membership function ranges**	**Linguistic terms**
External corrosion	Scale	Excellent (1–2) Good (1–3) Fair (2–4) Poor (3–5) Very poor (4–5)	External corrosion indicates the corrosion on external side of pipe walls
Performance module			
Input variables	**Unit**	**Membership function ranges**	**Linguistic terms**
Structural integrity	Scale	Excellent (1–2) Good (1–3) Fair (2–4) Poor (3–5) Very poor (4–5)	Structural integrity indicates the condition of pipe material.

(Continued)

Table 13.3 (Continued)

Performance module			
Input variables	**Units**	**Membership function ranges**	**Explanation**
Internal condition	Scale	Excellent (1–2) Good (1–3) Fair (2–4) Poor (3–5) Very poor (4–5)	Internal condition of the pipe indicates vulnerability to internal corrosion and pressure-related failures
External stress	Scale	Excellent (1–2) Good (1–3) Fair (2–4) Poor (3–5) Very poor (4–5)	External stress indicates the influencing factors that are outside the pipe boundary
External corrosion	Scale	Excellent (1–2) Good (1–3) Fair (2–4) Poor (3–5) Very poor (4–5)	External corrosion indicates the corrosion on external side of pipe walls
Output variables	**Unit**	**Membership function ranges**	**Linguistic terms**
Performance index	Scale	Excellent (1–2) Good (1–3) Fair (2–4) Poor (3–5) Very poor (4–5)	Performance index is a comprehensive index of pipe condition which incorporates structural integrity and internal and external conditions

13.4.4 Fuzzy Logic Methodology

The methodology for the performance index rating is robust and holistic. The aim is to build a system that combines both expert knowledge and available data. This model is applicable to all pipeline material and diameter ranges. The first strategy is to get information from human experts. It means that the knowledge of the fuzzy rule based system (FRBS) is defined manually by engineers, who interview human experts to extract and represent their knowledge.

13.4.4.1 Incorporating Expert Knowledge Model with Data

The previous sections introduced the expert knowledge performance model (i.e. weighted factor and fuzzy inference techniques). However, they are mainly based on expert knowledge rather than historical water pipe database.

As shown in Figure 13.4, the main idea is to use the Bayesian method to combine a knowledge-based model with data. The model structure and parameter prior distribution can be derived from the existing knowledge-based model (weighted factor model or FL model).

In order to obtain a customized model, the pipes are stratified into three groups. According to the feedback from cooperating water utilities, 16 inches of diameter are

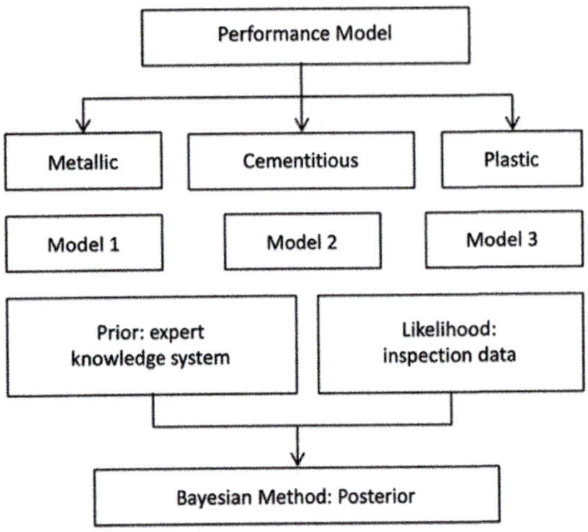

Figure 13.4 Framework of incorporating a knowledge-based model with data.

usually treated as a cutoff value between distribution pipelines and transmission pipelines. For small-diameter pipes, utilities most care about their failures or breaks rather than detailed condition assessments, so the data source of such kinds of pipes is a mainly historical work order. For large-diameter pipes, utilities usually prevent them from failure by maintaining them before they fail. For this reason, the break data in large-diameter pipes are rarely seen. So, the data source of large-diameter pipes is mainly from condition assessment data.

Bayesian inference is an important method to infer the underlying model function $g(\boldsymbol{x};\theta)$ for any empirical model that learns from data. Given the input vectors $\boldsymbol{X}_N = \boldsymbol{x}_i \forall i \in \{1,2,\ldots,N\}$ and the outputs $\mathbf{Y}_N = \boldsymbol{y}_i \forall i \in \{1,2,\ldots,N\}$, the posterior probability distribution of the function $g(\boldsymbol{x};\theta)$ can be expressed as

$$p(g(\boldsymbol{x};\theta)\,|\,\mathbf{Y}_N,\boldsymbol{X}_N) = \frac{p(\mathbf{Y}_N\,|\,g(\boldsymbol{x};\theta),\boldsymbol{X}_N)\,p(g(\boldsymbol{x};\theta))}{p(\mathbf{Y}_N\,|\,\boldsymbol{X}_N)} \tag{13.1}$$

where $g(\boldsymbol{x};\theta)$ is a generic representation for any model structure with parameters θ and variables $\boldsymbol{x}$.

For parametric models where the model structure of $g(\boldsymbol{x};\theta)$ is known, only the model parameters need to be inferred, and thus Eq. (13.1) can be expressed as

$$p(\theta\,|\,\mathbf{Y}_N,\boldsymbol{X}_N) = \frac{p(\mathbf{Y}_N\,|\,\theta,X_N)\,p(\theta)}{p(\mathbf{Y}_N\,|\,\boldsymbol{X}_N)} \tag{13.2}$$

In the case of fuzzy Bayesian learning [30], the function $g(\boldsymbol{x};\theta)$ is a fuzzy inference system with the rule base

$$\beta_k R_k : \mathit{if}\ A_1^k \oplus A_2^k \oplus \ldots \oplus A_{T_k}^k \text{ then } C_k \tag{13.3}$$

where β_k is a binary variable indicating whether to include the kth rule in the fuzzy inference system, A_i^k is a referential value of the ith antecedent attribute in the kth rule, T_k is the number of antecedent attributes used in the kth rule, C_k is the consequent in the kth rule, and $\oplus \in \{\vee,\wedge\}$ represents the logical operators (OR, AND) in the rules.

As triangular membership functions have been used in many previous studies [31–34], the membership functions for all the input and output variables in this study are also considered triangular, and each triangular membership function can be expressed as

$$f_i(u,\phi_i) = \begin{cases} \dfrac{u-\phi_{i1}}{\phi_{i2}-\phi_{i1}}, & \text{if } \phi_{i1} \le u \le \phi_{i2} \\ \dfrac{u-\phi_{i3}}{\phi_{i2}-\phi_{i3}}, & \text{if } \phi_{i2} \le u \le \phi_{i3} \\ 0, & \text{otherwise} \end{cases} \tag{13.4}$$

where u is the universe of discourse, and ϕ_{ij} is the parameter to be estimated for the jth corner point in the ith triangular function.

In this study, some rules and membership functions are considered uncertain, so the parameters (ϕ_{ij} and β_k) for those rules and membership functions are selected as the targets of inference. The posterior distributions can provide detailed distributions of parameters. A traditional statistical approach cannot replace reasoning, in that correlation does not suggest causation. The Bayesian method can solve this problem by incorporating the expert knowledge model as a prior structure.

13.4.4.2 Fuzzy Logic Performance Index Model

The FL framework of developing the performance index for water pipes can be upgraded. First, performance index systems for concrete and plastic water pipes do not exist. Although there are already comprehensive water pipe performance indices, they mostly concentrate on only one type of pipe. For example, St. Clair [35] developed a comprehensive framework to evaluate the performance index of metallic water pipes. Taking advantage of big data in PIPE*i*D, all kinds of pipes can be collected, such as metallic pipes, concrete pipes, and plastic pipes. Due to the relative lack of records for concrete and plastic pipes, there haven't been FL performance models for them. This study developed FL performance models for concrete pipes and plastic pipes [35]. In addition, the membership functions in previous models are mostly expert-based rather than data-driven. Using the vast number of data collected in PIPEiD, the membership functions can be calibrated. The importance of different variables can be analyzed and ranked. Then these variables can be included in the upgraded performance index development.

In order to obtain a customized model, the pipes are separated into three groups. According to the feedback from cooperating water utilities, 16 inches of diameter are usually treated as a cutoff value between distribution pipelines and transmission pipelines. For small-diameter pipes, utilities most care about their failures or breaks rather than any detailed condition assessments, so the data source of such kinds of pipes is usually an old work order. For large-diameter pipes, utilities usually prevent them from failure by maintaining them before they fail. For this reason, the break data in large-diameter pipes are rarely seen. So, the data source of large-diameter pipes is mainly condition assessment data. However, some utilities provide condition ratings for small-diameter pipes. The small- and large-diameter pipes can share one model to incorporate such a data source. In the fuzzy inference system, the distinction can be implemented by fuzzy rules such as, "If the diameter is small and the number of pipe breaks is 1, then the structural condition is fair" and, "If the diameter is large and the number of pipe breaks is 1, then the structural condition is poor." The interpretation of pipe breaks is different for small-diameter pipes than for large-diameter pipes. Within each of the six models, there are subgroups of material types, such as cast iron, ductile iron, and steel. However, these subgroups' performance features can be well handled by taking them as input variables or covariates.

13.4.5 Model Verification and Validation

Model verification and validation are important when checking how closely the model represents the real world. A working group was established with 25 participating water utilities along with a project advisory committee (PAC), who closely monitored the model's development and implementation. The overall framework for model verification and validation is shown in Figure 13.5.

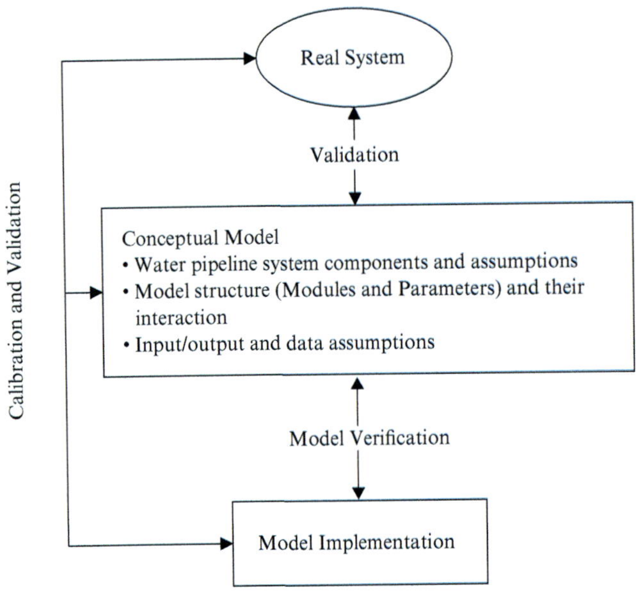

Figure 13.5 Model verification and validation steps.

13.4.5.1 Model Verification

Model verification is the task of determining if the implementation of the model was done correctly. The following techniques were used for model verification:

- Code verification: Software quality assurance (SQA) was performed by identifying and eliminating programming and implementation errors within the software. These were checked multiple times internally by the research team to ensure that the model was working as intended with the modules, parameters, and assumptions.
- Calculation verification: Model sensitivity to changing parameters and module results was assessed using correlation and sensitivity analysis. As part of the verification, the sensitivity of the model was tested based on the evaluation of three bands of input values classifying best, moderate, and worst-case values. The corresponding results from the model varied from 1 (excellent) to 5 (failed). The best values as inputs gave the best output, and correspondingly the moderate output gave *fair* results, and the worst values gave *failed* results. The model performed well when analyzing input data by the three-band scenarios. Test data were generated based on Monte Carlo simulations to statistically check the model's behavior.
- Boundary conditions were checked based on sample data from the collected datasets. The model algorithm was selected based on the following conditions:
  - Ability to incorporate knowledge into the framework: Given the uncertainty and missing parameters in the dataset, it is critical to ensure that the model was also able to include key assumptions based on knowledge.

- Ability to incorporate interrelationships between parameters: FL algorithm used is able to interpret parameter and module interrelationships using if–else rules.
- Ability to incorporate modules within the structure: Separate modules were developed within the overall model to incorporate specific areas (like structural integrity, internal stress, external stress, and corrosion), which is essential for model interpretability as well as result verification.

13.4.5.2 Model Validation Methods

The validation phase focuses on how well the real observed behavior of water pipeline systems can be mimicked within a model. This is critical for the applicability of any model in a real-world situation. Model validation is a rigorous process between the model developer and the intended users (in this case the water utility) to determine whether the model predictions are reliable and agree with the real world. The model can be adjusted based on feedback from the verification stage and preliminary validations. The validity of the models can be affected by the accuracy of the collected data, the usefulness and cost-effectiveness of implementing the model, the likelihood of its use, and its flexibility for further improvements by the water utility.

The research team had multiple rounds of validation with 25 participating water utilities within the model validation working group, which had high-quality data to support model validation and support the model results with ground truth information. The following techniques were used for model validation:

- Comparison to other models: The water utilities participating in the model validation stage shared their internal model results, which were compared with our results to check for potential discrepancy and variability in the result. These were rectified by calibrating model parameters and assumptions.
- Expert opinion: Around 40 meetings were conducted in total with participating water utilities in the validation stage, and results based on different scenarios were shared to discuss whether these results were reasonable. The application of the GIS (Geographical Information System) was discussed on GIS to pinpoint certain areas where the model might have predicted inaccurately. These errors were rectified by changing parameter assumptions and model rules.
- Subsystem validity: Some water utilities shared historical data, which were used to validate different modules within the performance model.

13.4.5.3 Model Validation with Ground Truth Data

There are many condition assessment techniques available to evaluate the current conditions of a water pipe. These can vary based on the level of details sought from the assessment, the type of material assessed, and the resource constraints with the water utility. The pipe condition values from these techniques can serve as the *ground truth* and can be used to validate the model performance in the real world. The following steps can be followed to validate the proposed performance model:

- Form a validation team comprising experts from a water utility: The utilities should have enough resources to enable the validation experiment. It is recommended to collaborate with larger water utilities for validation as they have enough resources, and the sheer size of their networks can support validations in a variety of contexts, like different soil corrosivity, pipe diameters, materials, etc.
- Collect ground truth data for the input parameters.
- Run the proposed FL model on the given input parameters.

- Share results with the validation team for blind testing.
- Compare results quantitatively (performance metrics like RMSE (root of the mean of the square of errors) MAE (mean of absolute value of errors)) and qualitatively (scenarios).

Given the resource intensiveness, as well as the complexity, of the real-world conditions, it is recommended that system level models such as the proposed performance model should not be evaluated in laboratory settings.

Based on verification with over 25 large water utilities who also participated in the PIPE*i*D project, the proposed technique can be summarized as a robust performance rating system capable of capturing performance accurately in a variety of scenarios. The proposed model can help water utilities in renewal decision-making, such as deciding when to replace a pipe.

13.5 Fuzzy Logic Performance Model Results

The current results seem to consolidate the performance index ratings mostly in the *good* and *fair* categories, which is due to favorable assumptions for the parameters that are missing in the database. This underlines the need to pilot the model in selected conditions with water utilities that can generate those datasets and help generate the results for validation. The percentage of pipe segments of different materials and diameters in different performance ratings are summarized in Table 13.4.

Table 13.4 Summary of pipe length as a percentage in different performance ratings.

Material type	Lengths of pipes in different performance ratings		Percentage in different performance ratings				
	Inches	**Miles**	**Excellent (%)**	**Good (%)**	**Fair (%)**	**Poor (%)**	**Very poor (%)**
DI	<16	50 310	0.00	75.74	24.26	0.01	0.00
	16–36	8137	0.00	76.41	23.59	0.00	0.00
	>36	772	0.00	76.38	23.62	0.00	0.00
CI pre-1930	<16	4527	0.00	47.08	52.91	0.00	0.00
	16–36	601	0.00	54.83	45.18	0.00	0.00
	>36	77	0.00	72.71	27.29	0.00	0.00
CI post-1930	<16	40 740	0.00	49.82	50.18	0.00	0.00
	16–36	3532	0.00	52.75	47.25	0.01	0.00
	>36	187	0.00	64.80	35.08	0.12	0.00
Steel	<16	7488	0.00	94.57	5.43	0.00	0.00
	16–36	1877	0.00	85.77	14.23	0.00	0.00
	>36	1903	0.00	81.89	18.11	0.00	0.00
GI	<16	19	0.00	62.84	37.16	0.00	0.00
PCCP	<16	576	0.00	62.68	37.32	0.00	0.00
	16–36	1943	0.00	86.25	13.75	0.00	0.00
	>36	1397	0.00	83.79	16.14	0.07	0.00

(Continued)

Table 13.4 (Continued)

Material type	Lengths of pipes in different performance ratings		Percentage in different performance ratings				
	Inches	Miles	Excellent (%)	Good (%)	Fair (%)	Poor (%)	Very poor (%)
RCCP	<16	35	0.00	93.64	6.36	0.00	0.00
	16–36	816	0.00	69.55	30.45	0.00	0.00
	>36	458	0.00	85.03	14.97	0.00	0.00
AC	<16	21 416	0.00	74.70	25.30	0.00	0.00
	16–36	2194	0.00	41.08	58.91	0.01	0.00
PVC	<16	26 271	13.00	86.97	0.03	0.00	0.00
	16–36	3293	25.55	74.45	0.00	0.00	0.00
HDPE	<16	399	14.33	85.67	0.00	0.00	0.00
	16–36	379	44.82	54.88	0.31	0.00	0.00

CI: cast iron; DI: ductile iron; GI: galvanized iron; PCCP: prestressed concrete cylinder pipe; RCCP: reinforced concrete cylinder pipe; AC: asbestos cement; PVC: polyvinyl chloride; HDPE: high density polyethylene.

The results are also summarized in Figures 13.6–13.8.

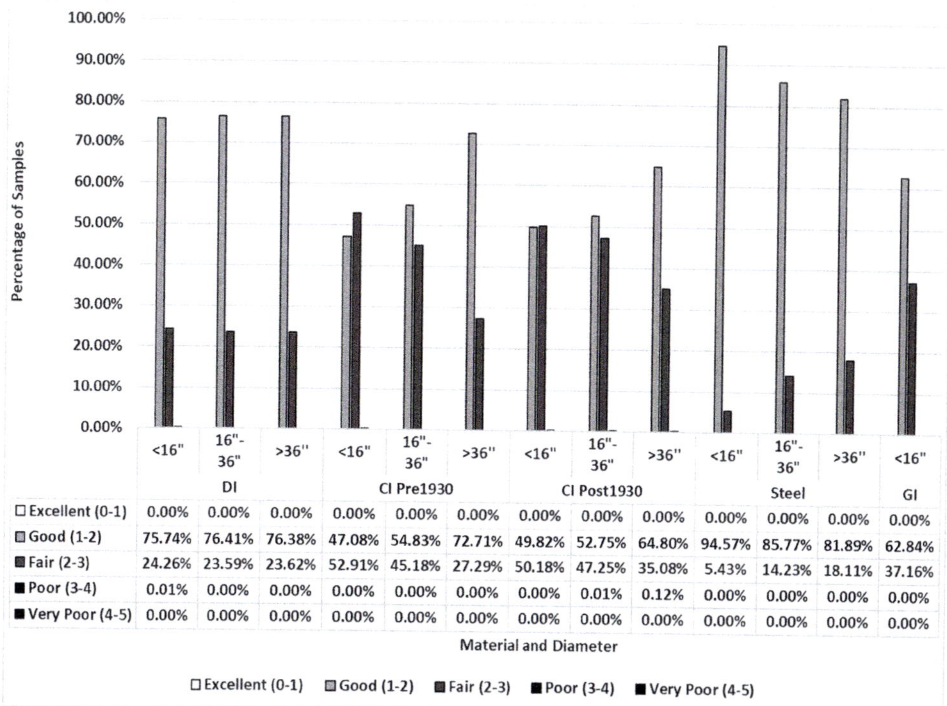

Figure 13.6 Percentage of metallic pipes in different performance ratings based on material and diameter.

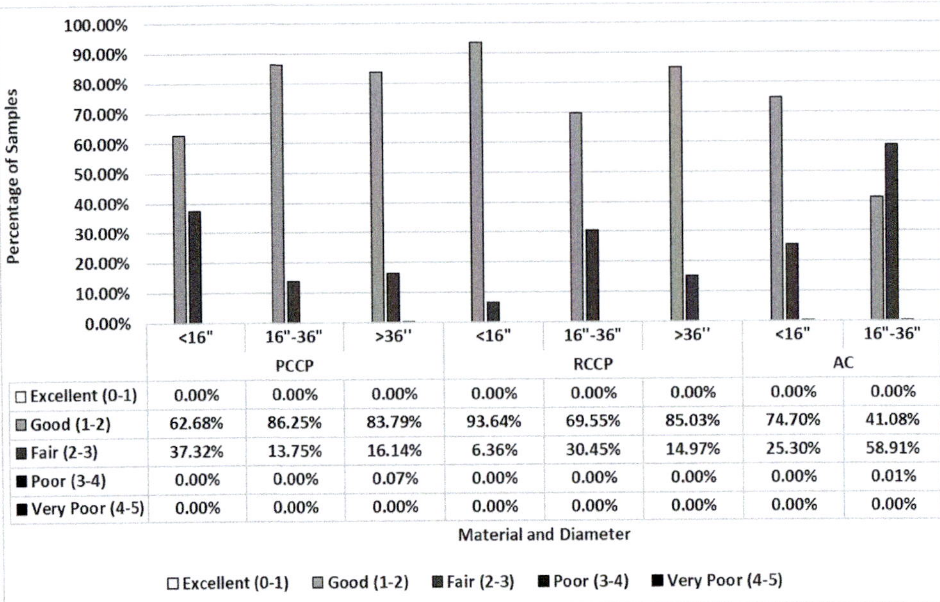

	PCCP			RCCP			AC	
	<16"	16"-36"	>36"	<16"	16"-36"	>36"	<16"	16"-36"
□ Excellent (0-1)	0.00%	0.00%	0.00%	0.00%	0.00%	0.00%	0.00%	0.00%
■ Good (1-2)	62.68%	86.25%	83.79%	93.64%	69.55%	85.03%	74.70%	41.08%
■ Fair (2-3)	37.32%	13.75%	16.14%	6.36%	30.45%	14.97%	25.30%	58.91%
■ Poor (3-4)	0.00%	0.00%	0.07%	0.00%	0.00%	0.00%	0.00%	0.01%
■ Very Poor (4-5)	0.00%	0.00%	0.00%	0.00%	0.00%	0.00%	0.00%	0.00%

Figure 13.7 Percentage of cementitious pipes in different performance ratings based on material and diameter.

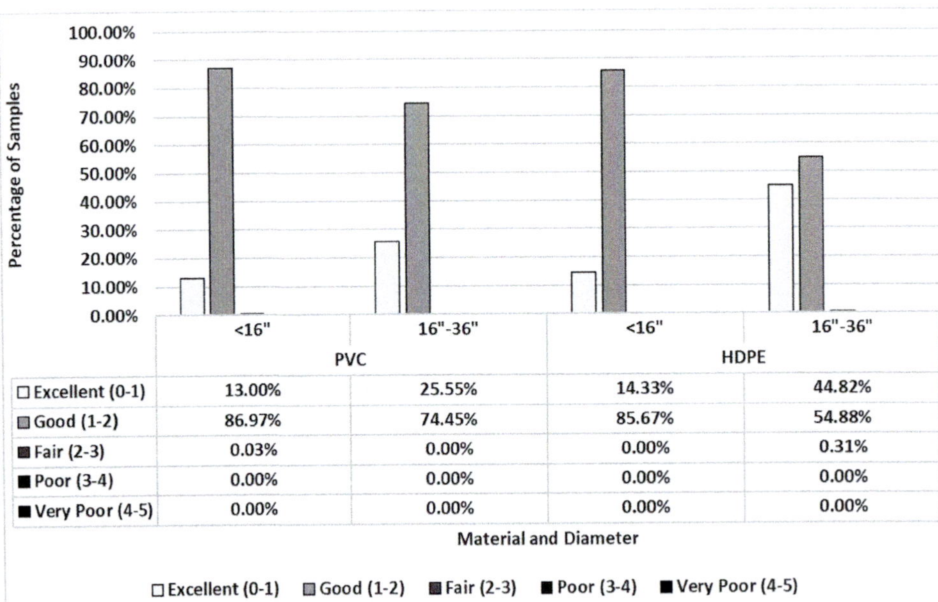

	PVC		HDPE	
	<16"	16"-36"	<16"	16"-36"
□ Excellent (0-1)	13.00%	25.55%	14.33%	44.82%
■ Good (1-2)	86.97%	74.45%	85.67%	54.88%
■ Fair (2-3)	0.03%	0.00%	0.00%	0.31%
■ Poor (3-4)	0.00%	0.00%	0.00%	0.00%
■ Very Poor (4-5)	0.00%	0.00%	0.00%	0.00%

Figure 13.8 Percentage of plastic pipes in different performance ratings based on material and diameter.

Figures 13.6–13.8 show that most of the pipes lie in the *good*, *fair*, and *poor* categories with fewer pipes lying in the *excellent* (brand new and without any problems) and *very poor* (almost failed) categories. The highest proportion of pipes in the excellent category is from HDPE pipes. A higher proportion of CI pipes are found in the *poor* category, which can be attributed to the fact that a large portion of these pipes are nearing the end of their performance lives.

13.6 Conclusions

We compared different water pipe performance modeling techniques and found that FIS are expert systems which can be built completely based on heuristic knowledge, in stark contrast to the other data-reliant, deterministic, probabilistic, and statistical techniques. The interpretability of a model is key for real-world application. As FL falls into explainable models category, it can have a wider acceptance among water utilities for predicting pipe performance and managing their water pipeline infrastructure systems. The key findings from this chapter are shown below:

- A standard performance index scale with clear definitions can be used to rate the water pipeline performance. The index scale can be decided based on the compatibility with other indices used for other assets as well as the granularity of the results.
- A fuzzy-based performance index model can capture data and knowledge and accurately represent water pipeline datasets.
- Consideration of the pipeline material and diameter is important in assessing the performance of the water pipelines.
- Based on the results from the proposed performance analysis, the current performance of plastic pipelines is relatively better than cementitious and metallic pipes since they are generally newly installed and have no failure records. Also, CI pre-1930 pipes perform worse in the US in arid conditions, which can be attributed to the influence of higher subsurface temperatures and stresses due to soil elasticity.
- Results from performance curves for different small (<16 in.) and medium (16–36 in.) material pipes show that CI prep1930 (16–36 in.) pipes have the highest performance life, which could be due to greater thickness which makes it more resistant to corrosion. On the other hand, HDPE (16–36 in.) has the lowest performance life, which can be attributed to the fact that these are newly installed and most failures are premature.
- Although the results from performance curves for large-diameter pipes in varying soil corrosivity showed that CI pipes have the highest performance life in high soil corrosivity, and DI has the lowest performance life in high soil corrosivity, these performance lives were found to vary based on the type of corrosion protection applied.
- It was also found from the application of machine learning techniques that the ANN model can learn more complexities and become better at generalization if more and better-quality data are fed to the algorithm.
- A performance index can be expanded in the future with the availability of more and better-quality data to improve granularity in the performance model results
- Most of the pipes in the US are in *good* and *fair* performance ratings. The performance of plastic pipes is relatively better than cementitious and metallic pipes since they are generally newly installed.
- Pipes in higher drainage class environments are subject to worse performance ratings. This can be attributed to higher failures due to corrosive environments.
- Pipes in higher flood frequency environments are subject to worse performance ratings. This can be attributed to a higher number of failures due to corrosive conditions.
- Unprotected metallic pipes in higher corrosive soil environments are subject to worse performance rating. Performance ratings can be influenced in this case by soil corrosivity and type of corrosion protection used (if any).

13.7 Recommendations

Some recommendations from this study are as follows:

- There should be more focus on collecting more reliable time-dependent data periodically in order to support stochastic and time-dependent analysis and more accurate water pipeline performance monitoring.
- More performance parameters need to be collected for water pipeline segments in the water utility dataset, which can support more comprehensive performance models. The water utility can also utilize datasets from other open-source databases like US Geological Survey (USGS), National Land Cover Database (NLCD), and the Natural Resources Conservation Service (NRCS) to enrich their dataset in case the data are unavailable, or collection is too costly.
- It is critical to use process-driven hybrid modeling or physics-based modeling, especially for the domain of water pipelines, because data reliability is a major issue. The hybrid model can then be evolved into a machine learning based model once data reliability is ensured, or the model is able to mimic the real-world situation accurately.
- The AI methodologies explored in this chapter discussed the application of different learning algorithms for analyzing water pipeline performance. This is an ongoing research and, as new and better-quality data are collected and shared with the research team, the learning can take place in a better way using more complex learning algorithms, including deep learning approaches.

References

1 Xu, H., Sinha, S.K., and Vishwakarma, A. (2020). Development of a fuzzy inference performance rating system for water pipelines using a comprehensive list of input variables. *Pipelines 2020*. https://ascelibrary.org/doi/10.1061/9780784483213.020 (accessed 4 October 2022). 178–188.

2 St. Clair, A.M. and Sinha, S. (2012). State-of-the-technology review on water pipe condition, deterioration and failure rate prediction models! *Urban Water Journal* 9 (2): 85–112.

3 Ge, S. and Sinha, S. (2014). Failure analysis, condition assessment technologies, and performance prediction of prestressed-concrete cylinder pipe: state-of-the-art literature review. *Journal of Performance of Constructed Facilities* 28 (3): 618–628.

4 Kahn, C., Damiani, A., and Ge, S. (2020). Validation of water main failure predictions: a 2-year case study. *AWWA Water Science* 2 (3): e1179.

5 Mazumder Ram, K., Salman Abdullahi, M., Li, Y., and Yu, X. (2018). Performance evaluation of water distribution systems and asset management. *Journal of Infrastructure Systems* 24 (3): 03118001.

6 Xu, H. (2017). *Seismic Fragility Assessment of Bridge and Its Network Using Response Surface Models*. Harbin Institute of Technology.

7 Xu, H. and Sinha, S.K. (2021). Modeling pipe break data using survival analysis with machine learning imputation methods. *Journal of Performance of Constructed Facilities* 35 (5): 04021071.

8 Opila Mary, C. and Attoh-Okine, N. (2011). Novel approach in pipe condition scoring. *Journal of Pipeline Systems Engineering and Practice* 2 (3): 82–90.

9 Wang, C., Niu, Z., Jia, H., and Zhang, H. (2010). An assessment model of water pipe condition using Bayesian inference. *Journal of Zhejiang University: Science A* 11 (7): 495–504.

10 Xu, H. and Sinha, S.K. (2020). Applying survival analysis to pipeline data: gaps and challenges. *Pipelines 2020*. https://ascelibrary.org/doi/10.1061/9780784483213.020 (accessed 4 October 2022). 148–158.

11 Carrión, A., Solano, H., Gamiz, M.L., and Debón, A. (2010). Evaluation of the reliability of a water supply network from right-censored and left-truncated break data. *Water Resources Management* 24 (12): 2917–2935.

12 Jenkins, L., Gokhale, S., and McDonald, M. (2014). Comparison of pipeline failure prediction models for water distribution networks with uncertain and limited data. *Journal of Pipeline Systems Engineering and Practice* 6 (2): 04014012.

13 Kleiner, Y. and Rajani, B. (2001). Comprehensive review of structural deterioration of water mains: statistical models. *Urban Water* 3 (3): 131–150.

14 Le Gat, Y. and Eisenbeis, P. (2000). Using maintenance records to forecast failures in water networks. *Urban Water* 2 (3): 173–181.

15 Pelletier, G., Mailhot, A., and Villeneuve, J.-P. (2003). Modeling water pipe breaks: three case studies. *Journal of Water Resources Planning and Management* 129 (2): 115–123.

16 Vladeanu, G.J. and Koo, D.D. (2015). A comparison study of water pipe failure prediction models using Weibull distribution and binary logistic regression. *Pipelines 2015*. 1590–1601.

17 Xu, H. and Sinha, S.K. (2019). A framework for statistical analysis of water pipeline field performance data. *Pipelines 2019*. https://ascelibrary.org/doi/10.1061/9780784482506.019 (accessed 4 October 2022). 180–189.

18 Yamijala, S., Guikema, S.D., and Brumbelow, K. (2009). Statistical models for the analysis of water distribution system pipe break data. *Reliability Engineering & System Safety* 94 (2): 282–293.

19 Zamenian, H., Abraham, D.M., and Mannering, F.L. (2016). Predicting water main breaks' frequency in water distribution systems. In: *Proceedings of the Construction Research Congress 2016* (ed. J.L. Perdomo-Riverahttps://ascelibrary.org/author/Perdomo-Rivera%2C%2BJos%C3%A9%2BL, A. Gonzáles-Quevedohttps://ascelibrary.org/author/Gonz%C3%A1les-Quevedo%2C%2BAntonio, C.L. del Puertohttps://ascelibrary.org/author/del%2BPuerto%2C%2BCarla%2BL%C3%B3pez, F. Maldonado-Fortunethttps://ascelibrary.org/author/Maldonado-Fortunet%2C%2BFrancisco, and O.I. Molina-Bashttps://ascelibrary.org/author/Molina-Bas%2C%2BOmar%2BI), 1599–1608. ASCE Library.

20 Zangenehmadar, Z. and Moselhi, O. (2016). Assessment of remaining useful life of pipelines using different artificial neural networks models. *Journal of Performance of Constructed Facilities* 30 (5): 04016032.

21 Hawari, A., Alkadour, F., Elmasry, M., and Zayed, T. (2020). A state of the art review on condition assessment models developed for sewer pipelines. *Engineering Applications of Artificial Intelligence* 93: 103721.

22 Al-Barqawi, H. and Zayed, T. (2006). Condition rating model for underground infrastructure sustainable water mains. *Journal of Performance of Constructed Facilities* 20 (2): 126–135.

23 Al-Barqawi, H. and Zayed, T. (2008). Infrastructure management: integrated AHP/ANN model to evaluate municipal water mains' performance. *Journal of Infrastructure Systems* 14 (4): 305–318.

24 Chen, T.Y.-J., Beekman, J.A., and Guikema, S.D. (2017). Drinking water distribution systems asset management: statistical modelling of pipe breaks. *Pipelines 2017*. 173–186.

25 El-Abbasy, M.S., Senouci, A., Zayed, T., Mirahadi, F., and Parvizsedghy, L. (2014). Artificial neural network models for predicting condition of offshore oil and gas pipelines. *Automation in Construction* 45: 50–65.

26 St. Clair, A.M. (2013). *Development of a Novel Performance Index and a Performance Prediction Model for Metallic Drinking Water Pipelines*. Virginia Tech.

27 Shi, F., Liu, Z., and Li, E. (2017) Prediction of pipe performance with ensemble machine learning based approaches. In: *Proc., 2017 International Conference on Sensing, Diagnostics, Prognostics, and Control (SDPC)*, IEEE, 408–414.

28 Shi, F., Liu, Y., Liu, Z., and Li, E. (2018). Prediction of pipe performance with stacking ensemble learning based approaches. *Journal of Intelligent & Fuzzy Systems* 34 (6): 3845–3855.

29 Angkasuwansiri, T. (2013). *Development of Wastewater Pipe Performance Index and Performance Prediction Model*. Virginia Tech.

30 Pan, I. and Bester, D. (2018). Fuzzy Bayesian learning. *IEEE Transactions on Fuzzy Systems* 26 (3): 1719–1731.

31 Angkasuwansiri, T. and Sinha, S. (2014). Development of a robust wastewater pipe performance index. *Journal of Performance of Constructed Facilities* 29 (1): 04014042.

32 St. Clair, A.M. and Sinha, S.K. (2014). Development of a fuzzy inference performance index for ferrous drinking water pipelines. *Journal of Pipeline Systems Engineering and Practice* 5 (3): 04014003.

33 Uslu, B. (2017). *Development of Protocols and Methods for Predicting the Remaining Economic Life of Wastewater Pipe Infrastructure Assets*. Virginia Tech.

34 Velayutham Kandasamy, V.P. and Sinha, S.K. (2018). Stormwater pipe performance index using fuzzy inference method. *Journal of Water Resources Planning and Management* 144 (10): 04018062.

35 St. Clair, A.M. (2013). *Development of a novel performance index and a performance prediction model for metallic drinking water pipelines*. Dissertation submitted to the faculty of the Virginia Polytechnic Institute and State University. https://vtechworks.lib.vt.edu/bitstream/handle/10919/19342/St._Clair_AM_D_2013.pdf?sequence=1&isAllowed=y (accessed 4 October 2022).

Part V

Using Fuzzy Logic for the Optimization of Water Treatment and Waste Management

14

Developing a Fuzzy-based Model for Regional Waste Management

Hirushie Karunathilake, Tharindu Prabatha, Kasun Hewage, and Rehan Sadiq

School of Engineering, University of British Columbia (Okanagan), Kelowna, BC, Canada

14.1 Introduction

With an increasing population, sustainable waste management has become a major issue, especially in urban environments throughout the world. Municipal solid waste (MSW) is produced through human activities, mainly as material disposed from urban households, which is then gathered and disposed of by the relevant municipalities [1]. While waste management primarily focuses on hygienic aspects, material recovery, reuse, and recycling are also becoming priorities. Therefore, at present the main goals of waste management are preventing harmful impacts on the environment and human health, and conserving limited resources [2, 3]. Improper waste management can result in contamination of environmental, media such as air, water, and soil, which in turn results in severe human health impacts and social issues [4]. Sustainable waste management focuses on reducing the environmental and social impacts of waste, while also curbing the use of virgin materials through recycling, and conserving energy and resources expended in waste disposal [5].

MSW is composed of a multitude of material types, both organic and inorganic, and is often toxic in nature. The composition of the waste mix varies significantly by region and is affected by a number of factors, including culture, economic development, industrialization, and climate [6]. The complex composition of waste matter makes waste disposal and treatment more challenging, requiring sophisticated collection and sorting mechanisms, and different treatment processes based on the type of waste [2, 6].

In a regional waste management strategy, waste processing options can be selected at different levels. In the hierarchy of waste management, prevention is the first step, followed by the three R's: reduce, reuse, and recycle. After these avenues of waste management, energy recovery is considered a preferred alternative to disposal [7]. Accordingly, while the conventional focus is on waste disposal, the next levels of waste processing include material recovery, resource extraction, recycling, and energy generation, with different technologies available at each level. Waste-to-energy conversion is used to recover energy from waste matter in its different forms, and thus can turn problematic waste into a resource. The energy recovered can be in the form of heat, electricity, or fuels [6].

Fuzzy Systems Modeling in Environmental and Health Risk Assessment, First Edition. Edited by Boris Faybishenko, Rehan Sadiq, and Ashok Deshpande.

In order to minimize the adverse economic, environmental, and health impacts of waste and convert it to a resource, it is necessary to investigate the potential effects of each treatment technology and define an optimal waste management system at the regional level. By using a mix of solutions instead of a single waste processing mechanism, it is possible to obtain the best outputs with regards to energy generation and economic returns while also minimizing the negative environmental and social effects. The objective of this chapter is to demonstrate the pathway toward developing an optimal mix of solutions in a regional waste management system, while considering cost, emissions, and health aspects, and minimizing leftover waste.

14.2 Methodology

While the most basic form of waste management is direct disposal through landfilling, the next steps in the waste management hierarchy focus on recovering materials and generating new resources from the waste. Accordingly, four different processing and treatment modes have been defined below.

- Direct disposal: landfilling
- Material recovery and recycling (MRR): Materials which can be reused (e.g. glass and metal) are extracted, and the rest is sent to landfill. Material is recovered and sorted, and recycling is done to generate new materials out of waste. The leftover waste matter is sent to landfill.
- Composting: Simple composting technologies are used to convert certain waste fractions into a resource in the form of fertilizer.
- Waste-to-energy conversion (WtE): Renewable electricity and heat are generated with a fraction of waste, using WtE technologies (i.e. thermochemical or biochemical processes).

At each level, different technologies are identified for each mode of waste treatment (Table 14.1).

Table 14.1 Waste treatment modes and technologies.

Waste treatment mode	Technologies
Composting	Windrows Aerated static piles Modular in-vessel containers (static) Modular in-vessel tunnels (static) In-vessel bays (mechanical agitation) In-vessel vertical silos Rotary drums
Material recovery and recycling	Source separated Dual Stream Single stream Mixed waste
Waste-to-energy (thermochemical processes)	Direct incineration Gasification Pyrolysis Refuse-derived fuel (RDF)
Waste-to-energy (biochemical processes)	Anaerobic digestion (AD) Fermentation Landfill gas Microbial fuel cell (MFC)

14.2.1 Selecting Best Technologies Under Different Modes of Processing

Out of the above identified technologies, the most suitable technologies need to be selected at each level. This selection is done based on a set of generalizable criteria, which represent the needs of the community and the priorities of its decision-makers and stakeholders. To enable this selection, technologies under each mode of processing (where multiple technology options are available) are ranked to select the most suitable technologies at each level, based on the indicators given below. A fuzzy multi-attribute decision-making (MADM)[1] problem is formulated with the available alternatives and performance indicators.

- Total project cost per tonne of waste processed.
- Value generation per tonne of waste processed.[2]
- Emissions per tonne of waste processed.
- Human health risk (HHR) resulting from the waste processing mechanism.

In the life cycle costing, all cost factors, including the initial capital cost, operation and maintenance (O&M) cost, and disposal costs are subject to uncertainties due to potential variations in external conditions and limited data availability. If data are available on a most likely value, a triangular fuzzy number (a, b, c) [8] and fuzzy arithmetic principles can be used when carrying out the algebraic computations for triangular fuzzy numbers [9] (Chapter 1).

Fuzzy TOPSIS is used to select the most suitable technologies at each level. It is a straightforward decision-making process that can represent human decision making [10, 11]. In this technique, the best and worst scenarios (negative and positive ideal solutions) are identified for each criterion. The alternative with the best performance is selected based on the distance from negative and positive ideal solutions [8, 12].

In the fuzzy TOPSIS decision matrix, the triangular fuzzy number for i^{th} technology under j^{th} performance indicator x_{ij} was defined as $x_{ij} = \left(a_{ij}, b_{ij}, c_{ij}\right)$.

The normalized value of r_{ij} is defined below for benefit criteria that need to be maximized and cost criteria that need to be minimized, by linear scale transformation [13].

$$\mathbf{r_{ij}} = \left(\frac{\mathbf{x_{ij}}}{\mathbf{c^{*}_{j}}}\right)$$

for benefit criteria, where

$$\mathbf{c^{*}_{j}} = \mathbf{max_{i}(c_{ij})} \tag{14.1}$$

i.e.

$$\mathbf{r_{ij}} = \left(\frac{\mathbf{a_{ij}}}{\mathbf{c^{*}_{j}}}, \frac{\mathbf{b_{ij}}}{\mathbf{c^{*}_{j}}}, \frac{\mathbf{c_{ij}}}{\mathbf{c^{*}_{j}}}\right)$$

1 MADM is a subdivision within the field of multicriteria decision-making (MCDM). MADM techniques can be used for selection and prioritization among a finite number of alternatives under multiple decision criteria [22, 23].

2 Value generation is defined as the economic value of the recovered and recycled material, generated fertilizer through composting, and produced renewable energy (electricity and/or heat).

$$\mathbf{r_{ij}} = \left(\frac{\mathbf{a^-_j}}{\mathbf{x_{ij}}}\right)$$

for cost criteria, where

$$\mathbf{a^-_j} = \mathbf{min_i(a_{ij})} \tag{14.2}$$

i.e.

$$r_{ij} = \left(\frac{a^-_j}{c_{ij}}, \frac{a^-_j}{b_{ij}}, \frac{a^-_j}{a_{ij}}\right)$$

The weighted normalized decision matrix is derived as followed, where $\mathbf{w_j}$ is the criterion weight [13].

$$\mathbf{v_{ij}} = \mathbf{r_{ij}}(.)\mathbf{w_j} \tag{14.3}$$

The positive and negative ideal solutions for all indicators are given below.

- Fuzzy positive ideal solution: $A^* = (v^*_1, v^*_2, \ldots, v^*_n)$; v^*_j is the maximum v_{ij} under each attribute.
- Fuzzy negative ideal solution: $A^- = (v^-_1, v^-_2, \ldots, v^-_n)$; v^-_j is the minimum v_{ij} under each attribute.

For each performance indicator, the v_{ij} with the largest d value is v^*_j and the one with the smallest a value is v^-_j. The distance from positive ideal solution and negative ideal solution are as follows [13].

$$\mathbf{d^*_i} = \sum\nolimits_{j=1}^{n} \mathbf{d(v_{ij},\ v^*_j)} \tag{14.4}$$

$$\mathbf{d^-_i} = \sum\nolimits_{j=1}^{n} \mathbf{d(v_{ij}, v^-_j)} \tag{14.5}$$

The equation given in Section 14.6 represents the Euclidean distance between two triangular fuzzy numbers, where $\mathbf{x_1} = (a_1, b_1, c_1)$ and $\mathbf{x_2} = (a_2, b_2, c_2)$. This is used to calculated the distance of each alternative technology from A* and A^- [13].

$$\mathbf{d(x_1, x_2)} = \sqrt{\frac{1}{3}\left[(\mathbf{a_1} - \mathbf{a_2})^2 + (\mathbf{b_1} - \mathbf{b_2})^2 + (\mathbf{c_1} - \mathbf{c_2})^2\right]} \tag{14.6}$$

The closeness coefficients (CC_i) are then determined [13].

$$CC_i = \frac{d^-_i}{d^*_i + d^-_i} \tag{14.7}$$

Based on the closeness coefficients (closer the better), the alternatives are ranked.

14.2.2 Defining Optimal Combination of Solutions for Waste Management

Different combinations of the treatment discussed in Section 14.2.1 and processing models can be defined for a regional waste management plan. The technologies selected during the previous stage are used in identifying the optimal mix of waste treatment options. The volume and composition of the available waste and the sorting mechanism are the main inputs to the model. The composition is defined by the mass of each type of waste, such as organics, metal, glass, textiles, plastics, paper, and wood.

Considering a continuous value range for plant capacities of waste management technologies does not make much sense from a practical engineering viewpoint. The reason for this is that the waste processing plant capacities vary in discrete values depending on the scenario-specific constraints (available standard equipment sizes and land area). Furthermore, employing a continuous optimization function increases the computational time. Therefore, this framework uses a discrete combinatorial optimization approach. The performance of alternative waste management plans is simulated by combining discrete, user-defined alternative capacities for each waste management strategy. The developed framework will simulate all possible combinations of alternative capacities of the selected technologies, considering external constraints such as waste composition to obtain the performance of each system based on performance evaluation criteria discussed in this chapter. For each waste management plan, the simulation algorithm will generate aggregated fuzzy numbers, which will be used to rank the overall performance of each combination.

14.2.3 Performance Objectives and Constraints

The performance objectives of the developed waste management system have to be defined to identify the best combination of technology solutions. The following objectives were defined based on the environmental, economic, and social considerations focused on similar waste management studies [14].

- Minimizing total project cost (TPC).
- Minimizing emissions.
- Minimizing HHR.
- Maximizing value generation (value is generated from the outputs of waste processing, in the form of recycled material, compost, or energy).

The waste management plan is constrained by the following factors:

- Maximum budget allocation: This is user-dependent and defined by the financial capabilities of the municipality or the relevant regional authority.
- Maximum land area which can be allocated for landfilling: Siting of a landfill is limited by provincial bylaws regarding the physical conditions of the site, the remoteness of the area, the existing and planned land uses in proximity to the site, the sensitivity of the receiving environment, etc., [15].

14.2.4 Models for Performance Objective Assessment

Some of the performance objective functions are made up of multiple components, which are all fuzzy in nature. For example, discounted net present value of TPC is selected as the economic evaluation criterion, and this needs to be minimized. This is made up of several variables, including investment cost, O&M costs, residual value, disposal costs, interest rate, and the rate of inflation. The present value (PV) of a one-time cash flow occurring at time t, and the PV of annually recurring equal cash flows for t years (PVA) are respectively given by Eqs. (14.8) and (14.9) [16].

$$\mathbf{PV} = \frac{\mathbf{FV}}{(1+\mathbf{r})^{\mathbf{t}}} \tag{14.8}$$

where FV is the future value and r is the annual discount rate.

$$\mathbf{PVA} = \frac{\mathbf{A}[(1+\mathbf{r})^{\mathbf{t}}-1]}{\mathbf{r}(1+\mathbf{r})^{\mathbf{t}}} \tag{14.9}$$

where A is the recurring cash flow.

$$\textbf{Total discounted project cost} = \sum\nolimits_{\mathbf{t}}\mathbf{IC} + \sum\nolimits_{\mathbf{t}}\mathbf{AOM} + \sum\nolimits_{\mathbf{t}}\mathbf{DC} - \sum\nolimits_{\mathbf{t}}\mathbf{RV} \tag{14.10}$$

where IC = investment cost, AOM = PV of annual O&M cost, DC = PV of disposal cost, and RV = PV of residual value.

The variables inside the discounted TPC objective function need to be aggregated as per Eq. (14.10). Each variable is defined as a triangular fuzzy number, based on the available data from literature and other sources.

14.2.5 Defining Weights for the Objectives

The relative importance of each performance objective is considered in assigning a linguistic qualifier to denote the weight of a particular objective. A fuzzy-weighting scheme (w) will be defined to represent the importance weight of each criterion based on the assigned linguistic qualifier, as shown in Table 14.2, based on the literature [17].

Table 14.2 Weighting scheme.

Linguistic term	Very low	Low	Medium	High	Very high
Membership function	(0, 0, 0.3)	(0, 0.3, 0.5)	(0.2, 0.5, 0.8)	(0.5, 0.7, 1)	(0.7, 1, 1)

Once the performance scores are identified for all optimization objectives under the selected technologies, the aggregated performance score of a technology combination (solution) is determined in the form of a fuzzy number. For this, the scores under criteria such as TPC and emissions are normalized and then multiplied with the fuzzy weights discussed in Table 14.2. Equation (14.11) was used in normalizing the performance scores.

$$p_{ij} = \begin{cases} \left(\dfrac{x_{ij1}-x_{min}}{x_{max}-x_{min}}, \dfrac{x_{ij2}-x_{min}}{x_{max}-x_{min}}, \dfrac{x_{ij3}-x_{min}}{x_{max}-x_{min}}\right); \text{for cost criteria} \\ \left(\dfrac{x_{max}-x_{ij1}}{Nx_{max}-x_{min}}, \dfrac{x_{max}-x_{ij2}}{x_{max}-x_{min}}, \dfrac{x_{max}-x_{ij3}}{x_{max}-x_{min}}\right); \text{for cost criteria} \end{cases} \tag{14.11}$$

where

$$x_{max} = \max(x_{ij3}) \text{ and } x_{min} = \min(x_{ij1})$$

The overall performance score (p) for each waste management plan will be calculated using the following equation, where a_{ij} is the normalized performance score under the jth

optimization criteria (performance objective) for the ith waste management plan, w_{ij} is the relative importance for j^{th} optimization criteria for the ith waste management plan, and p_i is the overall performance score of the ith waste management plan.

$$\mathbf{p_i} = \sum_{j=1}^{5} \mathbf{w_{ij}} * \mathbf{a_{ij}} \tag{14.12}$$

14.2.6 Ranking Fuzzy Numbers

The fuzzy numbers denoting the overall performance of a technology combination will be employed in identifying the best waste management plan for the given location using a fuzzy number ranking methodology. The maximizing set and minimizing set method is widely used in the literature and is relatively simple to apply [17, 18]. Therefore, that method was employed in this study to rank the fuzzy overall performance scores to identify the best waste management plan.

Fuzzy appropriateness index (overall performance score) of the m number of alternative waste management solutions will be defined as $F_i\ (i = 1,2,\ldots,m)$. Maximizing set M can be defined using the following equations, where $f_M(x)$ is the membership of x to M [18].

$$M = \{(x, f_M(x)) | x \in R\} \tag{14.13}$$

where

$$f_M(x) = \begin{cases} \dfrac{x - x_{min}}{x_{max} - x_{min}}, & x_{min} \le x \le x_{max} \\ 0 & , \text{otherwise} \end{cases}$$

Minimizing set G can be defined using the following equations, where $f_G(x)$ is the membership of x to G [18].

$$G = \{(x, f_G(x)) | x \in R\} \tag{14.14}$$

where

$$f_G(x) = \begin{cases} \dfrac{x - x_{max}}{x_{min} - x_{max}}, & x_{min} \le x \le x_{max} \\ 0 & , \text{otherwise} \end{cases}$$

where

$$x_{min} = \inf S;\ x_{max} = \sup S;\ S = \bigcup_{i=1}^{m} F_i;\ F_i = \{x | f_{F_i}(x) > 0\};\ i = 1,2,\ldots,m$$

For the ith waste management plan, the right utility value, left utility value, and the total utility can be represented as $U_M(F_i)$, $U_G(F_i)$, and $U_T(F_i)$. These parameters are defined as

$$U_M(F_i) = \sup(f_{F_i}(x) \cap f_M(x)),\ i = 1,2,\ldots,m \tag{14.15}$$

$$U_G(F_i) = \sup(f_{F_i}(x) \cap f_G(x)),\ i = 1,2,\ldots,m \tag{14.16}$$

$$U_T(F_i) = \frac{U_M(F_i) + 1 - U_G(F_i)}{2} \quad (14.17)$$

Figure 14.1 represents the general form of F_i, f_M, and f_G, if the fuzzy appropriateness index of the ith solution is denoted as $F_i = [a_i, b_i, c_i]$.

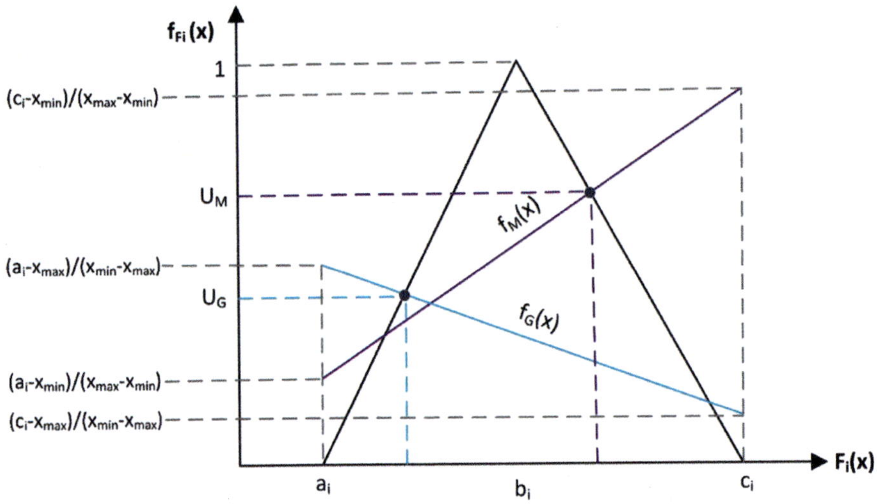

Figure 14.1 Fuzzy ranking based on maximizing set and minimizing set.

The waste management plan with the maximum total utility will be selected as the best waste management plan under the given conditions.

14.2.7 Performance of the Selected Combination of Solutions

The fuzzy extension principle is used in determining a fuzzy output based on fuzzy inputs [16]. The Dong, Shah, and Wong (DSW) algorithm is one technique used in research to simplify the application of the extension principle in fuzzy calculations using functions or variables with continuous values.

The DSW algorithm employs α-cut representation for fuzzy extension, and uses full α-cut intervals in standard interval analysis [16]. By using this technique, complex output membership functions for parameters made up of multiple variables can be calculated, such as the life cycle cost (LCC) of the final combined system. The method is:

1) Variables are represented as fuzzy numbers.
2) An "α" value is selected where $0 \leq \alpha \leq 1$.
3) The intervals in the input membership functions variables (i.e. interest rate, inflation, investment cost, O&M costs, disposal costs, salvage value) that correspond to the selected α-cut level are identified.
4) Standard binary interval operations are used to calculate the output membership function for the LCC at the selected α-cut level.
5) The above steps are repeated in an iterative process for different α-cut levels, at 25% intervals between (0, 1) membership.

A similar process can be taken when determining the membership function for health risk, fraction of leftover mass sent to landfill, and value generation.

14.3 Developing the Regional Waste Management Model

The waste management model was developed to identify the optimal waste management solutions for a given region. British Columbia (BC) in Canada was selected as the area of interest to demonstrate the developed method through a case study.

14.3.1 Inputs to the Decision Model

The throughput and composition of the waste that needs to be processed with the developed management model are dependent on the selected locality. The case study was conducted for a municipality with a population of 130 000. The waste generation data for the Okanagan region from 2010 was used to define the fuzzy number for per capita waste generation in tonnes per annum as (610, 650, 700) [19].

Organic waste composition was defined using the information provided in a waste composition study for Metro Vancouver, BC. The hazardous waste, electronic waste, building material, and hygiene waste components were excluded from the decision model. Accordingly, the composition of the waste mass that will be processed through the developed waste management mechanism is given in Table 14.3. A point to note in defining a waste management strategy is that some material components can only be processed through certain technologies. In Table 14.3, different types of waste matter are linked to the technologies which can process them.

Table 14.3 Waste material composition and processing technologies.

Waste matter		**Food**	**Yard**	**Plastics**	**Paper**	**Textile**	**Glass**	**Metal**	**Rubber**	**Wood**
Composition (%)		**44**	**4**	**15**	**20**	**7**	**4**	**4**	**1**	**2**
Technologies	Landfilling	✓	✓	✓	✓	✓	✓	✓	✓	✓
	Incineration	✓	✓	✓	✓	✓			✓	✓
	Windrows composting	✓	✓		✓					
	In-vessel composting	✓	✓		✓					✓
	Anaerobic digestion	✓	✓							
	Refuse-derived fuel			✓	✓	✓			✓	✓
	Dual stream recycling			✓	✓		✓	✓		

For the technology ranking and selection within a particular mode of waste processing, the information on total project cost, emissions, health risk, and value generation were compiled. An example of the ranking process has been carried out for the WtE technologies. As the purpose is to demonstrate the methodology for comparing technologies, the energy generation potential of each technology has been used as a representative parameter of the value generation, and the HHR of WtE technologies has been rated on a comparative scale based on expert judgement. The rating scale defined based on literature is given in Table 14.4 While this qualitative method can be used in comparative assessments, the users can input the quantified data into the decision model, if such data is available.

Table 14.4 Fuzzy linguistic ratings for human health risk (HHR).

Human health risk	Fuzzy number
Very low	(1, 1, 3)
Low	(1, 3, 5)
Medium	(3, 5, 7)
High	(5, 7, 9)
Very high	(7, 9, 9)

The values given in Table 14.5 were used in ranking the thermochemical WtE technologies. The quantitative values are given on a per-tonne basis. (The disposal costs and residual values were assumed to be zero for all technologies.) Table 14.6 presents a normalized decision matrix.

Table 14.5 Performance of thermochemical technologies[3,4].

Technology	Capital	O&M	Emissions[5]	HH risk	Energy potential
Units	\$/annual design tonne	\$/tonne	KgCO$_2$ eq/tonne	—	MWh/annual tonne
Incineration	(387.5, 775, 1162.5)	(45.5, 65, 84.5)	(495, 550, 605)	(5, 7, 9)	(0.5, 0.675, 0.85)
Gasification	(510, 850, 1190)	(33.55, 61, 88.45)	(915.3, 1017, 1118.7)	(5, 7, 9)	(0.4, 0.6, 0.8)
Pyrolysis	(307.23, 539, 770.770)	(24.48, 51, 77.52)	(915.3, 1017, 1118.7)	(5, 7, 9)	(0.5, 0.65, 0.8)
Refuse-derived fuel	(280, 400, 520)	(28, 40, 52)	(990, 1100, 1210)	(7, 9, 9)	(0.8, 1.7, 2.6)

Source: Adapted from [1].

After calculating the distances from the positive and negative ideal solutions and the corresponding closeness coefficients, the technology ranking revealed RDF to be the best-performing thermochemical WtE technology.

3 The values given here were approximated based on the literature to demonstrate the functionality of the mathematical model [24–28]. The provided data may not be representative of the exact up-to-date costing or the emissions factors for the given technologies.

4 The values given here were approximated based on the literature to demonstrate the functionality of the mathematical model [22–26]. The provided data may not be representative of the exact up-to-date costing or the emissions factors for the given technologies.

5 Some emission values were obtained from the ecoinvent 3 database using the SimaPro life cycle assessment software.

Table 14.6 Normalized decision matrix.

Parameter	NPV of TPC	Emissions	HHR	Energy potential
Weights	(0.36, 0.4, 0.44)	(0.135, 0.15, 0.165)	(0.135, 0.15, 0.165)	(0.27, 0.3, 0.33)
Incineration	(0.28, 0.39, 0.63)	(0.82, 0.9, 1)	(0.56, 0.71, 1.0)	(0.19, 0.26, 0.33)
Gasification	(0.27, 0.38, 0.67)	(0.44, 0.49, 0.54)	(0.56, 0.71, 1.0)	(0.15, 0.23, 0.31)
Pyrolysis	(0.35, 0.52, 1.0)	(0.44, 0.49, 0.54)	(0.56, 0.71, 1.0)	(0.19, 0.25, 0.31)
Refuse-derived fuel	(0.52, 0.67, 0.96)	(0.41, 0.45, 0.5)	(0.56, 0.56, 0.71)	(0.31, 0.65, 1.0)

NPV: net present value; TPC: total product cost.

In the next step, the top-ranking technology from each mode of treatment is considered in defining the optimal technology combination. To demonstrate the case study, the following technologies were selected for the optimization model:

- Landfilling.
- Material recovery and recycling (MRR): dual stream type.
- Composting: windrows.
- WtE (thermochemical): refuse-derived fuel.
- WtE (biochemical): anaerobic digestion.

The input data for the above technologies have been summarized in Table 14.7. For the WtE technologies, the value generation was defined by the expected sales revenue of generated renewable energy. The energy price ($/kWh) is defined as (0.042, 0.0936, 0.1872) [16]. For the MRR and composting technologies, the value generation was evaluated based on the market prices of produced compost and recycled material [20].

For the demonstration, the constraints were defined in the following manner. The maximum capital budget was set at $200 000. Based on the incoming annual solid waste mass, the size of the active area at a landfill site is limited as follows [15]. This information

Table 14.7 Input data for waste processing technologies of all modes.

Technology	Capital	O&M	Emissions	HH risk	Value generation
Units	$/annual design tonne	$/tonne	$KgCO_2$ eq/tonne	–	$/tonne
Windrows	(40, 100, 150)	(25, 40, 50)	(432, 480, 528)	(1, 1, 3)	(5.71, 15.912, 38.188)
Dual stream recycling	(200, 260, 300)	(30, 40, 50)	(985.5, 1095, 1204.5)	(3, 5, 7)	(71.51, 130.58, 193.84)
Refuse-derived fuel	(280, 400, 520)	(28, 40, 52)	(990, 1100, 1210)	(7, 9, 9)	(33.6, 159.12, 486.72)
Anaerobic digestion	(380, 610, 853)	(60, 90, 110)	(765, 850, 935)	(7, 9, 9)	(5.94, 24.99, 26.66)
Landfilling	(40, 70, 100)	(10, 20, 50)	(633.6, 704, 774.4)	(3, 5, 7)	(0, 0, 0)

was used in defining rules for the minimum area that needs to be available in compliance with siting requirements, to set up a landfilling facility to process the corresponding waste tonnage. If the mass of waste sent to landfill is m_{lf} (in tonnes) and the required landfill area is AR_{lf} (in hectares), the rules are

$$\text{If}\,(m_{lf} \leq 10000),\ \text{Then}\ AR_{lf} = 1\,\text{Ha}$$

$$\text{If}\,(10000 < m_{lf} \leq 20000),\ \text{Then}\ AR_{lf} = 2\,\text{Ha}$$

$$\text{If}\,(20000 < m_{lf} \leq 50000),\ \text{Then}\ AR_{lf} = 4\,\text{Ha}$$

$$\text{If}\,(50000 < m_{lf} \leq 100000),\ \text{Then}\ AR_{lf} = 6\,\text{Ha}$$

$$\text{If}\,(100000 < m_{lf} \leq 200000),\ \text{Then}\ AR_{lf} = 8\,\text{Ha}$$

$$\text{If}\,(200000 < m_{lf} \leq 500000),\ \text{Then}\ AR_{lf} = 10\,\text{Ha}$$

14.3.2 Defining Combined Solutions for Optimization

In order to define the possible set of solutions from which the optimal solution is selected, technology and waste mass combinations are defined as described in this section. This method operates on the assumption that waste material components are subjected to full source separation to send them to different processing technologies. In the developed decision model, the users are afforded the flexibility to define the solutions by deciding the resolution of partial processing for waste mass components. In partial processing, one waste material type is sent to multiple technologies in partial mass fractions. For "m" number of material types (j = 1, 2, ..., p) and "n" number of technologies (i = 1, 2, ..., q) the partial processing is as defined in Eq. (14.18).[6]

$$M_{MSW} = \sum_{i=1}^{p}\sum_{j=1}^{q} m_{ij} \tag{14.18}$$

where

M_{MSW} = Total MSW mass processed under the developed waste management plan

m_{ij} = Mass of a particular material component processed under a given technology

$$m_j = M_{MSW} x_j \tag{14.19}$$

where

m_j = Total mass of a particular material component in the processed MSW

x_j = Total mass fraction of a particular material component in the processed MSW

Table 14.3 links the material types to the technologies that can process them. The minimum mass from material "j" that can be sent to be processed via one technology is defined as

$$m_{j.min} = m_j.Pr_j \tag{14.20}$$

6 For this case study, p = 9, and q = 5.

where

$$Pr_j = \text{Resolution of partial processing for a particular material component}$$

$$m_{j.min} = \begin{array}{l}\text{Minimum mass of a particular material}\\ \text{which can be processed by one technology}\end{array} \tag{14.21}$$

This is done to avoid waste management plans where waste processing plants will be installed to process very small amounts of waste (i.e. if the waste mass of a particular material type is too small, it does not make practical sense to separate the waste streams and redirect them to separate plants to be processed under different technologies, as the value addition will be too low). Based on the same rationale, it was decided that materials that contribute less than 10% of the total MSW mass were to be processed by only one technology in a single combination of solutions (e.g. glass accounts for only 4% of the MSW mass, and therefore it was assumed to be either processed via recycling or by landfilling, without using partial mass combinations).

The resolutions of partial processing were defined using a rule-based method:

$$\text{If}\,(x_j \le 10\%),\ \text{Then}\ Pr_j = 1$$

$$\text{If}\,(10\% < x_j \le 20\%),\ \text{Then}\ Pr_j = 0.5$$

$$\text{If}\,(20\% < x_j \le 35\%),\ \text{Then}\ Pr_j = 0.3$$

$$\text{If}\,(35\% < x_j \le 50\%),\ \text{Then}\ Pr_j = 0.25$$

$$\text{If}\,(50\% < x_j \le 70\%),\ \text{Then}\ Pr_j = 0.2$$

$$\text{If}\,(70\% < x_j \le 100\%),\ \text{Then}\ Pr_j = 0.1$$

The relationship between total possible combinations for a material (C_j), the resolution of partial processing (Pr_j), and the number of technologies capable of processing that material (n_j) is given in Eq. (14.22).

$$C_j = \frac{\left(\dfrac{1}{Pr_j} + n_j - 1\right)!}{\left(\dfrac{1}{Pr_j}\right)! \times (n_j - 1)!} \tag{14.22}$$

The combination inputs for the case study are listed in Table 14.8. An example of how all the possible mass combinations were defined for food waste is provided in Table 14.9.

Once the individual combinations are defined for all waste types, the total number of possible combinations can be estimated. For "m" number of material types (j = 1, 2, ..., m), total number of solution combinations (C_T) is given by Eq. (14.23).

$$C_T = \prod_{i=1}^{j} C_j \tag{14.23}$$

For the case study, the total number of combinations (solutions) was 86 400. These combinations were then screened using the previously defined maximum capital budget and land area availability constraints. With this, the total number of viable solutions was reduced to

Table 14.8 Waste mass combination inputs.

Waste matter	Food	Yard	Plastics	Paper	Textile	Glass	Metal	Rubber	Wood
Material type number ($j = 1, 2, \ldots, m$)	1	2	3	4	5	6	7	8	9
Composition (x_j)	44%	4%	15%	20%	7%	4%	4%	1%	2%
Resolution of partial processing (Pr_j)	0.25	1	0.5	0.5	1	1	1	1	1
Number of technologies capable of processing the material	3	3	3	4	2	2	2	2	2
Number of possible combinations for a material	15	3	6	10	2	2	2	2	2

Table 14.9 Combination matrix for food waste vs. processing technologies.

Modes		Composting	MRR	WtE (thermo)	WtE (bio)	
Technologies		**Windrows**	**Dual stream**	**RDF**	**AD**	**Landfill**
Combinations	1	1	0	0	0	0
	2	0	0	0	1	0
	3	0	0	0	0	1
	4	0.25	0	0	0.25	0.5
	5	0.25	0.0	0.0	0.5	0.25
	6	0.5	0.0	0.0	0.25	0.25
	7	0.0	0.0	0.0	0.25	0.75
	8	0.25	0.0	0.0	0.0	0.75
	9	0.0	0.0	0.0	0.75	0.25
	10	0.25	0.0	0.0	0.75	0.0
	11	0.75	0.0	0.0	0.25	0.0
	12	0.75	0.0	0.0	0.0	0.25
	13	0.5	0.0	0.0	0.5	0.0
	14	0.5	0.0	0.0	0.0	0.5
	15	0.0	0.0	0.0	0.5	0.5

RDF: refuse-derived fuel; AD: anerobic digestion.

14 307. An overall performance score was calculated for these viable solutions. In calculating the total discounted project cost, the project life and discount rate were assumed to be crisp numbers (project life = 20 years; interest rate (i) = 5%; rate of inflation (f) = 2%; discount rate, $r = i - f = 3\%$).

14.3.3 Identifying Optimal Solutions

Fuzzy arithmetic operations were used to calculate the overall performance score by aggregating the normalized objective function values for every solution. For the aggregation, the weights for the optimization objectives were given as per the scheme defined in Table 14.2. The relative importance of the performance objectives was decided as follows: minimizing total project cost, *very high*; minimizing health risk, *medium*; minimizing emissions, *low*; maximizing value generation, *high*. The overall aggregated scores (for all viable solutions) are generated in the form of triangular fuzzy numbers. These fuzzy numbers, which represent the fuzzy appropriateness index (F_i), were ranked in the next step to find the best (optimal) solution according to the method discussed in Section 14.2.5.

The top five solutions that define the optimal plant capacities for each mode of waste processing are given in Table 14.10.

Table 14.10 Optimal waste management system configurations.

	Per day design capacity (tonnes/day)					
	Composting	MRR	WtE (thermo)	WtE (bio)		
Waste management systems	**Windrows**	**Dual stream**	**RDF**	**AD**	**Landfill**	**$U_T(F_i)$**
Combination 1	408	69	6	0	123	0.61532
Combination 2	408	90	0	0	108	0.61521
Combination 3	408	93	0	0	105	0.61511
Combination 4	408	69	12	0	117	0.61471
Combination 5	408	90	6	0	102	0.61460

The top-ranked solution from Table 14.10 was selected as the proposed waste management system, for the economic performance evaluation in the next phase.

14.3.4 Economic Evaluation of Waste Management Systems

The overall economic impact of the proposed waste management system is evaluated through the LCC of the final combination. The membership function for the LCC is generated using the fuzzy DSW algorithm explained in Section 14.2.5. The LCC is calculated using Eq. (14.24), considering net cash inflows and outflows during the project lifetime. The expected revenues (NPV of value generation, or VG) from renewable energy, compost, and recycled material were considered to be recurring annual incomes, while the project costs were the capital and O&M expenditure

$$\mathrm{LCC} = \sum\nolimits_t \mathrm{IC} + \sum\nolimits_t \mathrm{AOM} + \sum\nolimits_t \mathrm{DC} - \sum\nolimits_t \mathrm{VG} - \sum\nolimits_t \mathrm{RV} \tag{14.24}$$

The calculated LCC normalized decision matrix is given in Table 14.11, and a graphical presentation of membership functions is shown in Figure 14.2.

Table 14.11 LCC intervals for different α-cut levels.

α-cut level (membership)	NPV of total project cost ($)	NPV of value generation ($)	LCC ($)
0	207 878	95 492	526 013
0.25	259 569	134 885	344 423
0.5	316 826	179 682	169 652
0.75	379 891	230 152	−947
1	449 008	286 565	162 443
0.75	526 921	380 838	296 769
0.5	612 026	486 477	432 344
0.25	704 603	603 992	569 718
0	804 935	733 891	709 443

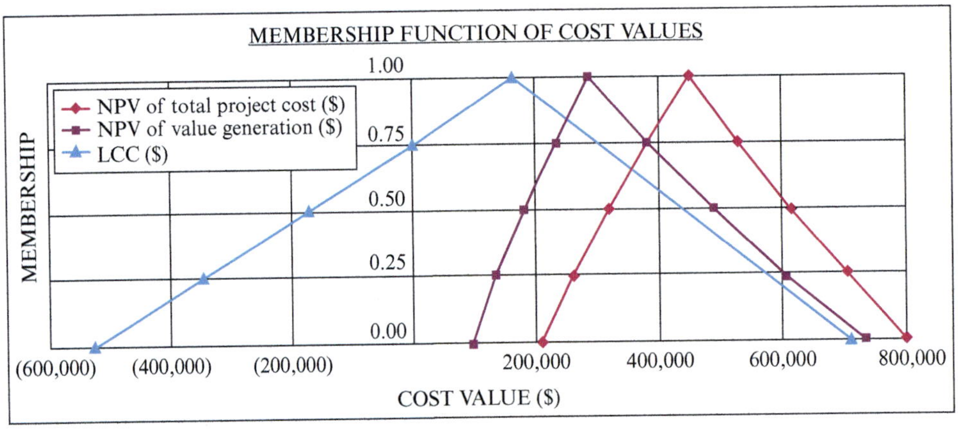

Figure 14.2 Membership function for total project cost, value generations, and LCC.

It can be seen that the LCC of the proposed waste management system varies across a range of values from negative to positive. This means that the profitability of the proposed project is fuzzy, and the final outcome can either be a profit or a loss based on the exact costs and revenues during operationalization. If a deterministic method was used for analyzing the LCC, this information would not have been captured as the LCC would be in the form of a crisp value (indicating either a profit or a loss based on the parameter values). The data on the whole range of possibilities captured through a fuzzy logic (FL) approach help decision-makers to make a more informed decision in comparison to a deterministic analysis.

14.4 Discussion and Conclusions

This type of fuzzy-based framework allows the qualitative information and the variation in human preferences, expert judgments, and imprecise data to be accounted for in the decision making. Where data are unavailable on the exact inputs, qualitative judgments which are converted to fuzzy rating scales and ruled-based methods can be used, as seen for the HHR assessment of technologies. Using FL can help decision-makers to capture a range of variations in different decision parameters and will help them to explore a range of possible outcomes.

While fuzzy arithmetic operations were used for calculating and aggregating the performance scores in the ranking and optimization steps, the fuzzy DSW algorithm was used to calculate the final LCC of the selected system. The DSW algorithm is capable of capturing the variation accurately during fuzzy calculations, due to the use of multiple α-cuts. Depending on the expected accuracy of the output membership function, the number of α-cuts can be increased for a higher resolution. The DSW principle can be used to apply the fuzzy extension principle for continuous valued and complex functions [16]. The simplified fuzzy arithmetic operations are limited to special types of fuzzy numbers, such as triangular and trapezoidal fuzzy numbers, where the upper and lower bounds have memberships of 0 and the most likely value (or value range) has a membership of 1. Since the discounting rates and project lifetime were taken as crisp numbers in the analysis, all parameters could be reasonably represented in the form of triangular fuzzy numbers in the technology selection and optimization analyses.

The fuzzy number ranking method based on total utility was used in this application. However, the centroid defuzzification method can also be used for type of ranking when the membership functions are known [21]. This method allows the entire fuzzy range to be reflected in the final decision. It compares the whole range of values in the set of numbers to be ranked by considering their closeness to the maximizing and minimizing set. It was decided that the above method results in a lower loss of information compared to the defuzzified ranking method, which converges the fuzzy range (in the appropriateness indices) to a single value without comparing the number to the entire range.

In this type of decision-making problem, the final outcome is heavily influenced by the current performance levels for technologies (i.e. cost factors, technology maturity, and capacity to process different types of waste, value generation potential), as well as the defined weights for performance criteria. If the values used for the above factors are changed, the obtained results (i.e. the optimal waste management solution) will vary significantly.

Resolution in defining possible system combinations is limited by the computational time taken for the optimization. As an example, the resolution of partial processing could have been further increased by separating the waste stream of a particular material into smaller mass packages. However, this would have significantly increased the computational time, and a compromise needs to be achieved between the decision resolution and the computational resource requirement in this type of decision model. When the resolution of partial processing is increased in an extreme manner (dividing waste masses into smaller and smaller blocks), the optimization technique can approximate a continuous optimization problem.

In this problem, the optimal system configuration of the waste management plan was defined by separating waste material types into blocks which can be separated by different technologies. In practical engineering applications, the plant design capacities also come in defined sizes (e.g. 50 000 or 100 000 tonnes per year of incineration facilities). These plant capacity steps also need to be considered in defining the combinations for the optimization algorithm to increase the practicality of the proposed waste management system planning model.

References

1 Cheng, H. and Hu, Y. (2010). Municipal solid waste (MSW) as a renewable source of energy: current and future practices in China. *Bioresource Technology* 101 (11): 3816–3824.

2 Brunner, P.H. and Rechberger, H. (2015). Waste to energy: key element for sustainable waste management. *Waste Management* 37: 3–12.

3 Bhada-Tata, P. and Hoornweg, D.A. (2012). *What a Waste: A Global Review of Solid Waste Management*. Washington, DC: World Bank Group.

4 Giusti, L. (2009). A review of waste management practices and their impact on human health. *Waste Management* 29 (8): 2227–2239.

5 Pearson, D. and Dale, A. (2013). Sustainable waste management: the value of recycling and composting programs. *Community Research Connection Case Studies*. https://www.crcresearch.org/community-research-connections/crc-case-studies/sustainable-waste-management-value-recycling-and (accessed 28 September 2022).

6 World Energy Council. (2013). *World Energy Resources: 2013 Survey*. London: World Energy Council.

7 Šomplák, R., Pavlas, M., Kropáč, J., Putna, O., and Procházka, V. (2014). Logistic model-based tool for policy-making towards sustainable waste management. *Clean Technologies and Environmental Policy* 16 (7): 1275–1286.

8 Kahraman, C. (2008). *Fuzzy Multi-Criteria Decision Making* (ed. C. Kahraman). Boston, MA: Springer.

9 Arithmetic, B.S. (2012). Operations on generalized trapezoidal fuzzy number and its applications. *Turkish Journal of Fuzzy Systems* 3 (1): 16–44.

10 Shih, H.-S., Shyur, H.-J., and Lee, E.S. (2007). An extension of TOPSIS for group decision making. *Mathematical and Computer Modelling* 45 (7–8): 801–813.

11 Yaakob, A.M. and Gegov, A. (2015). Fuzzy rule based approach with z-numbers for selection of alternatives using TOPSIS. *2015 IEEE International Conference on Fuzzy Systems (FUZZ-IEEE)*, (i), 1–8.

12 García-Cascales, M.S. and Lamata, M.T. (2012). On rank reversal and TOPSIS method. *Mathematical and Computer Modelling* 56 (5–6): 123–132.

13 Melin, P., Castillo, O., Ramirez, E.G., Kacprzyk, J., and Pedrycz, W. (eds.) (2007). *Analysis and Design of Intelligent Systems Using Soft Computing Techniques*. Berlin: Springer.

14 Cucchiella, F., D'Adamo, I., and Gastaldi, M. (2014). Sustainable management of waste-to-energy facilities. *Renewable and Sustainable Energy Reviews* 33: 719–728.

15 BC Ministry of Environment. (2016). *Landfill criteria for municipal solid waste*. https://www2.gov.bc.ca/assets/gov/environment/waste-management/garbage/landfill_criteria.pdf (accessed 28 September 2022).

16 Ruparathna, R., Hewage, K., and Sadiq, R. (2017). Economic evaluation of building energy retrofits: a fuzzy based approach. *Energy and Buildings* 139: 395–406.

17 Lin, H.Y., Hsu, P.Y., and Sheen, G.J. (2007). A fuzzy-based decision-making procedure for data warehouse system selection. *Expert Systems with Applications* 32 (3): 939–953.

18 Salahshour, S., Abbasbandy, S., and Allahviranloo, T. (2011). Ranking fuzzy numbers using fuzzy maximizing-minimizing points. *Proceedings of the 7th Conference of the European Society for Fuzzy Logic and Technology (EUSFLAT-11)*, Atlantis Press, 763–769.

19 Government of British Columbia, Canada. (2018). Municipal solid waste disposal in BC (1990–2016). *Environmental Reporting BC*. https://www2.gov.bc.ca/assets/gov/environment/research-monitoring-and-reporting/reporting/envreportbc/archived-indicators/sustainability/envreportbc_municipal_solid_waste_may2018.pdf (accessed 28 September 2022).

20 Recycling Council of Alberta. (2018). *Market updates*. https://recycle.ab.ca/newsyear/2018 (accessed 4 October 2022).

21 Wang, Y.M. (2009). Centroid defuzzification and the maximizing set and minimizing set ranking based on alpha level sets. *Computers and Industrial Engineering* 57 (1): 228–236.

22 San Cristóbal, J.R. (2011). Multi-criteria decision-making in the selection of a renewable energy project in Spain: the Vikor method. *Renewable Energy* 36 (2): 498–502.

23 Yoon, K. and Hwang, C.-L. (1995). *Multiple Attribute Decision Making*. Thousand Oaks, CA: Sage.

24 Pham, T.P.T., Kaushik, R., Parshetti, G.K., Mahmood, R., and Balasubramanian, R. (2015). Food waste-to-energy conversion technologies: current status and future directions. *Waste Management* 38 (1): 399–408.

25 Morrison Hershfield Ltd. (2016). *Residual waste management options for the regional district of Nanaimo*. Burnaby. https://www.rdn.bc.ca/sites/default/files/inline-files/2018%20SWMP%20Amendment_1.pdf (accessed 4 October 2022).

26 Stantec. (2010). *Waste to energy: a technical review of municipal solid waste thermal treatment practices*. Victoria, BC, Canada. https://www2.gov.bc.ca/assets/gov/environment/waste-management/garbage/bcmoewteemmissionsrevmar2011.pdf (accessed 29 September 2022).

27 Compost Council of Canada. (2016). *Best practices for operating an aerated windrow composting facility*. http://www.compost.org/wp-content/uploads/2021/12/FINAL-WINDROW-COMPOSTING-MANUAL-UPDATED-2016.pdf (accessed 4 October 2022).

28 Recycling Council of Alberta. (2006). *Municipal solid waste (MSW) options: integrating organics management and residual treatment/disposal*. Calgary, Canada. https://recycle.ab.ca/wp-content/uploads/2016/01/MSW_Options_Report.pdf (accessed 4 October 2022).

15

Development of a Fuzzy-based Risk Assessment Model for Process Engineering

Rachid Ouache[1,4], Muhammad Nomani Kabir[2], Husnain Haider[3], Nurdin Said[4], Farid Wajdi Akashah[5], Abdullah Ibrahim[6], Rajeev Ruparathna[7], Kasun Hewage[1], and Rehan Sadiq[1]

[1] *School of Engineering, University of British Columbia (Okanagan), Kelowna, BC, Canada*
[2] *Department of Computer Science & Engineering, Trust University, Barishal, Ruiya, Bangladesh*
[3] *Qassim University, Buraydah, Qassim, Saudi Arabia*
[4] *Faculty of Chemical Engineering and Natural Resources, University of Malaysia Pahang, Pahang, Malaysia*
[5] *Department of Building Surveying, Faculty of Built Environment, University of Malaya, Kuala Lumpur, Malaysia*
[6] *Faculty of Engineering Technology, University Malaysia Pahang, Pahang, Malaysia*
[7] *Department of Civil and Environmental Engineering, University of Windsor, Windsor, ON, Canada*

15.1 Introduction

A risk generally is a combination of a probability of an event and its consequence [1]. Lindhe defined risk as uncertainty, and the severity of the consequences of an activity [2], and Torstensson stated that risk is the combination of the probability of an event and the degree of an injury or damage that can arise from that hazardous incident [3]. Engineered systems pose an elevated risk to people, the environment, and physical properties. Hence, the identification of hazards and the determination of their risk level can help an organization prepare for hazardous events [4].

Risks can be characterized using several factors that need to be cautiously estimated for effective risk assessment [5–7]. Three scientific approaches (i.e. qualitative, semiquantitative, and quantitative) have been commonly adopted for risk assessment to reduce risk to an acceptable level or as low as reasonably practicable (ALARP). In general, qualitative methods are used for risk identification, and semiqualitative techniques are used for risk categorization, while quantitative risk assessment is considered the best approach [8]. However, a lack of data impels analysts to select qualitative and semiquantitative risk assessment techniques.

Many methods have been used by the engineering sector to reduce the risk to ALARP [9–11], including quantitative risk analysis (QRA) [12–14], checklist analysis, preliminary hazard analysis (PHA) [15], what if analysis, hazard and operability (HAZOP) studies [9, 16, 17], failure modes and effects analysis (FMEA), Bowtie analysis [10, 18, 19], fault tree analysis (FTA), event tree analysis (ETA) [20], cause–consequence analysis (CCA), human reliability analysis (HRA), layer of protection analysis (LOPA) [21, 22], and risk matrix [23–26]. Data requirements for these methods depend on objectives, limitations, strengths, and level of computational complexities. The risk analysis method is selected based on the type of industry and level of risk associated with the industrial processes. Risk matrix has been used by many researchers, particularly in engineering sectors.

In this study, a risk matrix model (RMM) was developed to handle uncertainties using a semiquantitative approach to assess the level of risk effectively and precisely in process

Fuzzy Systems Modeling in Environmental and Health Risk Assessment, First Edition. Edited by Boris Faybishenko, Rehan Sadiq, and Ashok Deshpande.

system engineering. All the factors used to compute the level of risk were included in the risk matrix analysis. To increase the reliability of the decision-making process in process system engineering, three mathematical equations were developed, and FL was used to compare the results of the developed equations. Finally, to facilitate the computations of RMM, a Simulink-MATLAB model was developed and implemented in a case study of a reboiler oven in the petroleum industry.

15.2 Risk Assessment

Risk can be measured by nonlinearities or by the variance of the probability distribution of possible gains and losses associated with particular alternatives. As well, many authors have described and defined risk from different perspectives in different areas [1, 27–44]. Risk assessment through the determination of risk and its potential damage likewise determines the actions needed to control that risk. A risk matrix is commonly used for this task.

A risk matrix was developed by the Electronic System Center of the United States Airforce in April 1995 to assess risk over the life cycle of a project. A risk matrix was also found to be applicable for assessing operational risk in different areas (e.g. engineering projects, commercial banks, enterprise resource planning, the weapons industry, etc.) [8]. Industries, such as manufacturing, prefer the risk matrix approach because the method is simple to use and easy to understand with graphical expression. A risk matrix is a qualitative risk assessment tool that rates the risk based on two main criteria: probability and consequence [45], where the probability and the consequence estimation are subjective [45].

The RMM is based on a 10×10 matrix with 100 risk cells. The rows indicate ten different levels of probabilities; probability in this study is based on two variables: (i) frequency of consequences and (ii) probability of failure on demand (PFD) of the barrier placed to decrease the frequency of the consequence [46]. The columns imply ten different levels of severity of consequences (impacts) on humans, the environment, properties, and reputation. In the chemical process industry, the quantitative risk has been defined as "a ratio of hazards to safeguards" and as "a combination of event, probability and consequences" [47]. The likelihood is the measure of expected probability or frequency of the event occurring, which can be qualitatively or quantitatively estimated [48, 49]. Probability is the expression of event occurrence as a number ranging from 0 to 1 [12, 47, 50]. The PFD is the indication probability that can predict failure of a system to perform a required function on demand [51], and impact is the severity of the effects of the undesirable event.

A risk matrix is developed to evaluate the risk following two essential steps: first, categorization and scaling of the severity of consequences and corresponding frequencies as inputs, and risk index as the output; second, establishing the risk-based rules knowledge to determine the level of risk based on the inputs. In traditional risk models, the risk is expressed by a crisp category based on the analysis. However, the uncertainty and imprecision of data produce some errors in the level of risk. In the traditional risk matrix approach, the risk is expressed by a crisp category (e.g. [1 3], [4 6], etc.). Different forms of risk matrix have been developed, such as 4×4 [52], 5×4 [53], and 7×5 [54]. Due to data limitations and imprecision, the results of the conventional risk matrix may be somewhat vague. For instance, the estimated level of risk is presented as a range (i.e. the scenario range is between levels 3 and 4 without an exact value). The real value of a risk level could be near the lower or upper probabilities. As a result, the conventional risk matrix still faces challenges to overcoming uncertainties for prioritizing scenarios in process system engineering due to the imprecision of risk level and absence of data to include all the factors that could affect the risk level. Therefore, different studies have been done to handle uncertainties for risk matrix, emerging from FL applications [45, 54]. However, the risk matrix still needs investigation in terms of the involved parameters, defined categories, and precision of the risk level.

15.3 Methodology

The RMM was developed in this study to conduct a risk assessment for process industries. The model had to be able to overcome the uncertainty of the risk level effectively and evaluate whether the risk was within the acceptable range of the industry or not. The RMM displayed the basic properties *frequency* and *impact*. Traditionally, risk has been defined as

$$Risk\ level_i = frequency\ of\ consequence_i \times severity\ of\ consequence_i \tag{15.1}$$

where i is the number of scenarios.

The framework of the proposed RMM is presented in Figure 15.1. An RMM analysis follows three main steps:

1) Analyze the probability of scenarios from different perspectives: event's probability, exposure, PFD, and vulnerability with weight.
2) Analyze the impact of scenarios on people, properties, environment, and reputation with weights
3) Determine risk level using developed equations and FL to precise risk level. All these steps are described in detail in this chapter.

15.3.1 Estimation of the Frequency of Risk Consequence

The frequency of risk depends on the frequency of risk events and the PFD for safeguards. The matrix shown in Table 15.1 is developed here to determine the frequency of risk. To determine the frequency of risk consequence in the developed matrix, the following formula is used

$$Fr_{RC} = Fr_{RE} \cdot PFD_S \tag{15.2}$$

where Fr_{RC} is the frequency of risk consequence, Fr_{RE} is the frequency of risk events, and PFD_S is the PFD of safeguards.

The proposed categories for PFD and the frequency of occurrence of the risk event are presented in Tables 15.2 and 15.3, and the classification of vulnerability and exposure levels are classified in Table 15.4. All the categories are linguistically defined with small intervals to facilitate the decision-making process for evaluating multiple scenarios and to precisely estimate the level of risk.

The inputs (i.e. PFD and Fr) are classified into ten levels ranging from *absolutely high* to *absolutely low*, as shown in Table 15.2, with the corresponding values of 10^{-5} and 10^{0} respectively. Impact of consequence (i.e. the second input of the RMM) distributed over ten categories is presented in Table 15.3, which helps to differentiate the levels of impact and prioritize the consequences. Nine levels of risk as outputs are presented in Table 15.2, varying from *absolutely acceptable* to *absolutely not acceptable*. The categories listed in Table 15.8 can evaluate different scenarios even if the difference in risk level is extremely small.

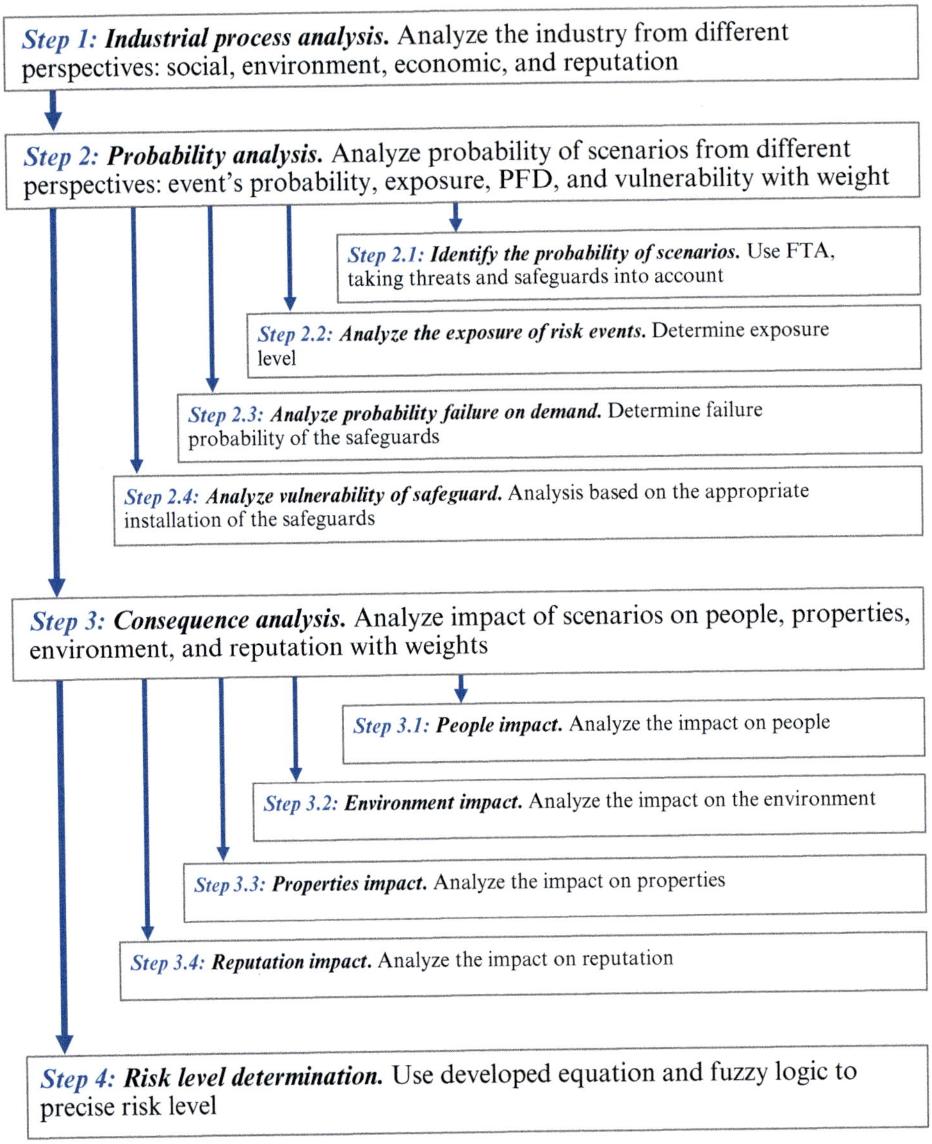

Figure 15.1 Framework for assessing risk level in the process industry.

15.3.2 Estimation of Impact Level

Four intermediate factors were included to determine the level of impact of a failure event, including the impacts on humans, the environment, properties, and reputation. Table 15.5 shows the matrix of risk impact developed in this study to determine the level of severity of an impact. The classification of consequence is presented in Table 15.6. The following equation was used to estimate the impact level in the proposed matrix using the weighted sum method

$$I_C = w_1\, I_H + w_2\, I_E + w_3\, I_P + w_4\, I_R \qquad (15.3)$$

where I_C is the impact of consequences, I_H is the impact on humans, I_E is the impact on the environment, I_P is the impact on properties, I_R is the impact on reputation, and w_i are the

Table 15.1 Frequency of risk consequence matrix (see Tables 15.2–15.4 for details of abbreviations).

		Probability of failure on demand of safeguards											
Frequency of risk events		AH	EH	QH	H	MH	ML	L	QL	EL	AL	Exposure level	
AL	[1 2]	MH	ML	ML	L	L	QL	QL	EL	EL	EL	[1 2]	AL
EL	[2 3]	MH	MH	ML	ML	L	L	QL	QL	EL	EL	[2 3]	EL
QL	[3 4]	H	MH	MH	ML	ML	L	L	QL	QL	EL	[3 4]	QL
L	[4 5]	H	H	MH	MH	ML	ML	L	L	QL	QL	[4 5]	L
ML	[5 6]	QH	H	H	MH	MH	ML	ML	L	L	QL	[5 6]	ML
MH	[6 7]	QH	QH	H	H	MH	MH	ML	ML	L	L	[6 7]	MH
H	[7 8]	EH	QH	QH	H	H	MH	MH	ML	ML	L	[7 8]	H
QH	[8 9]	EH	EH	QH	QH	H	H	MH	MH	ML	ML	[8 9]	QH
EH	[9 10]	AH	EH	EH	QH	QH	H	H	MH	MH	ML	[9 10]	EH
AH	[10 11]	AH	AH	EH	EH	QH	QH	H	H	MH	MH	[10 11]	AH
		[10 11]	[9 10]	[8 9]	[7 8]	[6 7]	[5 6]	[4 5]	[3 4]	[2 3]	[1 2]		
		AH	EH	QH	H	MH	ML	L	QL	EL	AL		
		Vulnerability											

Table 15.2 Classification of the probability of failure on demand (PFD) and frequency (Fr).

Category	Abbreviation	Range of PFD and Fr
Absolutely high	AH	[$5{\cdot}10^{-1}$ 10^{0}]
Extremely high	EH	[10^{-1} $5{\cdot}10^{-1}$]
Quite high	QH	[$5{\cdot}10^{-2}$ 10^{-1}]
High	H	[10^{-2} $5{\cdot}10^{-2}$]
Mildly high	MH	[$5{\cdot}10^{-3}$ 10^{-2}]
Mildly low	ML	[10^{-3} $5{\cdot}10^{-3}$]
Low	L	[$5{\cdot}10^{-4}$ 10^{-3}]
Quite low	QL	[10^{-4} $5{\cdot}10^{-4}$]
Extremely low	EL	[$5{\cdot}10^{-5}$ 10^{-4}]
Absolutely low	AL	[10^{-5} $5{\cdot}10^{-5}$]

Table 15.3 Classification of frequency of consequences.

Category	Abbreviation	Range
Absolutely high	AH	[10^{-1} 10^{0}]
Extremely high	EH	[10^{-2} 10^{-1}]
Quite high	QH	[10^{-3} 10^{-2}]
High	H	[10^{-4} 10^{-3}]
Mildly high	MH	[10^{-5} 10^{-4}]
Mildly low	ML	[10^{-6} 10^{-5}]
Low	L	[10^{-7} 10^{-6}]
Quite low	QL	[10^{-8} 10^{-7}]
Extremely low	EL	[10^{-9} 10^{-8}]
Absolutely low	AL	[10^{-10} 10^{-9}]

Table 15.4 Classification of vulnerability and exposure levels.

Category	Abbreviation	Range
Absolutely high	AH	[10 11]
Extremely high	EH	[9 10]
Quite high	QH	[8 9]
High	H	[7 8]
Mildly high	MH	[6 7]
Mildly low	ML	[5 6]
Low	L	[4 5]
Quite low	QL	[3 4]
Extremely low	EL	[2 3]
Absolutely low	AL	[1 2]

Table 15.5 Impact of consequence matrix.

		Impact on environment											
		AH	EH	QH	H	MH	ML	L	QL	EL	AL		
		[10 11]	[9 10]	[8 9]	[7 8]	[6 7]	[5 6]	[4 5]	[3 4]	[2 3]	[1 2]		
Impact on human													Impact on properties
AL	[1 2]	A	MA	MA	QA	QA	EA	EA	AA	AA	AA	[1 2]	AL
EL	[2 3]	A	A	MA	MA	QA	QA	EA	EA	AA	AA	[2 3]	EL
QL	[3 4]	MNA	A	A	MA	MA	QA	QA	EA	EA	AA	[3 4]	QL
L	[4 5]	MNA	MNA	A	A	MA	MA	QA	QA	EA	EA	[4 5]	L
ML	[5 6]	QNA	MNA	MNA	A	A	MA	MA	QA	QA	EA	[5 6]	ML
MH	[6 7]	QNA	QNA	MNA	MNA	A	A	MA	MA	QA	QA	[6 7]	MH
H	[7 8]	ENA	QNA	QNA	MNA	MNA	A	A	MA	MA	QA	[7 8]	H
QH	[8 9]	ENA	ENA	QNA	QNA	MNA	MNA	A	A	MA	MA	[8 9]	QH
EH	[9 10]	ANA	ENA	ENA	QNA	QNA	MNA	MNA	A	A	MA	[9 10]	EH
AH	[10 11]	ANA	ANA	ENA	ENA	QNA	QNA	MNA	MNA	A	A	[10 11]	AH
		[10 11]	[9 10]	[8 9]	[7 8]	[6 7]	[5 6]	[4 5]	[3 4]	[2 3]	[1 2]		
		AH	EH	QH	H	MH	ML	L	QL	EL	AL		
		Impact on reputation											

Table 15.6 Classification of consequence (impact).

Category	Abbreviation	Range	Consequence (impact) criteria
Slightly low	SLL	[0 1]	Slightly low impact
Slightly high	SLH	[1 2]	Slightly high impact
Minor low	MIL	[2 3]	Minor low impact
Minor high	MIH	[3 4]	Minor high impact
Moderately low	MOL	[4 5]	Moderately low impact
Moderately high	MOH	[5 6]	Moderately high impact
Critically low	CRL	[6 7]	Critically low impact
Critically high	CRH	[7 8]	Critically high impact
Catastrophic low	CAL	[8 9]	Catastrophic low impact
Catastrophic high	CAH	[9 10]	Catastrophic high impact

weights for the intermediate factors used to scale the factors for the computation of impact in the risk matrix I_R. Note that for the total weights the following condition holds true

$$\sum_{i=1}^{n} w_i = 1, i = 1,2,\ldots,n \tag{15.4}$$

15.3.3 Estimation of Risk Level

The template of the RMM based on two inputs is presented in Table 15.7. The inputs are distributed to ten levels of (i) frequency limited by 10^{-10} and 10^{0} as minimum and maximum values and (ii) impact limited by 1 as a minimum and 11 as a maximum value. Table 15.8 shows that the minimum value of frequency is 10^{-10}, which can be observed by multiplying the frequency of risk events with the PFD of safeguards. The output (i.e. risk level) is divided into nine levels ranging from 1 to 10.

To find the appropriate value of the risk level in the developed RMM, the following formula is deduced

$$R_L = \frac{1}{2}\left[(log(Fr_{RC}) + 11) + I_C \right] \tag{15.5}$$

whereFr_{RC} is the frequency of risk consequence, I_C is the impact of consequence, and R_L is the risk level.

15.3.4 Fuzzy Sets

FL was used to incorporate uncertainties associated with risk matrix parameters. FL is used to define specific linguistic variables for the risk that is based on the frequency and severity of the consequences. The FL system consists of the following elements:

- Fuzzify crisp inputs for each risk matrix component. Tables 15.9–15.11 show for the frequency, impact, and risk level, respectively, and Figure 15.2 depicts the structure of a typical FL set. The triangular fuzzification was considered in this study.
- The inference engine to set rules using if–then statements.
- Defuzzify outputs to one single output value.

Table 15.7 The risk matrix model (RMM) developed in this study.

		Impact of consequence									
		SLL	SLH	MIL	MIH	MOL	MOH	CRL	CRH	CAL	CAH
		[1 2]	[2 3]	[3 4]	[4 5]	[5 6]	[6 7]	[7 8]	[8 9]	[9 10]	[10 11]
Frequency of consequence	AH	[10^{-1} 10^{0}]	A	MNA	MNA	QNA	QNA	ENA	ENA	ANA	ANA	ANA
	EH	[10^{-2} 10^{-1}]	A	A	MNA	MNA	QNA	QNA	ENA	ENA	ANA	ANA
	QH	[10^{-3} 10^{-2}]	MA	A	A	MNA	MNA	QNA	QNA	ENA	ENA	ANA
	H	[10^{-4} 10^{-3}]	MA	MA	A	A	MNA	MNA	QNA	QNA	ENA	ENA
	MH	[10^{-5} 10^{-4}]	QA	MA	MA	A	A	MNA	MNA	QNA	QNA	ENA
	ML	[10^{-6} 10^{-5}]	QA	QA	MA	MA	A	A	MNA	MNA	QNA	QNA
	L	[10^{-7} 10^{-6}]	EA	QA	QA	MA	MA	A	A	MNA	MNA	QNA
	QL	[10^{-8} 10^{-7}]	EA	EA	QA	QA	MA	MA	A	A	MNA	MNA
	EL	[10^{-9} 10^{-8}]	AA	EA	EA	QA	QA	MA	MA	A	A	MNA
	AL	[10^{-10} 10^{-9}]	AA	AA	EA	EA	QA	QA	MA	MA	A	A

Table 15.8 Classification of the risk levels.

Category	Abbreviation	Risk level	Description
Absolutely acceptable	AA	[1 2]	Absolutely no mitigation required
Extremely acceptable	EA	[2 3]	Extremely no mitigation required
Quite acceptable	QA	[3 4]	Quite no mitigation required
Mildly acceptable	MA	[4 5]	Mildly no mitigation required
Acceptable	A	[5 6]	No mitigation required
Mildly not acceptable	MNA	[6 7]	Mildly mitigation required
Quite not acceptable	QNA	[7 8]	Quite mitigation required
Extremely not acceptable	ENA	[8 9]	Extremely mitigation required
Absolutely not acceptable	ANA	[9 10]	Absolutely mitigation required

Table 15.9 Fuzzy sets for frequency factor.

Linguistic variables of risk factors	Linguistic term (fuzzy set)	Description range of frequency
Frequency (F)	Absolutely high (AH)	AH[$5{\cdot}10^{-1}$ 1]
	Extremely high (EH)	EH[10^{-1} $5{\cdot}10^{-1}$]
	Quite high (QH)	QH[$5{\cdot}10^{-2}$ 10^{-1}]
	High (H)	H[10^{-2} $5{\cdot}10^{-2}$]
	Mildly high (MH)	MH[$5{\cdot}10^{-3}$ 10^{-2}]
	Mildly low (ML)	ML[10^{-3} $5{\cdot}10^{-3}$]
	Low (L)	L[$5{\cdot}10^{-4}$ 10^{-3}]
	Quite low (QL)	QL[10^{-4} $5{\cdot}10^{-4}$]
	Extremely low (EL)	EL[$5{\cdot}10^{-5}$ 10^{-4}]
	Absolutely low (AL)	AL[10^{-5} $5{\cdot}10^{-5}$]

Table 15.10 Fuzzy sets for impact of consequences factor.

Linguistic variables of risk factors	Linguistic term (fuzzy set)	Description range of severity	Universe of discourse (*X*)
Severity of consequences (*C*)	Slightly low (SLL)	SLL[0 1]	$X_C \in (0,10)$
	Slightly high (SLH)	SLH[1 2]	
	Minor low (MIL)	MIL[2 3]	
	Minor high (MIH)	MIH[3 4]	
	Moderately low (MOL)	MOL[4 5]	
	Moderately high (MOH)	MOH[5 6]	
	Critically low (CRL)	CRL[6 7]	
	Critically high (CRH)	CRH[7 8]	
	Catastrophic low (CAL)	CAL[8 9]	
	Catastrophic high (CAH)	CAH[9 10]	

Table 15.11 Fuzzy sets for risk factor

Linguistic variables of risk factors	Linguistic term (fuzzy set)	Description range of risk	Universe of discourse (*X*)
Risk category (*R*)	Absolutely acceptable (AA)	AA[1 2]	$X_R \in$
	Extremely acceptable (EA)	EA[2 3]	
	Quite acceptable (QA)	QA[3 4]	
	Mildly acceptable (MA)	MA[4 5]	
	Acceptable (A)	A[5 6]	
	Mildly not acceptable (MNA)	MNA[6 7]	
	Quite not acceptable (QNA)	QNA[7 8]	
	Extremely not acceptable (ENA)	ENA[8 9]	
	Absolutely not Acceptable (ANA)	ANA[9 10]	

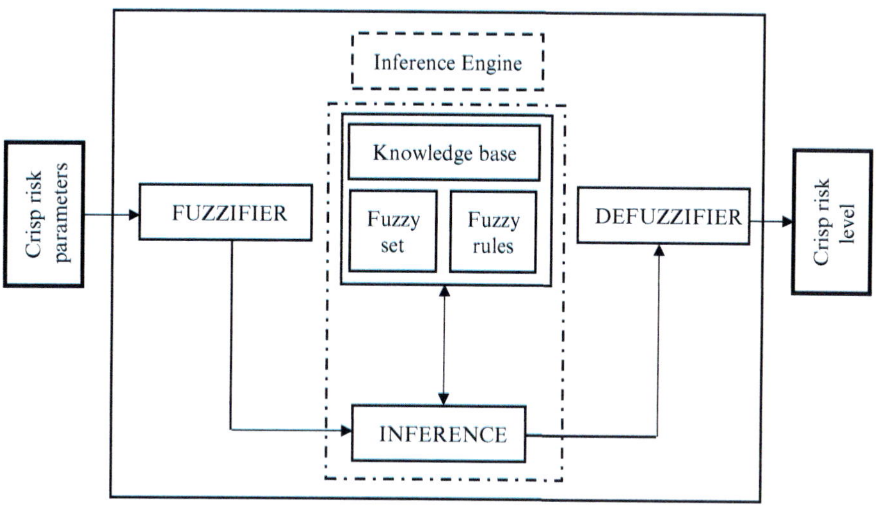

Figure 15.2 The structure of a typical fuzzy logic set.

15.4 Case Study of the Proposed Method

15.4.1 System Description

The production division of Sonatrach in Hassi R'Mel, Algeria is a compressor station that recovers heavy hydrocarbons in the forms of condensate and liquefied petroleum gas (LPG). The specific process evaluated in this research is the flow diagram of the reboiler oven H101 operating in the industry (Figure 15.3). Raw gas gathered from numerous wells is used to produce treated gas, which is known as the sales gas. A reboiler is a heat exchanger that is used to generate the vapor supplied to the bottom tray of a distillation column. The oven reboiler produces lighter fuel gas (sales gas), which is essentially composed of methane and ethane gases. The risk assessment framework developed in this study was implemented to evaluate the practicality of the above process.

The process of producing lighter fuel gas starts with the pre-separation of the crude gas from wells and then compression at the pump station at a pressure of 117 kg /cm^2 and a temperature of 62°C. In the high-pressure separation section, the recovered liquid hydrocarbons are separated as LPG and condensate in the deethanizer C101 of the fractionation section. Once the light components are extracted to the deethanizer C101, the accumulator separates the LPG condensate. To prevent hydrate formation in the upper part of the column, a glycol solution is injected. The separated liquid hydrocarbons are sent toward the C101. Part of these hydrocarbons is sent through pump 31-P101A and B to the reboiler H101 oven for heating at 150°C. The hydrocarbon flow rate is controlled by the motorized control valve FICA 136. The fluid exiting the heater to about 180°C is fed to the column C101 of the light to extract combustible gases (sales gas).

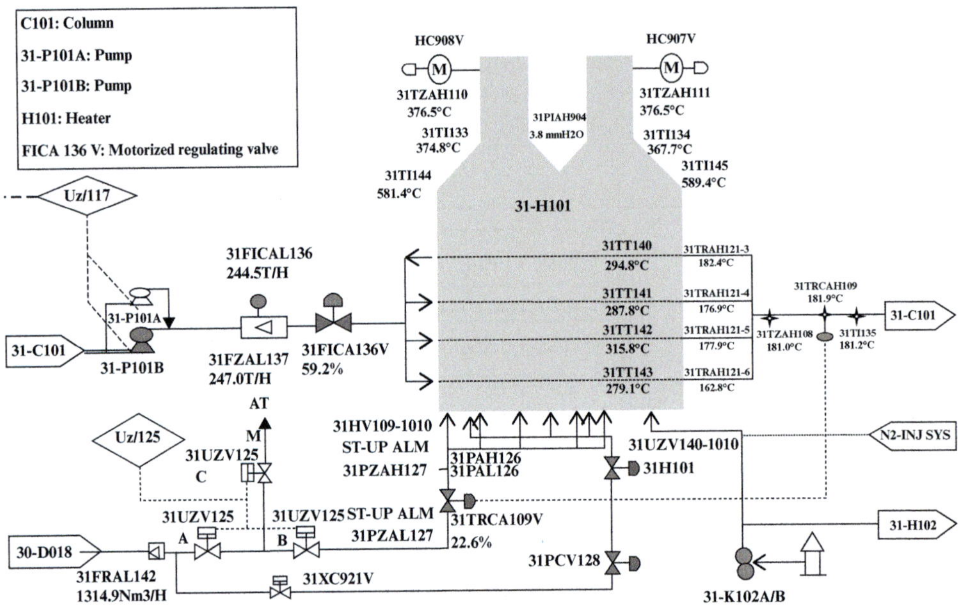

Figure 15.3 Process flow diagram of the reboiler oven H101. *Source*: [55]/With permission of Elsevier.

15.4.2 Simulink-MATLAB: Model for Risk Matrix Model

The Simulink-MATLAB model was used to evaluate eight possible undesirable consequences, which are boiling liquid expanding vapor explosion (BLEVE), jet fire, asphyxia, potential hazard, unconfined vapor cloud explosion (UVCE), flash fire, impacts on population, and environmental impacts. The values of risk level are determined based on frequency and impact of consequences with the help of Simulink. Frequency (indicated by u_0) and severity (indicated by u_1) are used as two inputs for applying S-function. Eight scenarios are presented in Table 15.12 showing minimum, median, and maximum values obtained from the simulations to overcome the issue of uncertainty. Figure 15.4 shows the integration of S-Function with the Simulink model to simulate Eq. (15.5).

Four scenarios out of the eight given in Table 15.12 were selected for detailed analysis for simplicity in the presentation of results. Figure 15.5 shows the risk level variation with 100 possible cases to show the possible level that each scenario could reach. The figure shows the risk level for four scenarios; it can be clearly seen that the highest risk level was observed for scenario 4 and the lowest for scenario 6. The results of 100 simulations show the range of risk level for each scenario. For instance, for scenario 6 (i.e. flash fire) the range of risk level varies between 0.98 and 8.85, whereas the figure shows that the risk level rarely approaches the low extreme due to the use of average data. For risk managers in the process industry, the

Table 15.12 The minimum, median, and maximum values of the risk level for eight scenarios obtained from the Simulink model.

		Risk level		
No.	**Scenario**	**Min**	**Median**	**Max**
1	BLEVE	0.18	5.96	7.90
2	Jet fire	0.09	6.61	8.66
3	Asphyxia	0.0	5.75	8.55
4	Potential hazard	0.0	4.88	7.43
5	UVCE	0.54	5.81	7.57
6	Flash fire	0.98	6.99	8.85
7	Impacts on population	0.0	3.90	6.53
8	Impacts on environment	0.0	4.71	7.09

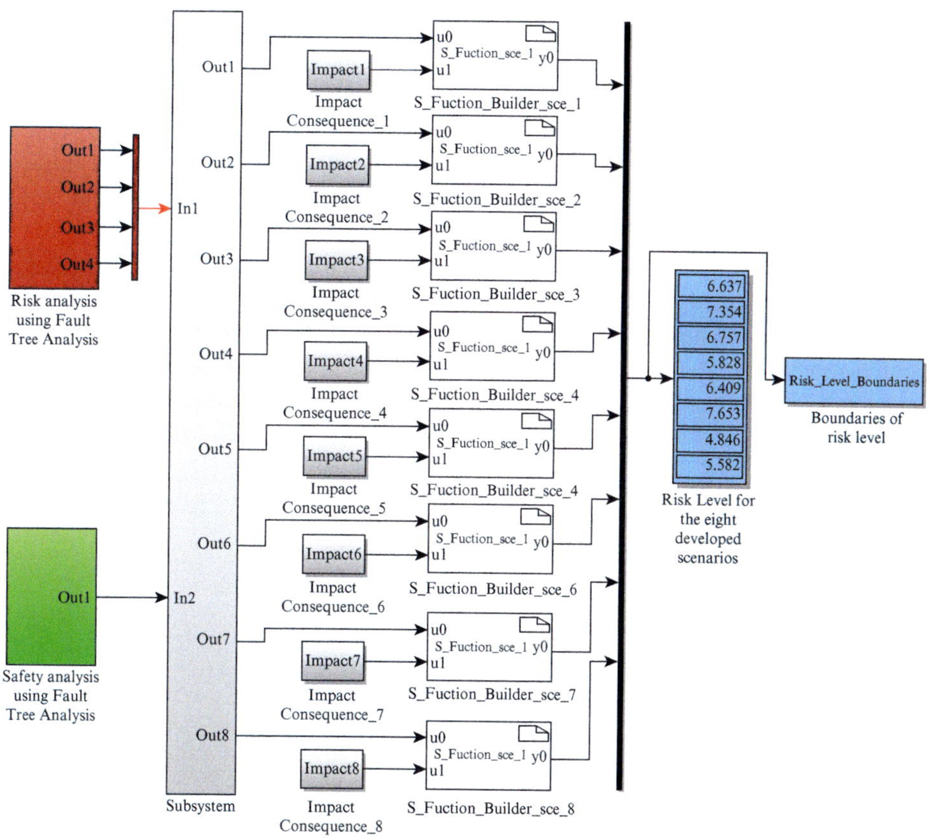

Figure 15.4 Simulink-MATLAB model to determine level of risk for eight scenarios.

highest risk values are important for improving safeguards. The values of Table 15.12 are slightly different from those in Figure 15.5, because the figure presents the entire range of data, whereas in Table 15.12 the average of minimum, median, and maximum values are presented.

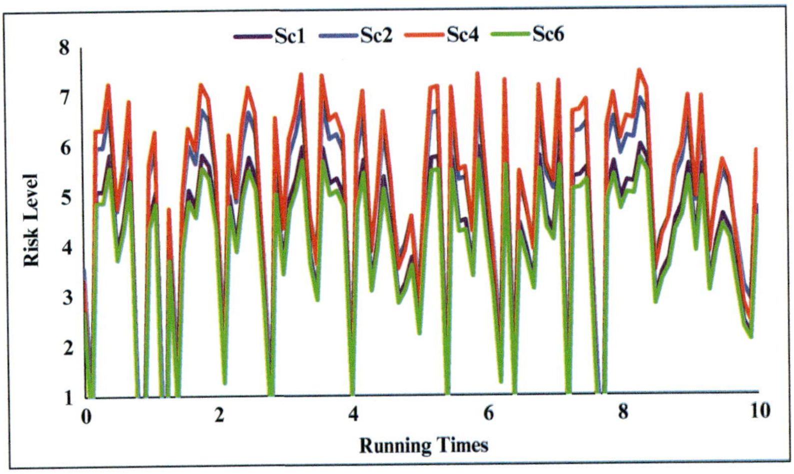

Figure 15.5 Simulink results showing variation of risk level for four scenarios.

Table 15.13 shows the average risk levels of eight scenarios and prioritizes the scenarios that could happen in the reboiler oven. The colors used in Table 15.13 correspond to the colors of the RMM (presented in Table 15.7). Simulink results of new RMM are more conservative and precise than the conventional model (Table 15.13). A *t*-test between the median risk values given in Table 15.13 and the mean values of risk level revealed that the *p*-value was found to be much lower than 0.05. Furthermore, results revealed that the mean difference between the mean risk values obtained from the conventional approach and RMM ranged between 0.95 and 1.3, with a 95% confidence interval. These results show that the risk can be underestimated using a conventional approach and that the RMM gives more precise results for estimating the risk level in the petrochemical industry.

Figure 15.6 shows a comparison of calculated risk levels using the fuzzy-based RMM and the classical risk model. Managers cannot make decisions based on minimum risk values to secure their systems, while mitigation actions for maximum values can be expensive for the industry. Results of the fuzzy-based RMM in terms of crisp risk level are certainly more desirable for effective decision-making. Figure 15.6 also shows the robustness of the proposed RMM when the risk levels are close to the maximum values obtained from the classical risk model and

Table 15.13 Comparison between a new RMM and a classical risk model for eight risk scenarios in the petrochemical industry.

No.	Scenario	Risk level using RMM	Risk level using classical risk model
1	BLEVE	6.93	[6 7]
2	Jet fire	7.63	[7 8]
3	Asphyxia	7.15	[7 8]
4	Potential hazard	6.15	[6 7]
5	UVCE	6.69	[6 7]
6	Flash fire	7.92	[7 8]
7	Population can be affected	5.21	[5 6]
8	Environment can be affected	5.90	[5 6]

close to the minimum for scenarios 3 and 4. In scenarios 2 and 5, the risk levels obtained by the RMM are almost equal to the average values of the ranges estimated by the classical risk model. These results show the practicality of the fuzzy-based RMM with more reliable results.

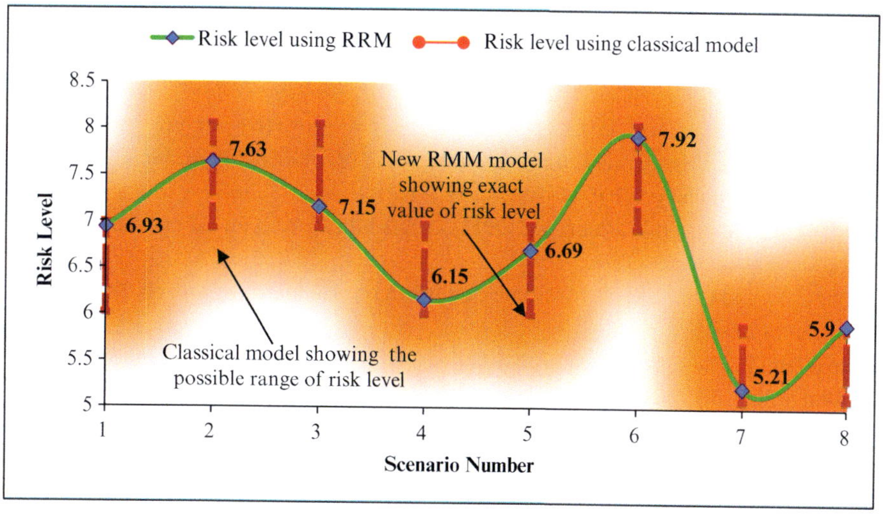

Figure 15.6 Variation of risk level using the RMM and the classical risk model.

The application of the RMM developed in this study manifests the more realistic estimation of the process risk levels. The RMM is useful for determining the risk level of each scenario using the frequency of risk events and the PFD of the safeguards, as well as based on the severity of the impacts on humans, the environment, properties, and reputation. The developed model will help ensure effective risk management and minimize the cost of safeguards. The proposed model effectively showed the level of risk for each scenario, which will help safety managers in risk mitigation planning.

15.5 Conclusions and Recommendations

The fuzzy-based RMM more realistically estimates the risk level of process industries than the classical risk model. The model scales and categorizes the frequency of risk events and PFD to determine the frequency and consequence of risks, to evaluate the impacts on humans, the environment, and properties. The fuzzy-based RMM also scales and categorizes the level of risk and calculates the precise level of risk while handling the uncertainties. The developed model effectively determines the level of risk for each scenario, which provides information to prioritize and identify scenarios that need emergency analysis to avoid possible catastrophic consequences. The overall results of the fuzzy-based RMM are more precise and reliable than the traditional risk matrices models.

References

1 Aven, T. (2010). On how to define, understand and describe risk. *Reliability Engineering and System Safety* 95 (6): 623–631.

2 Lindhe, A. (2010). *Risk Assessment and Decision Support for Managing Drinking Water Systems.* PhD thesis. Gothenburg, Sweden: Chalmers University of Technology.

3 Torstensson, E. (2012). *Risk analysis: the key to safe machinery*. http://ri.diva-portal.org/smash/get/diva2:962682/FULLTEXT01.pdf (accessed 4 October 2022).

4 Guanquan, C. and Jinhui, W. (2012). Study on probability distribution of fire scenarios in risk assessment to emergency evacuation. *Reliability Engineering and System Safety* 99: 24–32.

5 Aqlan, F. and Mustafa Ali, E. (2014). Integrating lean principles and fuzzy bow-tie analysis for risk assessment in chemical industry. *Journal of Loss Prevention in the Process Industries* 29: 39–48.

6 Ouache, R., Adham, A.A., and Azizan, N.A. (2014). Integrate fault tree analysis and fuzzy sets in quantitative risk assessment. *International Journal of Advanced Research in Engineering and Technology* 5 (10): 12–20.

7 Ouache, R., Kabir, M.N., and Adham, A.J (2015). A reliability model for safety instrumented system. *Safety Science* 80: 264–273.

8 Ni, H., Chen, A., and Chen, N. (2010). Some extensions on risk matrix approach. *Safety Science* 48 (10): 1269–1278.

9 Mohammadfam, I., Mahmoudi, S., and Kianfar, A. (2012). Comparative safety assessment of chlorination unit in Tehran treatment plants with HAZOP and ETBA techniques. *Procedia Engineering* 45: 27–30.

10 Ferdous, R., Khan, F., Sadiq, R., Amyotte, P., and Veitch, B. (2012). Handling and updating uncertain information in bow-tie analysis. *Journal of Loss Prevention in the Process Industries* 25 (1): 8–19.

11 Ouache, R., Adham, A., and Rasydan, A. (2014). Technical methods for the risk assessment at an industry system: review paper. *International Journal of Engineering Research & Technology* 3 (2): 1602–1611.

12 American Institute of Chemical Engineers. (1989). *Guidelines for process equipment reliability data with data tables*. https://www.aiche.org/resources/publications/books/guidelines-process-equipment-reliability-data-data-tables (accessed 4 October 2022).

13 Wu, W., Cheng, G., Hu, H., and Zhou, Q. (2013). Risk analysis of corrosion failures of equipment in refining and petrochemical plants based on fuzzy set theory. *Engineering Failure Analysis* 32: 23–34.

14 Fu, M., Weng, W.G., and Yuan, H.Y. (2014). Quantitative assessment of the relationship between radiant heat exposure and protective performance of multilayer thermal protective clothing during dry and wet conditions. *Journal of Hazardous Materials* 276: 383–392.

15 Ozog, H. (1985). Hazard identification, analysis and control: a systematic way to assess potential hazards helps promote safer design and operation of new and existing plants. *Chemical Engineering* 92 (4): 161–170.

16 Dunjó, J., Fthenakis, V., Vílchez, J.A., and Arnaldos, J. (2010). Hazard and operability (HAZOP) analysis: a literature review. *Journal of Hazardous Materials* 173 (1–3): 19–32.

17 Bendixen, L. and O'Neill, J.K. (1984). Chemical plant risk assessment using HAZOP and fault tree methods. *Plant/Operations Progress* 3 (3): 179–184.

18 Khakzad, N., Khan, F., and Amyotte, P. (2012). Dynamic risk analysis using bow-tie approach. *Reliability Engineering & System Safety* 104: 36–44.

19 Mokhtari, K., Ren, J., Roberts, C., and Wang, J. (2011). Application of a generic bow-tie based risk analysis framework on risk management of sea ports and offshore terminals. *Journal of Hazardous Materials* 192 (2): 465–475.

20 Ouache, R. and Adham, A.A.J. (2014). Reliability quantitative risk assessment in engineering system using fuzzy event tree analysis. *International Journal of Current Engineering and Technology* 4 (2): 1117–1123.

21 Summers, A. (2003). Introduction to layers of protection analysis. *Journal of Hazardous Materials* 104 (1-3): 163–168.

22 Ouazraoui, N., Nait-Said, R., Bourareche, M., and Sellami, I. (2013). Layers of protection analysis in the framework of possibility theory. *Journal of Hazardous Materials* 262: 168–178.

23 Jianfen, P. and Heng, L. (2013). Fuzzy comprehensive risk assessment method based on consistent matrix and information entropy. *The Open Automation and Control Systems Journal* 5: 139–149. https://benthamopen.com/TOAUTOCJ/home/.

24 Ale, B., Burnap, P., and Slater, D. (2015). On the origin of PCDS (probability consequence diagrams). *Safety Science* 72: 229–239.

25 Qinqin, C., Jia, Q., Yuan, Z., and Huang, L. (2014). Environmental risk source management system for the petrochemical industry. *Process Safety and Environmental Protection* 92 (3): 251–260.

26 Pinto, A. (2014). QRAM a qualitative occupational safety risk assessment model for the construction industry that incorporate uncertainties by the use of fuzzy sets. *Safety Science* 63: 57–76.

27 Duijm, N.J. (2015). Recommendations on the use and design of risk matrices. *Safety Science* 76: 21–31.

28 Netjasov, F. and Janic, M. (2008). A review of research on risk and safety modelling in civil aviation. *Journal of Air Transport Management* 14 (4): 213–220.

29 Aven, T. and Reniers, G. (2013). How to define and interpret a probability in a risk and safety setting. *Safety Science* 51 (1): 223–231.

30 Steen, R. and Aven, T. (2011). A risk perspective suitable for resilience engineering. *Safety Science* 49 (2): 292–297.

31 FDA/CFSAN. (2012). *Draft qualitative risk assessment: risk of activity/food combinations for activities (outside the farm definition) conducted in a facility co-located on a farm*. https://www.fda.gov/food. (accessed 4 October 2022).

32 Cascos, I. and Molchanov, I. (2013). Choosing a random distribution with prescribed risks. *Insurance: Mathematics and Economics* 52 (3): 599–605.

33 Medina-Herrera, N., Jiménez-Gutiérrez, A., and Grossmann, I.E. (2014). A mathematical programming model for optimal layout considering quantitative risk analysis. *Computers and Chemical Engineering* 68: 165–181.

34 Rockwell Automation. (2009). *Safety related control systems for machinery*. https://www.controleng.com/articles/machine-safety-advice-think-beyond-the-safety-device-rockwell-automation-says/ (accessed 4 October 2022).

35 Nieto-Morote, A. and Ruz-Vila, F. (2011). A fuzzy approach to construction project risk assessment. *International Journal of Project Management* 29 (2): 220–231.

36 Peng, Z., Lyu, Y., Miller, A., Johnson, C., and Zhao, T. (2019). *Risk assessment of railway transportation systems using timed fault trees*. http://www.dcs.gla.ac.uk/~alice/papers/qreiRisk.pdf (accessed 28 September 2022).

37 Refauld, F. (2011). *Quantitative risk analysis in an uncertain and dynamic environment*. PhD thesis, Memorial University of Newfoundland.

38 Vaidogas, E.R., Kisežauskienė, L., and Girnienė, I. (2016). The risk to structures built near roads and rails used for moving hazardous materials. *Journal of Civil Engineering and Management* 22 (3): 442–455.

39 Argenti, F., Landucci, G., Spadoni, G., and Cozzani, V. (2015). The assessment of the attractiveness of process facilities to terrorist attacks. *Safety Science* 77: 169–181.

40 EFSA (2010). Guidance on the environmental risk assessment of genetically modified plants. *EFSA Journal* 8 (11): 1–111.

41 Ho, V. (2010). *The risk of using risk matrix in assessing safety risk*. https://www.hkarms.org/web_resources/20101116_risk_matrix_hkieb_print.pdf (accessed 28 September 2022).

42 Canadian Society for Chemical Engineering. (2004). *Risk assessment: recommended practices for municipalities and industry*. https://www.cheminst.ca/wp-content/uploads/2019/04/

Risk20Assessment20E2809320Recommended20Practices20for20Municipalities20and20 Industry.pdf (accessed 28 September 2022).

43 Andersen, S. and Mostue, B.A. (2012). Risk analysis and risk management approaches applied to the petroleum industry and their applicability to IO concepts. *Safety Science* 50 (10): 2010–2019.

44 Aven, T. (2011). *Quantitative Risk Assessment: The Scientific Platform*. Norway: Cambridge University Press.

45 Bao, C., Li, J., and Wu, D. (2016). A fuzzy mapping framework for risk aggregation based on risk matrices. *Journal of Risk Research* 9877: 1–23.

46 Karanikas, N. (2015). An introduction of accidents' classification based on their outcome control. *Safety Science* 72: 182–189.

47 Center for Chemical Process Safety. (2000). *Guidelines for chemical process quantitative risk analysis*.

48 International Association of Oil & Gas Producers. (2013). *Oil spill risk assessment and response planning for offshore installations*. https://www.ospri.online/site/assets/files/1135/jip_finding_6_oilspill_risk_assessment.pdf (accessed 28 September 2022).

49 American Petroleum Institute. (2009). Risk-based inspection. *API Publishing Services*. http://www.irantpm.ir/wp-content/uploads/2013/08/API-RP-580-Risk-Based-Inspection-2009.pdf (accessed 28 September 2022).

50 Joy, J. and Griffiths, D. (2007). *National minerals industry safety and health risk assessment*. https://www.nost.edu.au/icms_docs/286339_National_Minerals_Industry_Safety_and_Health_Risk_Assessment_Guideline_-_Jim_Joy.pdf (accessed 28 September 2022).

51 American Institute of Chemical Engineers. (2003). *Investigating Chemical Process Incidents*. Wiley.

52 Ozog, H. (2009). Designing an effective risk matrix. *An ioMosaic Corporation Whitepaper*. https://iomosaic.com/docs/default-source/papers/designing-an-effective-risk-matrix-2009.pdf?sfvrsn=222fcdd4_8 (accessed 28 September 2022).

53 Sanders, E. (2017). *Risk matrix as a tool for risk assessment in the chemical process industry* https://silo.tips/download/risk-matrix-as-a-tool-for-risk-assessment-in-the-chemical-process-industry (accessed 4 October 2022).

54 Markowski, A.S. and Mannan, M.S. (2008). Fuzzy risk matrix. *Journal of Hazardous Materials* 159 (1): 152–157.

55 Ouache, R. and Kabir, M.N. (2016). Models of probability of failure on demand for safety instrumented system using atmospheric elements. *Safety Science* 87: 38–46.

16

Application of Fuzzy Theory to Investigate the Effect of Innovation Power in the Emergence of an Advanced Reusable Packaging System

Péter Böröcz, Adrienn Buruzs, Ádám Bukovics, Péter Földesi, and László T. Kóczy

Széchenyi István University, Győr, Hungary

16.1 Introduction

The need for modern and environmentally friendly packaging systems raises the question of whether this is related to the development of a given industrial area. Namely, based on the already existing mainly production factors within the given territorial boundaries, the developed and environmentally effective information and factors can be explained by the causal relations in connection with the design of packaging systems.

Businesses in industrial parks are concentrated in one place, usually because they prefer to be close to the right infrastructure and existing historical knowledge and experience. However, industrial parks nowadays have to comply not only with settling-in companies but also with society [1], as they try to operate in the spirit of sustainable development. This includes the use of a packaging system in their logistic activities that allows them to do so.

However, the development path of industrial parks, and thus the design of a modern and environmentally friendly packaging system, is largely determined by the unique features of the country, region, and city in which they are based. However, in addition to the basic economic and social conditions already known for investment, the skills, development capacity, and willingness to innovate of the workforce are also essential in the long run. Measuring the latter two characteristics is a more difficult task.

In this chapter, the authors attempt to analyze the role of these two subjective factors, using fuzzy theory, to provide more accurate predictions for the engineer on the design of a reusable packaging system that is more environmentally friendly.

16.2 Establishing the Territorial Potential for an Environmentally Effective Packaging System

A provider of a modern packaging system must consider certain criteria when deciding whether to relocate to a different area or develop their existing place of business.

There is a wide literature on site selection, in which researchers mainly approach the issue from a corporate perspective and ask how a site is chosen by a company. The generally accepted strategy is characterized by the following phases: sorting, site selection, and then physical planning.

This method basically examines the microeconomic, macroeconomic, and social situation of a given area (or region). Nowadays, this method is complemented by various market

Fuzzy Systems Modeling in Environmental and Health Risk Assessment, First Edition. Edited by Boris Faybishenko, Rehan Sadiq, and Ashok Deshpande.

estimation methods and the requirement for a skilled workforce. What has changed in recent decades is the method of analysis [2]. However, the decision to choose the most appropriate place to provide a pool service requires other criteria.

Such existing capabilities include, in particular, the propensity to innovate or the existence of an environmentally friendly business philosophy, neither of which can be clearly assessed objectively.

In order to address the problem as a whole, we need to look at existing local conditions, such as the social, economic, and infrastructural situation in a given area. However, these analytical methods will not be able to make more subjective judgments about a desire to innovate or be more environmentally aware.

However, fuzzy theory [3] can be a suitable way of studying such factors. In this chapter, the authors' intention is to investigate a region with an objective-based potential to accommodate a modern packaging system and decide, using fuzzy methodologies, whether this region has a so-called subjective-based potential for a modern packaging system.

16.3 The Model

The purpose of the model is to determine the potential of a particular region. This model can be used as a decision-making tool for a specific installation planning task. This tool applies multicriteria evaluation in order to analyze the suitability of different areas as locations for new industrial parks.

The model requires three distinct phases of analysis. In the first phase, the area is geographically sorted and selected. This analysis is still widely appreciated by the region. The purpose is to evaluate the new potential development area and to determine which location would be best suited for geographical purposes, such as the introduction of a pool system or the launch of a pool service. This process gives the potential of a given area to compare the characteristic value of the two areas to one another in the future. The given region is denoted by P_i, where i denotes the ith region.

In the second phase, objective factors that have a clear impact on the potential of the region or directly provide the most appropriate choice should be delimited. These factors are predefined variables at hierarchical levels that can be grouped into categories and subcategories. Categories and subcategories represent the intermediate level in the decision-making and analytical work of the model. The variables test the existence of the necessary requirements that are essential for possible improvement. Ultimately, they can be used to evaluate available resources. The appropriate categories must be defined and grouped in a clear, understandable way, and then the basic potential of a given area can be determined.

For example, the following basic potential categories can be formed: social (S_i), economic (E_i), infrastructure (I_i), research and development (R&D) (K_i), and, for example, the local rule set (R_i) (grouping is shown in Table 16.1). Each factor can be evaluated as a parameter, or if its mere existence is a question, we can decide whether it exists in the region.

It is easy to see that the potential of a region can be derived as a function of the various variables described in Table 16.1. Of course, for an accurate result, it is necessary to expand the categories as widely as possible. If the appropriate categories are considered in their entirety, then we may assume that $P_i = f(S_i, E_i, K_i, I_i, R_i)$.

It is clear that different factors can be specified with different extensions and units. Therefore, it is necessary to examine some of their comparisons by category. Thus, for example, one possible solution is to compare the value of the factors of the other regions to the region with the *most favorable values* in some factors of the studied regions. The most

Table 16.1 Categories and some possible metrics to measure spatial potential.

Categories	Some possible metrics
Social	Demographic metrics (births and death rates, migration balance, etc.) Education (proportions by education, etc.) Labor force (rate of unemployment, etc.)
Economic	Economic activity (existence of industrial and service sectors, etc.) Costs (resources, manpower, waste disposal, etc.)
Infrastructure	Transport network (road, rail, air, etc.) Energy, water, etc. Recycling (waste management, recycling facilities, etc.) Communication, IT (internet, telephone, etc.)
Knowledge	Number of R&D centers Proportion of R&D expenditure in GPD
Rule set	Legal framework Taxes State aids

favorable value may also be the minimum or maximum value within a given group, depending on whether the upper or lower limit of that value is logically the most favorable.

The following factors were selected for uploading the model:

- s_i: unemployment rate in the region.
- e_i: Gross domestic product (GDP) per capita in the region.
- k_i: R&D expenditure as a percentage of GDP in the region.
- i_i: the value of the motorways per square kilometer in a given region.

The rule set was omitted from the model because it was almost impossible to determine a real, exact value without a thorough legal knowledge. So, we say that the values for the categories in a given territorial unit can be given as

$$s_i = \frac{s\,\underset{i}{\min}}{s_i e}; e_i = \frac{E_i}{E\,\underset{i}{\max}}; k_i = \frac{K_i}{K\,\underset{i}{\max}}; j = i_i = \frac{I_i}{I\,\underset{i}{\max}}$$

A possible form of estimating the potential of a region is Eq. (16.1). This specifies how the values defined by the independent variables are added together

$$P_i = a_s s_i + a_E e_i + a_K k_i + a_I i_i \tag{16.1}$$

The notation a shows the weight of the given objective parameter in the given region. It will be appreciated that additional parameters may be added to the equation that, in some cases, may be required for a proper comparison of two or more regions. It should be mentioned here that the equation contains only existing possible objective parameters, without being exhaustive, but does not provide satisfactory information on the characteristics that can be interpreted as the subjective parameters of the region.

The third phase of the model is the delineation of subjective factors that may directly affect the potential of the region or modify the value of the potential. Such a suitable parameter is, for example, the willingness to innovate or the innovation environment, which may have a direct impact on our basic potential.

In order to analyze such a parameter, we can use the logic of fuzzy theory, which extends classical logic beyond the logic of real and false values [3]. As a result, we can interpret this task as looking at the degree of true and false meaning. This already allows the use of mathematical formulas to handle and analyze such information. In this way, we can interpret subjective parameters and concepts that cannot be precisely defined. Fuzzy logic (FL) is very useful for dealing with real-world phenomena, which is characterized by a great deal of uncertainty and complexity [4].

For example, if there is a subjective factor that affects the region's potential as a venue for an advanced packaging system, we can use FL to analyze this as an extended function of our base function. This is called P_i*, where * denotes the modified potential that has a fuzzy exponent [5]. The exponent can influence the slope of our potential function in different ways (Figure 16.1). This fuzzy exponent is denoted by (*b*).

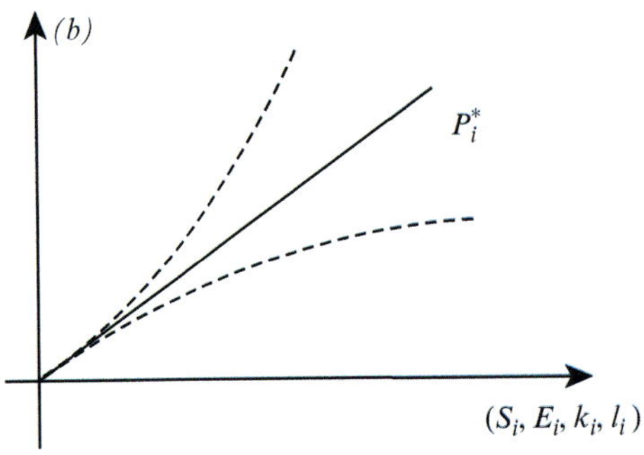

Figure 16.1 Changing the potential curve with the fuzzy exponent.

If Eq. (16.1) is extended with the fuzzy exponent presented, the next step can be determined. If P_i exists, the modified potential (P_i*) will show the *real potential* extended by the subjective parameters. Thus, a region can be found where the potential for advanced packaging may be lower or higher

$$p_i^{\ *} = a_s s_i^{\ b_s} + a_E e_i^{\ bE} + a_K k_i^{\ b_k} + a_I l_i^{\ b_l} \tag{16.2}$$

Equation (16.2) is a quadratic equation that is not easily solved by conventional methods of analysis and therefore requires the use of a search engine algorithm. Also, since the value of the exponent is uncertain, it is also advisable to include FL. In practice, the value of the fuzzy exponent is determined by the power of the innovation effect. The appropriate information must be collected about this. For example, we can look at how many patents have been filed in a given region on packaging topics, or how many scientific publications have been published in a given area on packaging topics, etc. Of course, we need to give the appropriate weighting values for each category for each test. In the developed model, a partially asymmetric and symmetric fuzzy solution can be applied to analyze previously presented independent variables [5]. Asymmetric and symmetric representations can handle the uncertainty of the power of innovation. See Figure 16.2.

Thus

$$\begin{aligned} P = {} & \frac{1}{1+(1+\lambda)(1-\alpha)}\left(s^{b_c}\right) + \frac{1-\alpha}{1+(1+\lambda)(1-\alpha)}\left(s^{b_c-(1-\alpha)(b_c-b_L)}\right) \\ & + \frac{(1-\alpha)\lambda}{1+(1+\lambda)(1-\alpha)}\left(s^{b_c+(1-\alpha)(b_R-b_C)}\right) \end{aligned} \tag{16.3}$$

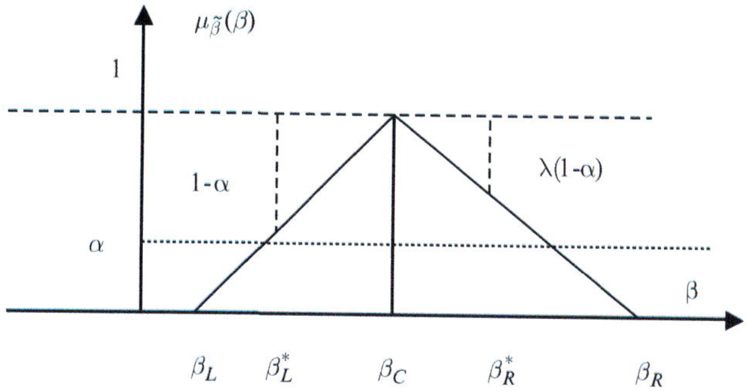

Figure 16.2 General asymmetric representation of any exponent beta with different α-cuts.

where $\lambda = 1$ and P is the product of s_i, e_i, k_i, and i_i, plus the results of factors extended by the fuzzy exponent. By choosing the value of α we can look at how the power of the innovation effect can be modeled as a fuzzy value. The b indicates where the lower (L), middle (C), upper (R) boundaries are. Of course, C will be different in symmetric and asymmetric cases.

16.4 Applying the Fuzzy Component Through a Numerical Example

To test the model, the advanced packaging system discussed here is investigated in three countries. These countries are Germany, Portugal, and Hungary. Definition of factors per category:

- s_i: unemployment rate in the given country (%), 2016.
- e_i: GDP per capita in the given country (US dollars/capita), 2016.
- k_i: R&D expenditure as a percentage of GDP in the given country, 2016.
- i_i: value of highways per km^2 in the given country (km/km^2), 2016.
- b: fuzzy exponent, power of innovation effect.
- a: importance of objective factor (weight).

It is readily apparent that the value of a has to be determined for each factor such that it modifies the magnitude of each factor in the model. In our case, we concluded that the unemployment rate is the least influential in terms of packaging technology innovation and development (as it is almost independent of the development of packaging technology and only indirectly in the case of a difference of 1 or 2%) and R&D costs in GDP proportion (since the research must be highly targeted and the research subject must be significant in terms of packaging technology), which are thus given a lower weighting. We count with the $a_s = 0.1$, $a_K = 0.2$ values. It should be noted here that, since weights are involved, weights do not necessarily have to be 1 in total. Per capita GDP was considered more important than the above two factors, as it assumes a high level of industrial output in a given area, which also implies a high demand for packaging technology. Thus, its value was taken as 0.3. The infrastructure factor is considered the strongest indicator in this area. There are two reasons for this: on the one hand, the density of the transport network shows a significant number of industrial areas, thus the size of the industrial volume, and assumes that the transport demand is significant depending on the significant freight tonne kilometers. Its value was thus determined with a weight of 0.4. The period considered is 2016, as data for this year were fully available. Data were collected from the statistical database of the International Monetary Fund (IMF), the Organisation for Economic Co-operation and Development (OECD), and

EUROSTAT. The values of the subjective factors of the studied area and the values of the weights used in this study are summarized in Table 16.2.

Table 16.2 The values of the subjective factors of the studied area and the values of the weights.

	a_S	s_i	a_E	e_i	a_K	k_i	a_I	i_i	S_i %	E_i GDP per capita	K_i %/GDP	I_i km/km^2
Portugal	0.1	0.96	0.3	0.525	0.2	0.32	0.4	0.785	7.8	18 464	0.8	2.8
Germany		0.77		1.0		1.0		1.0	9.8	35 203	2.5	3.5
Hungary		1.0		0.322		0.4		0.24	7.5	11 340	1.0	0.8

The determination of the fuzzy exponent b also needs to be considered logically, that is what effect the innovation has on the given factor, and if it is sufficiently strong or weak, and how it actually influences it. Based on these, it can be concluded that, in terms of unemployment rate or R&D expenditure, the power of innovation increases with the strength of that innovation (Figure 16.3), whereas it increases to only a given value in the case of infrastructure conditions and GDP, and has no effect thereafter. Therefore, the value of the exponent b of fuzzy function is evaluated between 1.01 and 1.4 (asymmetrically), at value a_S between 1.00 and 1.5 (asymmetrically), at value a_K between 0.5 and 0.9 (symmetrically), and at value a_E between 0.6 and 1.0 (symmetrically).

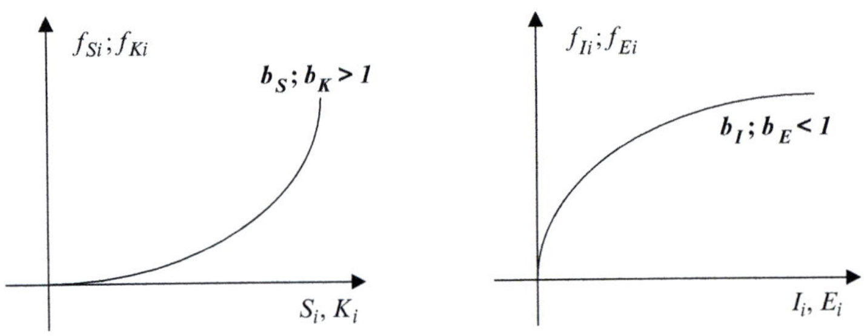

Figure 16.3 The value and effect of choosing the fuzzy exponent.

Using Eq. (16.2), based on the data presented in Tables 16.3 and 16.4, and after correcting for the given weights and summing up the values for each category, the following modified potential values were obtained:

- Portugal 0.509
- Germany 0.973
- Hungary 0.502

Assuming that the power of the innovation effect is fuzzy-like, the potential for the emergence of advanced packaging would be highest in Germany, while Portugal and Hungary would have similar potential.

Assuming that the power of the innovation effect is less fuzzy-like, we get:

- Portugal 0.505
- Germany 0.974
- Hungary 0.498

Based on the results of calculations the following conclusion can be drawn. When the strength of the innovation effect is less fuzzy-like, there is the highest potential in

Table 16.3 Value of certain objective factors calculated with fuzzy index supposing weak fuzzy circumstances (low α).

	s, e, k, i	y	x	y-fuzzy	β_L	β_C	β_R	alfa	cent	left	right
Portugal S	0.96	0.95414	0.96	0.9527517	1.15	1.01	1.4	0.01	0.3201811	0.3187778	0.3137928
Germany	0.77	0.7404	0.77	0.7340626	1.15	1.01	1.4	0.01	0.2484552	0.2550443	0.2305631
Hungary	1.0	1.0	1.0	1.0	1.15	1.01	1.4	0.01	0.3355705	0.3322148	0.3322148
Portugal E	0.525	0.59721	0.525	0.6004444	0.8	0.6	1.0	0.01	0.2004062	0.2254005	0.1746377
Germany	1.0	1.0	1.0	1.0	0.8	0.6	1.0	0.01	0.3355705	0.3322148	0.3322148
Hungary	0.322	0.40391	0.322	0.4106939	0.8	0.6	1.0	0.01	0.1355403	0.1679377	0.1072159
Portugal K	0.32	0.25479	0.32	0.2519095	1.2	1.0	1.5	0.01	0.0854995	0.1060667	0.0603432
Germany	1.0	1.0	1.0	1.0	1.2	1.0	1.5	0.01	0.3355705	0.3322148	0.3322148
Hungary	0.4	0.33302	0.4	0.3286705	1.2	1.0	1.5	0.01	0.1117521	0.1326426	0.0842758
Portugal I	0.32	0.45041	0.32	0.4580553	0.7	0.5	0.9	0.01	0.1511432	0.1875013	0.1194108
Germany	1.0	1.0	1.0	1.0	0.7	0.5	0.9	0.01	0.3355705	0.3322148	0.3322148
Hungary	0.4	0.52655	0.4	0.5323265	0.7	0.5	0.9	0.01	0.1766956	0.2097264	0.1459045

Table 16.4 Value of certain objective factors calculated with fuzzy index supposing weak fuzzy circumstances (high α).

	s, e, k, i	y	x	y-fuzzy	β_L	β_C	β_R	alfa	cent	left	right
Portugal S	0.96	0.95414	0.96	0.9541039	1.15	1.01	1.4	0.9	0.7951163	0.0795571	0.0794305
Germany	0.77	0.7404	0.77	0.7402208	1.15	1.01	1.4	0.9	0.6169971	0.0619259	0.0612979
Hungary	1.0	1.0	1.0	1.0	1.15	1.01	1.4	0.9	0.8333333	0.0833333	0.0833333
Portugal E	0.525	0.59721	0.525	0.5972188	0.8	0.6	1.0	0.9	0.4976754	0.0504131	0.0491303
Germany	1.0	1.0	1.0	1.0	0.8	0.6	1.0	0.9	0.8333333	0.0833333	0.0833333
Hungary	0.322	0.40391	0.322	0.4039274	0.8	0.6	1.0	0.9	0.3365917	0.0344307	0.0329049
Portugal K	0.32	0.25479	0.32	0.2545645	1.2	1.0	1.5	0.9	0.2123238	0.0217218	0.0205189
Germany	1.0	1.0	1.0	1.0	1.2	1.0	1.5	0.9	0.8333333	0.0833333	0.0833333
Hungary	0.4	0.33302	0.4	0.3327821	1.2	1.0	1.5	0.9	0.2775177	0.028265	0.0269993
Portugal I	0.32	0.45041	0.32	0.4504263	0.7	0.5	0.9	0.9	0.375339	0.0383991	0.0366882
Germany	1.0	1.0	1.0	1.0	0.7	0.5	0.9	0.9	0.8333333	0.0833333	0.0833333
Hungary	0.4	0.52655	0.4	0.5265676	0.7	0.5	0.9	0.9	0.4387941	0.0446909	0.0430826

Germany for the design of the advanced packaging. Furthermore, Portugal and Hungary would also have nearly the same potential. The conclusion can also be drawn from calculations that, while the final result is strongly determined by the objective influencing factors, a fuzzy index (force of innovation effect) cannot be considered a strong influencing factor.

16.5 Example of the CHEP Packaging System

If we want to check the control of the introduced model in respect of the packaging system occurring in real life, too, the CHEP multi-way packaging system would be a very good example to use. Currently, the CHEP system is available in more than 45 countries on various continents (Figure 16.4). It includes the different number of warehouses and service centers in various areas (regions/countries) of the world.

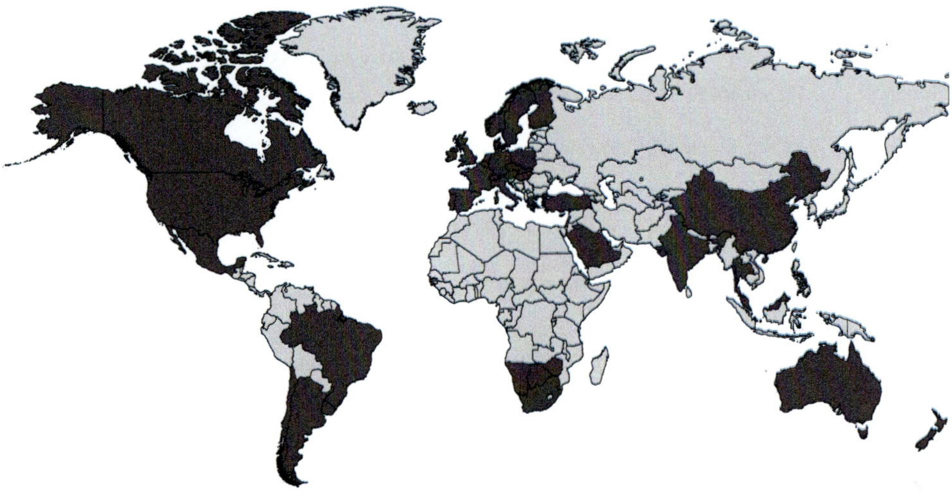

Figure 16.4 Availability of the CHEP system in the world.

The examined countries have the following number of centers within the given country: 6 in Portugal, 16 in Germany, 1 in Hungary. Result of the model relative to Hungary, further corrected by the number of inhabitants is shown in Table 16.5.

Table 16.5 Results of the model corrected by the number of inhabitants.

Result of the model	Corrected by the number of inhabitants
Portugal (0.509); 1.01	Portugal (10.5 million people); 1.06
Germany (0.973); 1.93	Germany (82.0 million people); 15.82
Hungary (0.502); 1.00	Hungary (10.0 million people); 1.00

By comparing the preliminary data obtained as the final result of the model with the data occurring in real life, we can see that data are only roughly correlate. While in the model Germany had the highest potential value, it was the same in industrial practice, too, yet practically the moduli are the same. At the same time, in the case of Portugal and Hungary, the model generated nearly the same results, although real values showed significant deviations. The reason for this could be that in the model several objective factors should have been taken into account, like the transport of goods by sea and the impact of transit, which can have a significant impact in Portugal.

Certainly, it cannot be unequivocally stated that the selected objective factors and the applied weighted numbers accurately reflect the reality. Obviously, that was not the aim of the model. The aim was to provide a proposal for a possible solution and offer the opportunity for the packaging experts to be interested in advanced packaging and be able to make reliable decisions considering a given time horizon.

16.6 Conclusion

In summary, it can be stated that an innovative milieu is an indispensable requirement for the introduction and design of advanced packaging, but most of the time the decisions depend on the synergic effect of local production factors and regional peculiarities. Modeling it is not as easy as it seems at first sight. That is determining certain subjective factors and interpreting their interrelations is practically impossible using traditional analytical methods. The model discussed in this chapter goes beyond the traditional method of basing the choice of a region on macroeconomic and microeconomic factors alone. And it also goes a long way to show that knowledge and the innovation that it generates both play a significant role in the creation and development of commercial systems.

Acknowledgments

This work was supported by Széchenyi István University, Győr (Hungary).

References

1 Gerstlberger, W. (2004). Regional innovation systems and sustainability: selected examples of international discussion. *Technovation* 24 (9): 749–758.
2 Fernández, I. and Ruiz, M.C. (2009). Descriptive model and evaluation system to locate sustainable industrial areas. *Journal of Cleaner Production* 17: 87–100.
3 Bojadziev, G. and Bojadziev, M. (1997). *Fuzzy Logic for Business, Finance and Management. World Science.*
4 Boclin Campos, A.S. and de Mello, R. (2006). A decision support method for environmental impact assessment using a fuzzy logic approach. *Ecological Economics* 58: 170–181.
5 Földesi, P., Botzheim, J., and Kóczy, L.T. (2008). Fuzzy exponents for heuristic based applications. *Acta Technica Jaurinensis Series Intelligentia Computatorica* 1 (3): 423–435.

Index

Fuzzy Systems Modeling in Environmental and Health Risk Assessment, First Edition. Edited by Boris Faybishenko, Rehan Sadiq, and Ashok Deshpande.

g

h

i

l

m

n

o

p

Printed and bound by CPI Group (UK) Ltd, Croydon, CR0 4YY

17/09/2026

14964314-0003